PROCEEDINGS OF SYMPOSIA
IN PURE MATHEMATICS
Volume XVIII, Part 1

Symposium on Nonlinear Functional Analysis, Chicago, 1968

NONLINEAR FUNCTIONAL ANALYSIS

AMERICAN MATHEMATICAL SOCIETY
Providence, Rhode Island
1970

Proceedings of the Symposium in Pure Mathematics
of the American Mathematical Society
Held in Chicago, Illinois
April 16–19, 1968

Prepared by the American Mathematical Society
under National Science Foundation Grant GP-8462

Edited by
FELIX E. BROWDER

AMS 1970 Subject Classifications.
Primary 47H05, 35J60, 47H15, 58E05, 35K55, 46E35, 49A30, 58B15, 47H10,
54H20, 93C20, 55F10, 76R10, 49A40, 57D25
Secondary 26A51, 35F20, 35F25, 25F30, 35G20, 35G25, 35G30, 35J45, 35J50,
35J60, 35J65, 35K25, 35K40. 35P10, 35R20, 47D05, 52A05, 52A40

Library of Congress Catalog Number 73-91392
International Standard Book Number 0-8218-0243-7

Printed in the United States of America

CONTENTS

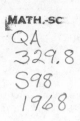

PREFACE

The present volume contains Part I of the Proceedings of the Symposium on Nonlinear Functional Analysis held at the April meeting of the American Mathematical Society in Chicago in April 1968 under the sponsorship of the A.M.S. and with financial support from the National Science Foundation. All of the speakers at that Symposium with the exception of George Minty have contributed papers to these Proceedings. Part I contains all of these papers, except that of the undersigned. The latter paper entitled "Nonlinear operators and nonlinear equations of evolution in Banach spaces" appears in Part II.

Felix E. Browder

A MONOTONE CONVERGENCE THEOREM FOR SEQUENCES OF NONLINEAR MAPPINGS

Edgar Asplund

In this paper we prove a theorem generalizing the elementary theorem on convergence of bounded, monotone sequences of real numbers, and also the theorem of Vigier and Nagy, cf. [**2**, Appendice II] on the convergence of certain sequences of symmetric linear operators on Hilbert space.

The paper consists of two sections. In the first we prove the main monotone convergence theorem (Theorem 1) and apply it to prove a decomposition for monotone operators which generalizes the decomposition of a linear operator into symmetric and antisymmetric parts. In the second section we apply Theorem 1 to linear operators. Some rather messy linear algebra computations have to be performed in order to get Theorem 4, which is a natural generalization of the above mentioned theorem of Vigier and Nagy. In the process, we arrive at a characterization of n-monotone linear operators by their numerical ranges (Theorem 3) which shows that even for linear mappings on a space of two real dimensions, the classes of n-monotone operators are all distinct. This settles a question of Rockafellar [**3**, p. 500].

1. **The convergence theorem.** Let X be a Banach space and let D be a subset of the dual Banach space X^*, such that D contains 0 in X^* and is weak* dense in some norm neighborhood of 0. We will consider the set X^D of all functions from D to X. On X^D, Rockafellar [**3**] has defined a set of relations $f \geq_n 0$ by the condition that

$$\langle f(x_1), x_1 - x_n \rangle + \sum_{k=2}^{n} \langle f(x_k), x_k - x_{k-1} \rangle \geq 0 \tag{1}$$

be valid for all n-tuples $(x_1, \ldots, x_n)$ in D^n. The relation $f \geq_2 0$ is the usual "monotonicity" relation. If one defines "$f \geq_c 0$" to mean "$f \geq_n 0$ for all n" then Rockafellar [**3**, Theorem 1] has proved that $f \geq_c 0$ if and only if f is pointwise the subgradient of some convex function defined on conv D. Generalizing (1), we say, given an $n \times n$ matrix $A = [a_{ik}]$, that $f \geq_A 0$ if and only if

$$\sum_{i,k=1}^{n} a_{ik} \langle f(x_i), x_k \rangle \geq 0 \tag{2}$$

holds for all n-tuples $(x_1, \ldots, x_n)$ in D^n. All these relations define orderings on

X^D, compatible with its vector space structure, in the usual way. Indeed, the "positive cones" of these orderings are defined by

$$P_A = \{f \in X^D : f \geq_A 0\}.$$

We will now put one restriction on the matrices A that we consider, namely we assume that all subgradient functions are in P_A, i.e. that $f \geq_c 0$ implies $f \geq_A 0$. By definition, the matrices A that correspond to the relations $f \geq_n 0$ defined by (1), in other words $A = I - P$ where P is a permutation matrix whose longest cycle has length n, all have this property. The class of all matrices A of this kind can be given an equivalent characterization.

PROPOSITION 1. *The relation $f \geq_c 0$ implies $f \geq_A 0$ if and only if A is (up to a positive scalar factor) of the form $A = I - S$ where I is the identity matrix and S is a doubly stochastic matrix, i.e. a matrix with positive elements whose rows and columns all sum to* 1.

DEFINITION 1. *A matrix A satisfying the conditions of Proposition* 1 *is called a monotonicity matrix and $f \geq_A 0$ is called "f is A-monotone".*

PROOF OF PROPOSITION 1. Let A be an $n \times n$ matrix such that $f \geq_A 0$ whenever $f \geq_c 0$. Consider the function f defined by

$$f(y) = g(\langle x, y\rangle)x \quad \text{for all } y \text{ in } D$$

where x is an element of X and g is a monotone, nondecreasing function of one real variable. It is clear that f is a subgradient function of the convex function $F(y) = G(\langle x, y\rangle)$, where G is an indefinite integral of g. Introducing this f into (2) we find that

$$\sum_{i,k=1}^{n} a_{ik} g(t_i) t_k \geq 0$$

must hold for all $\{t_k\}$ contained in some dense subset of a neighborhood of zero in R, and by taking limits and making elementary homotheties we may as well assume that this subset contains the interval $[-1, 1]$, with g finite valued and nondecreasing on it.

Now let g be such that $g(0) = 0$ and $g(1) = 1$ and let $t_k = 1$, $t_j = 0$ for $j \neq k$. It follows that all diagonal elements of A are nonnegative. Next, choose a g such that $g(x) = 0$ for $-1 \leq x < 0$, but $g(0) = 1$, and put $t_i = 0$, $t_k < 0$ for $k \neq i$; moreover, for some fixed $j \neq i$ choose $t_j = -1$ and let the other t_k be close to zero. It follows that the element a_{ij} is nonpositive, i.e. all nondiagonal elements of A are nonpositive. Finally, by choosing g to assume constant value $+1$ and constant value -1 and letting the t_k assume values 0 and 1 in each case, we find that the columns of A sum to zero, whereas using the function $g(-1) = -1$, $g(x) = 0$ for $-1 < x < 1$, and $g(1) = 1$ and choosing just one of the t_i's to be 1 and the others to be close, we find that the rows have nonnegative sums, and manipulating similarly near the discontinuity at -1 we find that these sums are also nonpositive. Hence all rows as well as all columns of A sum to zero. Let λ be larger than any of the diagonal elements of A. Then it follows from what we have found that $\lambda I - A = \lambda S$ where S is a doubly stochastic matrix, for the rows and columns of the matrix $\lambda I - A$ (which by construction has all its elements nonnegative) all sum to $\lambda > 0$.

Conversely, if $A = I - S$, then by Birkhoff [**2**, Theorem 1], A is a convex combination of matrices of the type $I - P$, and it follows from the definition that $f \geq_c 0$ implies $f \geq_A 0$. Thus Proposition 1 is proved.

COROLLARY. *If A is a monotonicity matrix, then $f \geq_A 0$ implies $f \geq_2 0$ (except in the trivial case $A = 0$, which we exclude now and henceforth).*

PROOF. This is immediate in case A is a 2×2 matrix, and the general case is reduced to this by taking for some k such that $a_{kk} > 0$, $x_k = x$ and $x_j = y$, for $j \neq k$ in (2).

To summarize: the object of our study are such orderings, defined on X^D by matrices in the sense of (2), that are finer than $\geq_2$ but coarser than $\geq_c$, and the corresponding matrices are characterised in Proposition 1. We must also define a subclass of these matrices.

DEFINITION 2. *A monotonicity matrix A is called sectorial if for some pair of indices i and k, $a_{ik} = 0$ but $a_{ki} \neq 0$.*

The following is the monotone convergence theorem mentioned in the title of this paper.

THEOREM 1. *Let A be a sectorial monotonicity matrix, $\{f_\alpha\} \subset X^D$ a net with $f_\alpha(0) = 0$ for all α, g a function in X^D which is norm to norm continuous at 0 in D and satisfies $g(0) = 0$, and suppose that*

$$0 \leq_A f_\alpha \leq_A f_\beta \leq_2 g \quad \textit{whenever } \alpha \leq \beta.$$

Then there exists a function f in X^D such that, in the norm on X,

$$\lim f_\alpha(x) = f(x) \quad \textit{for all } x \textit{ in } D.$$

REMARK. Obviously, the function f whose existence is asserted in Theorem 1 satisfies $f(0) = 0$ and $f_\alpha \leq_A f \leq_2 g$ for all α. Also, the theorem remains valid with $\leq_c$ instead of $\leq_A$.

PROOF of THEOREM 1. Using the notation $f_{\beta\alpha} = f_\beta - f_\alpha$ for $\alpha \leq \beta$ we have with $x_k = x$, $x_i = y$, and $x_j = 0$ for $j \notin \{i, k\}$.

$$a_{kk}\langle f_{\beta\alpha}(x), x\rangle + a_{ki}\langle f_{\beta\alpha}(x), y\rangle + a_{ii}\langle f_{\beta\alpha}(y), y\rangle \geq 0 \tag{3}$$

for all x, y in D. Furthermore, because $0 \leq_2 f_\alpha \leq_2 f_\beta \leq_2 g$ if $\alpha \leq \beta$,

$$0 \leq \langle f_\alpha(y), y\rangle \leq \langle f_\beta(y), y\rangle \quad \text{for all } y \text{ in } D. \tag{4}$$

Let $\varepsilon > 0$ be given. Since g is continuous at 0 there is a $\delta > 0$ such that

$$\langle g(y), y\rangle \leq \varepsilon\delta \quad \text{if } \|y\| \leq \delta. \tag{5}$$

Suppose that x in a fixed element of D. By (4), it is possible to find an index γ such that

$$0 \leq \langle f_{\beta\alpha}(x), x\rangle \leq \varepsilon\delta \quad \text{provided } \gamma \leq \alpha \leq \beta.$$

Combining this with (3), (4), and (5) one obtains

$$-a_{ki}\langle f_{\beta\alpha}(x), y\rangle \leq (a_{kk} + a_{ii})\varepsilon\delta \quad \text{if } \|y\| \leq \delta \text{ and } \gamma \leq \alpha \leq \beta.$$

Assuming as we may that D is dense in $\{y\colon \|y\| \leq \delta\}$ we conclude that

$$\|f_{\beta\alpha}(x)\| \leq \varepsilon(a_{kk} + a_{ii})/|a_{ki}|.$$

Therefore $\{f_\alpha(x)\}$ is a Cauchy net, and Theorem 1 follows.

As an application of Theorem 1 we prove the following theorem.

THEOREM 2. *Suppose that f is a monotone operator which is defined on some set in X^* whose weak* closure has a norm interior, and is norm to norm continuous at one point of this interior. Then there exists an operator g such that $0 \leq_c g \leq_2 f$ and such that if g_1 satisfies $0 \leq_c g_1 \leq_2 f - g$, then g_1 reduces to a constant.*

REMARK. It is well known that a linear monotone operator is precisely one with positive symmetric part, in which case the decomposition $f = g + (f - g)$ is just the elementary decomposition into symmetric and antisymmetric parts.

PROOF OF THEOREM 2. By translation and subtraction of a constant, we may assume that 0 is the indicated point of continuity for f, and that $f(0) = 0$, i.e. the domain D of f contains 0 and is weak* dense in some norm neighborhood of 0. Consider the family

$$G = \{g \in X^D\colon 0 \leq_c g \leq_2 f, g(0) = 0\}.$$

Let $\{g_\alpha\} \subset G$ be a linearly $\leq_c$-ordered net. By Theorem 1 this net has a $\leq_c$-majorant in G. It follows from Zorn's lemma that G contains a maximal element g which in view of the normalizations made satisfies the condition of Theorem 2.

One can under more restrictive conditions on X and D replace the continuity requirement on f by a boundedness requirement.

THEOREM 2′. *If f is a monotone operator defined and bounded on an open subset of a reflexive Banach space, then the same decomposition statement as in Theorem 2 holds.*

This theorem follows (in the same way as Theorem 2 follows from Theorem 1) from

THEOREM 1′. *If X is reflexive, D is a norm neighborhood of O in X^*, g is bounded on D (but not a priori continuous anywhere), and the other conditions are as in Theorem 1, then the same conclusion holds with weak instead of norm convergence.*

PROOF IN OUTLINE. One may assume $D = \{y\colon \|y\| \leq \delta\}$ and that $\varepsilon > 0$ is the bound of $\|g(y)\|$ on D. The proof then follows that of Theorem 1 except that one has to regard y as fixed, concluding for each y that

$$\lim \langle f_\alpha(x), y\rangle = \langle f(x), y\rangle \quad \text{with } \|f(x)\| \le \varepsilon a_{ii} / |a_{ik}|.$$

2. **Linear A-monotone operators.** In the important special case of linear operators one can characterize $(I - P)$-monotonicity (where P stands for a permutation matrix) by means of the numerical range of the operator or rather its complexification. Let T be a linear operator with values in X, defined on D, which we now assume is a (weak* dense) subspace of X^*. We also introduce the complexification D_c of D

$$D_c = \{z = x + iy \mid x, y \in D,\ \|z\| = (\|x\|^2 + \|y\|^2)^{1/2}\}$$

and, correspondingly, the complexification X_c of X. The complexification T_c of T is a linear operator from D_c to X_c defined by

$$T_c(x + iy) = T(x) + iT(y) \quad \text{for } x, y \text{ in } D.$$

Moreover, we assume that X_c and D_c are paired by the "sesquilinear" form

$$(x + iy \mid u + iv) = \langle x, u\rangle + \langle y, v\rangle + i(\langle y, u\rangle - \langle x, v\rangle)$$

for x, y in X and u, v in D. Finally, the numerical range of T_0 is as usual defined and denoted by

$$W(T_c) = \{(T_c z \mid z) \mid z \in D_c,\ \|z\| = 1\}.$$

We state now the characterization alluded to above.

THEOREM 3. *The linear operator T is $(I - P)$-monotone if and only if $W(T_c) \subset \{w \in C \mid |\arg w| \le \pi/n\}$, where n is the length of the longest cycle in P.*

PROOF. We first observe that if $A = I - P$ then $T \ge_A 0$ means precisely $T \ge_n 0$, for the latter relation implies $T \ge_m 0$ for all $m < n$ (this observation is true also for nonlinear operators). Also, we may assume $n \ge 3$, for $n = 1$ gives the trivial case $A = 0$ which we have excluded, and the case $n = 2$ follows by immediate use of the definition.

Assume then that $T \ge_n 0$. We will write this relation as

$$\sum_r \langle T(x_r), x_r - x_{r-1}\rangle \ge 0 \tag{6}$$

where the index r may be thought of as describing $\mathbf{Z}/n\mathbf{Z}$, the cyclic group with n elements. Let $z = x + iy$ be an arbitrary element of D_c and let

$$\omega = \cos 2\pi/n + i \sin 2\pi/n$$

be the "standard" nth root of unity. Consider the set $\{x_r\}$ of n elements in D defined by

$$x_r = \operatorname{Re} \omega^r z = \tfrac{1}{2}(\omega^r z + \omega^{-r}\bar{z})$$

and substitute this into (6).

$$\begin{aligned} 0 &\le \sum_r \langle T(x_r), x_r - x_{r-1}\rangle \\ &= \frac{1}{4}\sum_r (T_c(\omega^r z + \omega^{-r}\bar{z}) \mid (\omega^r - \omega^{r-1})z + (\omega^{-r} - \omega^{-r+1})\bar{z}) \\ &= \frac{n}{2}\operatorname{Re}(1-\omega)(T_c z \mid z). \end{aligned}$$

Replacing ω by $\bar{\omega} = \omega^{-1}$ in the definition of $\{x_r\}$ yields another inequality which combined with the one above implies that

$$W(T_c) \subset S_{\pi/n} = \{w \mid |\arg w| \le \pi/n\} \tag{7}$$

in view of the fact that

$$\arg(1-\omega) = (-\tfrac{1}{2} + 1/n)\pi = -\arg(1-\bar{\omega}).$$

Conversely, suppose that (7) holds and that $\{x_r\}$ is a set of n elements of D. The set $\{\alpha_s\}$, $s \in \mathbf{Z}/n\mathbf{Z}$ defined by

$$\alpha_s = \left(T_c\left(\sum_r \omega^{-rs}x_r\right) \middle| \sum_\beta \omega^{-ps}x_p\right) = \sum_r\sum_p \omega^{(p-r)s}\langle T(x_r), x_p\rangle$$

is obviously contained in $S_{\pi/n}$ hence

$$\operatorname{Re}\sum_{s=0}^{n-1}(1-\omega^s)\alpha_s \ge 0. \tag{8}$$

Indeed, each term in the above sum lies in the right half plane of the complex plane. But

$$\sum_s \omega^s\alpha_s = \sum_r\sum_p\sum_s \omega^{(p-r+1)s}\langle T(x_r), x_p\rangle = n\sum_r \langle T(x_r), x_{r-1}\rangle$$

so that, computing $\sum \alpha_s$ in the same way, we get

$$\sum_s (1-\omega^s)\alpha_s = n\sum_r \langle T(x_r), x_r - x_{r-1}\rangle.$$

In conjunction with (8), this proves that $T \ge_n 0$.

Having now proved Theorem 3 we remark that for $n \ge 3$ the matrix $A = I - P$ is indeed sectorial according to Definition 2, and the choice of the word sectorial is motivated by the appearance of a proper subsector of the right complex half plane, containing the set $W(T_c)$. Indeed, in the linear case one may in Theorem 1 replace the relations $\le_A$, with A sectorial, by $\le_\varphi$, which for a fixed φ, $0 \le \varphi < \pi/2$, has the following meaning: $T \ge_\varphi 0$ if and only if

$$W(T_c) \subset S_\varphi = \{w \mid |\arg w| \le \varphi\}. \tag{9}$$

To see this, we note that (9) means

$$\sin \varphi(\langle T(x), x\rangle + \langle T(y), y\rangle) + \cos \varphi(\langle T(x), y\rangle - \langle T(y), x\rangle) \geq 0$$

for all x, y in D, i.e. $T \geq_\varphi 0$ is the same as $T \geq_A 0$ with the matrix A given by

$$A = \begin{bmatrix} \sin \varphi & \cos \varphi \\ -\cos \varphi & \sin \varphi \end{bmatrix}.$$

This is not a sectorial matrix nor even a monotonicity matrix. However, since we now deal exclusively with linear mappings we have the fact that any $n \times n$ monotonicity matrix is equivalent with any one of its principal $(n-1) \times (n-1)$ minors as seen by replacing x_k by $x_k - x_n$ for $k = 1, 2, \ldots, n-1$. Also by substitution it is clear that A is equivalent (with respect to ordering) to any matrix of the type BAB', where B is nonsingular and B' is its transpose. Thus, if we can choose B so that one of the nondiagonal elements of BAB' is zero and the other is negative and of absolute value not greater than either of the two diagonal elements (of BAB'), then we can imbed BAB' as a 2×2 principal minor in a 3×3 sectorial matrix, and this is possible if $\pi/3 \leq \varphi < \pi/2$; a suitable choice for B is

$$B = \begin{bmatrix} 2 & 0 \\ -1 & \tan \varphi \end{bmatrix}.$$

Since the case $0 \leq \varphi \leq \pi/3$ is covered by Theorem 3, we could now announce as a corollary to Theorem 1 a monotone convergence theorem for linear mappings from the real normed linear space D to the real Banach space X. However, the use of numerical ranges makes it more natural to consider *complex* linear mappings from a complex normed linear space D into complex Banach space X, where D is a weak* dense subspace of the complex dual X^*. As usual we assume then that X and D are paired by a sesquilinear form $(x \mid y)$ for all x in X and y in D, and this is used to define the numerical range of the operator T:

$$W(T) = \{(T(z) \mid z) \mid z \in D, \|z\| = 1\}.$$

DEFINITION 3. *If D and X are as above and T is a linear mapping from D to X, then we denote*

$$W(T) \subset S_\varphi = \{w \mid |\arg w| \leq \varphi\}$$

by $T \geq_\varphi 0$

$$(0 \leq \varphi \leq \pi/2).$$

THEOREM 4. *If $\{T_n\}$ is a sequence of (complex) linear mappings from D to X, and R is a continuous linear mapping from D to X such that*

$$0 \leq_\varphi T_n \leq_\varphi T_{n+1} \leq_{\pi/2} R \quad \text{for all } n$$

with some fixed $\varphi < \pi/2$, then there exists a linear mapping T from D to X such that $T_n(z) \to T(z)$ in norm for all z in D.

REMARK. The limiting operator T clearly satisfies $0 \leq_\varphi T \leq_{\pi/2} R$. Also, though this fact is not explicitly used, all the operators T_n and T are actually bounded. Indeed, $(1 + \tan \varphi) \|R\|$ is a common bound for all of the norms.

The proof of Theorem 4 will fill up the rest of this paper. As we have already remarked, the proof is already complete in the case when T is the "complexification" of an originally real linear operator. The general case will be reduced to this by first regarding the given complex linear operator T as a real linear operator between the real linear spaces "underlying" D and X, and then "recomplexifying". Since the underlying spaces are identical as sets with the original ones, we will not use any special notation for them, nor for the "real" operator T. The difference between real and complex space comes out in the real inner product, defined from the sesquilinear complex pairing between D and X by

$$\langle x, y \rangle = \operatorname{Re}\,(x \mid y) \quad \text{for all } x \text{ in } X \text{ and } y \text{ in } D.$$

Conversely, the sesquilinear form can be regained by the formula

$$(x \mid y) = \langle x, y \rangle + i\langle x, iy \rangle.$$

Here i acting on the real vectors in X (or D) has to be regarded as a (real) linear transformation from X into X (D into D) satisfying $i^2 = -I$ and

$$\langle ix, y \rangle = -\langle x, iy \rangle \quad \text{for all } x \text{ in } X,\ y \text{ in } D.$$

With these conventions we may express the numerical range of the complex transformation T by

$$W(T) = \{\langle T(z), z \rangle + i\langle T(z), iz \rangle \mid z \in D,\ \|z\| = 1\},$$

whereas the numerical range of the "recomplexified" operator T_c is given by

$$W(T_c)$$
$$= \{\langle T(x), x \rangle + \langle T(y), y \rangle + i[\langle T(y), x \rangle - \langle T(x), y \rangle] \mid x, y \in D,\ \|x\|^2 + \|y\|^2 = 1\}.$$

Since the conclusion of Theorem 4 for the respective recomplexified operators obviously implies the same for the original ones, if suffices to show that the order relationships on these imply the same on the original ones. In other words, what we have to prove is precisely the following lemma:

LEMMA. *$W(T) \subset S_\varphi$ implies $W(T_c) \subset S_\varphi$ for $\varphi \leq \pi/2$.*

PROOF. The case $\varphi = \pi/2$ is obvious, so we may assume $\varphi < \pi/2$. Since we are taking an arbitrary point in $W(T_c)$ which is determined by two elements x and y in D, we may also regard D as having two complex dimensions, in fact as the real linear span of x, y, ix, and iy. Moreover, we may assume that the maximum of the expression

$$(10) \qquad (\langle T(y), x \rangle - \langle T(x), y \rangle)/(\langle T(x), x \rangle + \langle T(y), y \rangle), \quad \|x\|^2 + \|y\|^2 = 1$$

which we denote by M, is attained just for x and y. For if $\langle T(z), z\rangle = 0$ for some z on the unit sphere $\|z\| = 1$—which we may as well assume to be Hilbertian—then since $W(T) \subset S_\varphi$ and $\varphi < \pi/2$ we can use the well-known fact that in a two-dimensional space the numerical range is an elliptical disc, which in this case must degenerate to a line segment with one end point at 0, to indicate that $T(z) = T^*(z) = 0$ which effectively reduces the problem to the trivial one-dimensional case.

We will finish the proof of the lemma by showing, with a contradiction argument, that the vectors x and y for which the maximum in (10) is attained, are actually linearly dependent, thus reducing the problem again to the one-dimensional case. Suppose therefore that x, y is a complex basis for D and let u, v be a dual basis, i.e.

$$(u \mid x) = (v \mid y) = 1, \quad (u \mid y) = (v \mid x) = 0.$$

An elementary differentiation argument yields the following relations (note that we may dispense with the condition $\|x\|^2 + \|y\|^2 = 1$ in (10), by homogeneity).

$$\langle T(y), z\rangle - \langle T(z), y\rangle = M(\langle T(x), z\rangle + \langle T(z), x\rangle)$$
$$\langle T(z), x\rangle - \langle T(x), z\rangle = M(\langle T(y), z\rangle + \langle T(z), y\rangle),$$

valid for all z in D. Choosing successively z equal to x, y, ix, and iy in these two equations one gets

$$\langle T(y), x\rangle = -\langle T(x), y\rangle = M\langle T(x), x\rangle = M\langle T(y), y\rangle = AM$$
$$-M\langle T(y), ix\rangle = M\langle T(x), iy\rangle = \langle T(x), ix\rangle = \langle T(y), iy\rangle = BM,$$

where A and B denote two real quantities. Using these relations one can determine the matrix of T with respect to the basis x, y and u, v.

$$T(x) = (A + iBM)u - (AM - iB)v$$
$$T(y) = (AM - iB)u + (A + iBM)v.$$

An elementary computation (the matrix is normal; hence its numerical range is just the line segment bounded by its eigenvalues) now finishes the proof of the lemma and Theorem 4.

Bibliography

1. G. Birkhoff, *Tres observaciones sobre el algebra lineal*, Univ. Nac. Tucuman, Revista A. **5** (1946), 147–151.

2. J. Dixmier, *Les algèbres d'opérateurs dans l'espace hilbertien* (*Algèbres de von Neumann*), Cahiers scientifiques, Fascicule XXV, Gauthier-Villars, Paris, 1957, 376 pp.

3. R. T. Rockafellar, *Characterization of the subdifferentials of convex functions*, Pacific J. Math. **17** (1966), 497–510.

University of Stockholm

MULTIPLE SOLUTIONS OF NONLINEAR OPERATOR EQUATIONS ARISING FROM THE CALCULUS OF VARIATIONS

Melvyn S. Berger[1]

What can be said about the totality of real solutions of nonlinear elliptic partial differential equations defined on manifolds (or satisfying boundary conditions) using the methods of functional analysis? In this article we intend to discuss an important aspect of this problem as well as its converse, namely do the known results of nonlinear elliptic partial differential equations lead to new ideas and methods in functional analysis?

A. **Formulation of problems.** In order to put the above questions in a *general* setting, we begin by considering operator equations defined on a Hilbert space over the real numbers. (Extensions to more general reflexive Banach spaces will be mentioned where appropriate.) Let H denote a separable Hilbert space over the reals and T a locally Lipschitz continuous mapping of H into itself with $T(0) = 0$. We wish to study the real zeros of T in H without assuming the condition $(u, Tu) \to \infty$ as $\|u\| \to \infty$ (in which case the zeros of T would necessarily all be located in some bounded set in H).

We shall consider the following five general problems concerning the zeros of T:

(1) Local Problem. The structure of the zeros of T near $u \equiv 0$.

(2) Global Problem. The structure of the zeros of T without regard to norm.

(3) Continuation Problem. The relation between (1) and (2) above.

(4) Linearization Problems. What facts concerning the zeros of T can be decided by linearization?

(5) Problem of Nonlinear Effects. What qualitative features of the nonlinearity are relevant to the structure of the zeros of T?

The above problems can be reformulated if we assume T to be a gradient mapping, i.e. there exists a real-valued functional $Q(u)$ defined on H such that

$$\lim_{s \to 0} \frac{Q(u + sv) - Q(u)}{s} = (v, Tu)$$

for all u, v in H. If T is a gradient mapping, we may assume without loss of

[1] Research partially supported by NSF GP-7041X.

generality that

$$Q(u) = \int_0^1 (u, T(su))\, ds.$$

(See [**2**].) Thus the zeros of T are in one-to-one correspondence with the critical points of $Q(u)$ in H. Furthermore the questions above concerning the structure of the zeros of T can be phrased in terms of the structure of the critical points of $Q(u)$.

All the results stated in this article will pertain only to gradient mappings T. Indeed counterexamples will show that such results are false for nongradient mappings T. In terms of nonlinear elliptic partial differential equations, the restriction to gradient mappings implies that these equations are the Euler-Lagrange equations of some real-valued functional, and we say the equation $Tu = 0$ arises from the calculus of variations. The following example clarifies the last statement.

B. **A first example.** Let G be a bounded domain in the plane with boundary ∂G. We consider the following Dirichlet problem defined on G for p a nonnegative integer:

$$\text{(1)} \qquad \begin{aligned} \Delta u + u^{2p+1} &= 0 \quad \text{in } G \\ u &= 0 \quad \text{on } \partial G. \end{aligned}$$

Here Δ denotes the Laplace operator. To study the totality of the solutions of (1), we distinguish two cases:

Case I, $p = 0$ (*the linear case*). The number of solutions depends on the *size of* G and *shape of* ∂G, i.e. whether or not 1 is an eigenvalue of Δ with respect to G and the null boundary conditions. However the set of zeros always forms a finite-dimensional subspace.

Case II, $p \neq 0$ (*the nonlinear case*). In this case the following result is a *consequence of the nonlinearity* of the equation.

THEOREM 1. *If $p \neq 0$, (1) has a countably infinite number of distinct solutions $\{u_n\}$ $(n = 1, 2, \ldots)$ independent of the size of G or the shape of ∂G. The solutions are smooth in G and at all sufficiently smooth portions of ∂G.*

The proof of this result is a direct consequence of the Nonlinear Sturm-Liouville Theorem 2 discussed below. (See §E.) Furthermore in this case the linearized equation about $u \equiv 0$ is

$$\text{(1')} \qquad \begin{aligned} \Delta u &= 0 \quad \text{in } G \\ u &= 0 \quad \text{on } \partial G. \end{aligned}$$

As this equation has the unique solution $u \equiv 0$ we note that linearization in this case gives no hint as to the validity of Theorem 1.

In the Hilbert space $H = W_{1,2}(G)$, consisting of all functions $u(x)$ such that

$u(x)$ and $|\nabla u|$ are square integrable over G, and vanishing on ∂G in the generalized sense, it is easily shown that the operator equation corresponding to (1) is $Tu = 0$ where T is a gradient mapping with the associated functional

$$Q(u) = \tfrac{1}{2}D[u] - \frac{1}{2p+2}\int_G u^{2p+2},$$

where $D[u] = \int_G |\nabla u|^2 = \|u\|_H^2$.

C. **A basic idea.** A program for studying the problems mentioned in §A can be achieved by focussing attention on nonlinear invariants (of a topological nature) for the operator T which possess the following properties:

(a) The invariant is a qualitative measure of the zeros of T.

(b) The invariant is stable under restricted small perturbations.

(c) The invariant can be calculated locally by linearization.

What are possible nonlinear invariants that satisfy (a)–(c)? First there is the Leray-Schauder degree of T, see [**11**], defined irrespective of the fact that T is a gradient mapping. But, as we shall see, this invariant is not sufficiently delicate to prove results on multiple solutions of equations such as Theorem 1 above. A second possibility is the Morse Theory of nondegenerate critical points of functionals in a Hilbert space (cf. [**16**], [**19**]). The difficulty with the Morse-type invariants is that they have been found useful, to date, only under the assumption that all critical points of a given variational problem are nondegenerate. For problems in elliptic partial differential equations (i.e. multiple integral problems in the calculus of variations) such an assumption cannot be made. Of course a given problem often may be approximated by problems which possess nondegenerate critical points, but this fact is usually not sufficient for global problems in analysis [cf. for example the results of M. Morse on closed geodesics on a manifold $\mathcal{M}$ homeomorphic to a sphere, where the so-called Morse metric restriction must be imposed on $\mathcal{M}$ to insure that closed geodesics are "distinct" (see [**14**])].

In this article we suggest another invariant:

Critical values calculated by Minimax principles

where the appropriate Minimax principle is calculated over classes of subsets of a manifold $\mathcal{M}$ in H. Subsequently the classes of subsets of $\mathcal{M}$ will be calculated by the Ljusternik-Schnirelmann category (cf. [**16**], [**18**]). A simple finite-dimensional example of this idea can be seen from the following statement. Let $F(x)$ be a C' real-valued function defined on $\boldsymbol{R}^n$ with $F(x) \to \infty$ as $|x| \to \infty$. If $F(x)$ has two distinct isolated relative minima at $x = x_0$ and $x = x_1$, then $F(x)$ must have another distinct critical point at $x = x_2$ and

$$F(x_2) = \min_{[C]} \max_{C} F(x)$$

where C is a compact connected set in $\boldsymbol{R}^n$ joining x_0 and x_1, and $[C]$ is the class of all such sets. (Note that none of the critical points in this example is assumed to be nondegenerate.) See [**26**, p. 59].

D. Some difficulties and their resolution.

1. The functional

$$Q(u) = \int_0^1 (u, T(su))\, ds$$

may not be bounded from above or from below in H. Then, for example, $\min_H Q(u)$ or $\max_H Q(u)$ would not define a critical value of $Q(u)$. In particular, in the example of §B, denote by $D[u]$ the integral of $|\text{grad } u|^2$ over G, then

$$Q(u) = \tfrac{1}{2} D[u] - \frac{1}{2p+2} \int_G u^{2p+2}$$

is neither bounded from above nor below in $\mathring{W}_{1,2}(G)$. To surmount this difficulty we write $T = A - B$ where B is completely continuous in H (i.e. B maps weakly convergent sequences into strongly convergent sequences) and $(Au, u) \to \infty$ as $\|u\| \to \infty$, and we try to find critical points of $Q(u)$ on bounded subsets of H.

2. The critical points of $Q(u)$ may be degenerate. Then as we wish to study the number of *distinct* critical points, we focus attention on the critical point theory of Ljusternik-Schnirelmann instead of that of Morse.

3. From the fundamental studies of S. Bernstein [**6**] the existence and multiplicity of critical points of multiple integrals defined over a bounded domain G in $\boldsymbol{R}^n$ with boundary ∂G of the form

$$\int_G F(x, u, Du, \ldots, D^m u)$$

subject to null Dirichlet boundary conditions, where F is a real-valued C^1 function, are known to depend crucially on

(i) the size of G,

(ii) the shape of ∂G,

(iii) the growth of the integrand F with respect to its arguments.

We wish to state results independent of the shape of ∂G, and this requires us, by virtue of the results of [**6**], to restrict the growth of the integrand F. In particular, we shall assume that the leading part of the associated Euler-Lagrange equation satisfies certain coerciveness conditions and is strongly elliptic. If $T = A - B$, coerciveness means $(Au, u) \to \infty$ as $\|u\| \to \infty$.

To study the influence of the size of G on the solutions, we introduce a positive parameter into the problem by setting $x = \lambda^{1/2} x'$ (i.e. make a change of scale) and require, for example, that the diameter of G' be unity. Then the equation $T(u) = 0$ is transformed into $\tilde{T}(u, \lambda) = 0$ and we study the real solutions of this equation *as a function of* λ. In the example of §B such a transformation gives the following boundary value problem:

$$(1'') \qquad \begin{aligned} \Delta u + \lambda u^{2p+1} &= 0 \quad \text{in } G' \\ u &= 0 \quad \text{on } \partial G'. \end{aligned}$$

We shall assume (as is the case in (1″)) that $\tilde{T}(u, \lambda)$ depends *linearly on* λ. Furthermore, in §F, we show that this is an important case in applications. Thus we shall study the solutions of the operator equation $\tilde{T}(u, \lambda) = Au - \lambda Bu = 0$ as a function of λ.

4. The equation $\tilde{T}(u, \lambda) = 0$ always has the trivial solution $u \equiv 0$. In order to insure that the solutions of $\tilde{T}(u, \lambda) = 0$ are nontrivial (i.e. $u \not\equiv 0$), we note that if $\tilde{T}(u, \lambda) = Au - \lambda Bu$, and $Au = 0$ implies $u = 0$, then for a large class of pairs (A, B) the critical points of

$$b(u) = \int_0^1 (u, B(su))\, ds$$

on the hypersurface

$$\partial A_R = \left\{ u \mid a(u) = \int_0^1 (u, A(su))\, ds = R \right\}$$

as R varies from $-\infty$ to $+\infty$ are in (1-1) correspondence with the zeros of $\tilde{T}(u, \lambda) = 0$ in H. Hence if $R \neq 0$, the critical points on ∂A_R (if such exist) correspond to nontrivial solutions of $\tilde{T}(u, \lambda) = 0$.

E. **Statement of results for the equation** $Au = \lambda Bu$ **in** H. We shall state results here: First a global result (a Sturm-Liouville Theorem) describing solutions without regard to their norm, secondly a local result (a Bifurcation Theorem) describing the behavior of solutions near $u \equiv 0$ with respect to the parameter λ, and thirdly a Continuation Theorem describing the connection between the local and global results. The fourth result describes the effect of a nonlinear perturbation on a multiple eigenvalue λ_n. A counterexample at the end of the section then shows that *all* of the results may be false if one of the operators A or B is not a gradient mapping.

In Theorem 2 below (the global existence theorem) we shall assume

(i) $\dim H = \infty$ (as the finite dimensional case is well known);

(ii) the functionals $a(u)$, $b(u)$ corresponding to the operators A and B respectively are invariant under the antipodal mapping $J: u \to -u$, or equivalently, A and B are odd mappings; and

(iii) the set ∂A_R is homeomorphic to the sphere $\partial S_R = \{u \mid \frac{1}{2}\|u\| = R\}$ by rays through the origin. Then identifying the antipodal points on ∂A_R we obtain a set $\partial \tilde{A}_R = \partial A_R / J$ homeomorphic to the infinite-dimensional projective space $P^\infty(H)$ over H, along rays through the origin. If we denote the Ljusternik-Schnirelmann category of a subset V of X relative to X by cat (V, X), then on $\partial \tilde{A}_R$ we can consider the classes of sets

$$V_{n,R} = \{V \mid \operatorname{cat}(V, \partial \tilde{A}_R) \geq n\}.$$

Note that (i) $V_{1,R} \supset V_{2,R} \supset V_{3,R} \supset \cdots \supset V_{n,R} \supset \cdots$ is a strictly decreasing sequence, (ii) the class $V_{n,R}$ is invariant under continuous deformations.

(iv) A and B are locally Lipschitzian.

THEOREM 2 (A STURM-LIOUVILLE THEOREM) [**2**], [**7**]. *Suppose* dim $H = \infty$ *and the operators A and B are odd, gradient mappings of H into itself. Then for a large class of pairs (A, B) (see below) and every fixed $R > 0$ the equation $Au = \lambda Bu$ has a countably infinite number of distinct solutions $(u_n(R), \lambda_n(R))$ where $u_n(R) \in \partial A_R$ and $\lambda_n(R) \to \infty$ as $n \to \infty$. Furthermore $u_n(R)$ is characterized as a critical point of the variational problem:*

$$c_n(R) = \sup_{V_{n,R}} \min_{V} b(u).$$

The simplest hypotheses on A, B to insure the validity of Theorem 2 are (i)–(iv) above and

For A. (1) Coerciveness. $a(u) \to \infty$ as $\|u\| \to \infty$. This implies ∂A_R is a bounded set.

(2) $(u, Au) > 0$ for $u \neq 0$. This implies ∂A_R is starlike, for each $R > 0$.

(3) If $u_n \to u$ weakly, and $Au_n \to Au$ strongly, then $u_n \to u$ strongly. This condition implies the Palais-Smale condition C [**17**] together with (1), (2) below.

For B. (1) B is completely continuous in the sense that B maps weakly convergent sequences in H into strongly convergent sequences.

(2) $(u, Bu) > 0$ for $u \neq 0$.

Theorem 2 can be extended to hold in uniformly convex Banach spaces as well as under considerably more general hypotheses on the operator A [**2**], [**7**].

The next two results pertain to equations of the form

$$u + Mu = \lambda(Lu + Nu)$$

where M, L, N are completely continuous gradient mappings with L linear and positive and M, N, higher order mappings in the sense that

$$\|Nu - Nn\| \leq k(\|u\|, \|v\|)\, \|u - v\|$$

with $k(s, t) = o(|s| + |t|)$. We set $B = L + N$.

THEOREM 3 (A BIFURCATION THEOREM). *Let λ_n be an eigenvalue of $u = \lambda Lu$. Then for sufficiently small R (depending on n) if*

$$\partial A_R = \left\{ u \mid \tfrac{1}{2} \|u\|^2 + \int_0^1 (u, M(su))\, ds = R \right\}$$

is sphere-like, there is a one parameter family of solutions $(u_n(R), \lambda_n(R))$ of $u + Mu = \lambda(Lu + Nu)$ such that as $R \to 0$, $(u_n(R), \lambda_n(R)) \to (0, \lambda_n)$. Conversely, the numbers λ_n are the only real numbers at which such a local result holds.

Here sphere-like with respect to the family ∂A_R means that for all R, $0 < R < R_0$ (R_0 sufficiently small), the hypersurfaces ∂A_R form a nested family of nonintersecting starlike surfaces converging uniformly to the origin. For example, ∂A_R is starlike if $(Mu, u) \geq 0$.

The following graph of the norm of solutions u of $u + Mu = \lambda(Lu + Nu)$ versus λ summarizes the contents of the above result, Theorem 3.

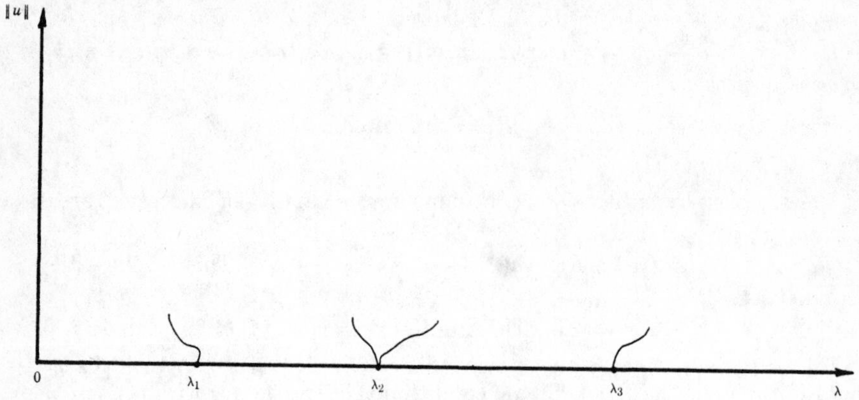

FIGURE 1. Illustrating the Bifurcation Theorem 3.

If $M \equiv 0$, Theorem 3 was proved by Krasnoselskii [**10**].

THEOREM 4 (A CONTINUATION THEOREM). *If the operators M and N are odd mappings with ∂A_R sphere-like and $(u, Bu) > 0$ for $0 < R \leq R_0$, then the one parameter families of solutions $(u_n(R), \lambda_n(R))$ mentioned in Theorem 2 have the property that $(u_n(R), \lambda_n(R)) \to (0, \lambda_n)$ as $R \to 0$ where λ_n is the nth eigenvalue of the linearized equation $u = \lambda Lu$ (the eigenvalues are ordered by magnitude and counted according to multiplicity).*

COROLLARY 5. *In addition to the hypotheses of Theorem 4 suppose λ_n is an eigenvalue of multiplicity p for $u = \lambda Lu$, then there are at least p distinct nontrivial one parameter families of solutions $(u_i(R), \lambda_i(R))$ $(i = 1, 2, \ldots, p)$ of $u + Mu = \lambda(Lu + Nu)$ bifurcating from $(0, \lambda_n)$.*

We end this section with the following counterexample showing that if A or B is not a gradient mapping all the results of this section may be false.

Consider the system defined on a bounded domain G in the plane with boundary ∂G

(a) $$\Delta u + \lambda(u + v^3) = 0 \quad \text{in } G,$$

(b) $$\Delta v + \lambda(v - u^3) = 0 \quad \text{in } G,$$

$$u = v = 0 \quad \text{on } \partial G.$$

We assert that this system has only the trivial solution. Indeed multiplying (a) by v, (b) by u, and subtracting we obtain by means of integration by parts

$$\int_G (v^4 + u^4) = 0$$

i.e. $u \equiv v \equiv 0$. On the other hand in the Hilbert space $H = \dot{W}_{1,2}(G) \times \dot{W}_{1,2}(G)$ the system (a), (b) can be written

$$w = \lambda(Lw + Nw)$$

where L and N are completely continuous mappings of H into H with L linear and N nonlinear. Note however that the operator N is not a gradient mapping because the vector (v^3, u^3) is not the gradient of any real-valued function $f(u, v)$.

F. **Applications.** Before indicating the proofs of the results of §E we wish to indicate some applications of these results to various specific problems in mathematics and mathematical physics.

We begin by considering a "one-dimensional" problem.

(i) *Periodic solutions of autonomous Hamiltonian systems.* Consider the system of ordinary differential equations

$$\ddot{x} + Lx + f(x) = 0 \tag{2}$$

where differentiation is taken with respect to t, x is an n-vector, L is a positive definite selfadjoint constant matrix with eigenvalues $a_1^2, a_2^2, \ldots, a_n^2$ and $f(x)$ is a locally Lipschitz continuous vector-valued function with

$$|f(x)| = o(|x|) \quad \text{and} \quad f(x) = \text{grad } F(x)$$

where $F(x)$ is a real-valued analytic function.

A difficulty with studying periodic solutions of (2) is that the periods of solutions are unknown a priori. However the following result due essentially to Liapunov is valid.

(THEOREM L.) *If $a_i/a_j \neq$ integer for $i \neq j$ $(i, j = 1, \ldots, n)$, then the system* (2) *has n distinct one parameter families of periodic solutions $x_i(R)$ $(i = 1, \ldots, n)$ for sufficiently small R and the period of the ith family $T_i(R) \to 2\pi/a_i$ as $R \to 0$.*

A basic problem with this result is to determine circumstances under which (i) the "irrationality" hypothesis $a_i/a_j \neq$ integer can be removed and (ii) the one parameter families $x_i(R)$ exist for large R. This can be accomplished by virtue of Corollary 5. Indeed by making a change of variables $t = \lambda s$ in (2) we obtain the system

$$\ddot{x} + \lambda^2[Ax + f(x)] = 0$$

where differentiation is now taken with respect to s. Furthermore the linearized problem (about $x \equiv 0$) over the class of 2π periodic vector functions x has eigenvalues N^2/a_i^2 $(N = 1, 2, \ldots$ and $i = 1, \ldots, n)$. Thus $a_i/a_j = N$ means that the eigenvalue $1/a_j^2$ is of multiplicity greater than one for the linearized system, so we are concerned with describing nonlinear perturbations of degenerate linear eigenvalue problems. Indeed by applying Theorem 4 and Corollary 5 in this situation we obtain the following

THEOREM 6. *If $F(x)$ is an even function of x, the conclusions of Theorem* L *hold. Furthermore, if $x - f(x) \geq 0$ for all x, then the n one parameter families $x_i(R)$ of periodic solutions exist for all positive R.*

A proof of this result will be found in [**22**]. In this case the appropriate Hilbert space H is the direct product of n copies of the Hilbert space of odd 2π periodic functions which possess a square integrable derivative.

A difficult geometrical problem related to the study of periodic solutions of Hamiltonian systems is the study of closed geodesics on a manifold $\mathscr{M}$ homeomorphic to a sphere. Indeed the results of [**12**], [**13**] on this problem were among the first important applications of the general minimax principles noted above.

Let us turn now to the two-dimensional problem stated in Theorem 1 (which is an application of Theorem 2).

(ii) *Sketch of Proof of Theorem 1.* First, the solutions of

$$\Delta u + \lambda u^{2p+1} = 0 \qquad (3)$$
$$\mu|_{\partial G} = 0$$

can be considered as the solutions of the operator equation $u = \lambda N u$ in the Hilbert space $\dot{W}_{1,2}(G)$. By virtue of the Sobolev imbedding theorem, the operator N is a completely continuous gradient mapping with

$$Nu = \text{grad} \int_G \frac{u^{2p+2}}{2p+2}.$$

Hence, by Theorem 2, on the sphere $\frac{1}{2}\|u\|^2 = R$ the operator equation $u = \lambda N u$ has a countably infinite number of distinct solutions $w_n(R)$ with associated $\lambda_n(R) \to \infty$. Now because of the nonlinearity and homogeneity of the operator N, $u_n = k_n w_n$ satisfies $u_n = N u_n$ where $k_n = (\lambda_n(R))^{1/2p}$; and as the $w_n(R)$ are distinct and $\lambda_n(R) \to \infty$, the $u_n(R)$ are all distinct. The smoothness of the solutions of (3) is a consequence of the L_p regularity theory for linear elliptic partial differential equations (see [**1**]).

This result and its proof can be generalized to n dimensions as follows:

THEOREM 1′. *Let G be a bounded domain in $\boldsymbol{R}^n$, then provided $p \neq 0$ and* $2p + 1 < (n+2)/(n-2)$, (1) *has a countably infinite number of distinct solutions.*

Note that if $2p + 1 \geq (n+2)/(n-2)$, Pohozahev has shown (by a counterexample) that Theorem 1′ may be false. Thus the assumption of complete continuity of B cannot in general be removed from Theorem 2.

We now pass to applications in differential geometry.

(iii) *Conformal deformation of Riemannian metrics on a compact manifold $\mathscr{M}$ of dimension* 2. The Uniformization Theorem for compact Riemann surfaces implies that any (C^∞) compact orientable simply connected 2 dimensional manifold is conformally equivalent to the sphere. This result (which we shall denote by (*)) can also be obtained by the methods discussed here. Indeed suppose $ds^2 = g_{ij}dx_i dx_j$ defines a C^∞ metric on $\mathscr{M}$ with Gaussian curvature $R(x)$. Then a conformally equivalent metric $\bar{g}_{ij}$ with Gaussian curvature $\bar{R}(x)$ on $\mathscr{M}$ is defined by setting $\bar{g}_{ij} = e^{2\sigma}g_{ij}$ where $\sigma \in C^\infty(\mathscr{M})$. Here $\bar{R}(x) = e^{-2\sigma}\{R(x) - \Delta\sigma\}$ where Δ denotes the Laplace-Beltrami operator with respect to g_{ij}. (To derive this formula recall that in terms of isothermal coordinates (u, v) with $g_{11} = g_{22} = \lambda$, $g_{12} = 0$;

$$R(x) = -\frac{1}{2\lambda}\{(\log \lambda)_{uu} + (\log \lambda)_{vv}\}$$

so that the desired formula follows by setting $\lambda' = (\exp 2\sigma)\lambda$.) Then since a sphere is the only orientable compact 2-manifold with constant positive curvature our result $(*)$ follows if we can determine a function $\sigma \in C^\infty(\mathscr{M})$ that satisfies the equation

$$(**) \qquad \Delta\sigma - R(x) + \bar{R}e^{2\sigma} = 0$$

where $\bar{R}$ is some positive constant. Now regarding $\bar{R}$ as a parameter we note that the C^∞ solutions of $(**)$ coincide with the set $\mathscr{S}$ of C^∞ solutions of the variational problem: Find the critical points of the functional $F(\sigma) = D[\sigma] + \int_{\mathscr{M}} R(x)\sigma$ on the hypersurface $\int_{\mathscr{M}} e^{2\sigma} = \text{const}$. The admissible functions are the elements $\sigma(x) \in W_{1,2}(\mathscr{M})$, i.e. the Hilbert space of functions $\sigma(x) \in L_2(\mathscr{M})$ which possess generalized derivatives $\in L_2(\mathscr{M})$. The interested reader will find the proof of $(*)$ and generalizations to nonsimply connected $\mathscr{M}$ in the author's paper [**23**]. This result for $n = 2$ prompts the question: to what extent can the above arguments be extended to n-dimensional compact manifolds with $n > 2$?

(iii)′ *Conformal deformation of Riemannian metrics on a compact manifold $\mathscr{M}$ of dimension $n > 2$.* Yamabe [**21**] considered the problem of deforming a given smooth Riemannian metric g_{ij} on $\mathscr{M}$ to a new metric with constant scalar curvature $\bar{R}$ by means of a conformal deformation. This problem can be reduced to the existence of a real-valued nonnegative solution of the equation

$$(4) \qquad \Delta u - R(x)u + \bar{R}u^{(n+2)/(n-2)} = 0$$

defined on $\mathscr{M}$ where $R(x)$ is the original scalar curvature of $\mathscr{M}$ with respect to g_{ij} and Δ is the associated Laplace-Beltrami operator. In the Sobolev space $W_{1,2}(\mathscr{M})$ the equation can be written in the form $u = \bar{R}Nu$, where N is a gradient mapping but not completely continuous. By virtue of the results mentioned in (ii) above, the existence of such nonnegative solutions of (4) is in doubt. Nonetheless for compact manifolds $(\mathscr{M})$ without boundary, quite a different situation can occur.

THEOREM [**20**]. *The equation* (4) *has a nonnegative smooth solution defined on $\mathscr{M}$, provided $\int_{\mathscr{M}} R(x) \leq 0$.*

For the final two applications we turn to problems of mathematical physics.

(iv) *Von Karman's equation for the buckling on thin elastic shells.* Von Karman's equations for thin elastic shells are a pair of nonlinear fourth order elliptic partial differential equations which determine the vertical displacement u and the stresses produced in a thin elastic body (which is possibly curved in its undeformed state) by forces acting on the boundary of the body. (We measure the magnitude of forces acting by the scalar λ.) Under appropriate boundary conditions the problem can be formulated in a closed linear subspace H of the Hilbert space $W_{2,2}(G)$, where G is a bounded domain (see [**4**], [**5**]) in $\boldsymbol{R}^2$, as an operator equation of the form

$$u + Nu = \lambda Lu$$

where N and L are completely continuous gradient mappings of $H \to H$ with L a linear, positive operator and N nonlinear with $N(0) = 0$. There are essentially two cases.

(1) A Plate (i.e. zero initial curvature). In this case linearization is only locally valid (that is, nontrivial solutions exist for $\lambda > \lambda_1$ where λ_1 is the smallest eigenvalue of $u = \lambda Lu$). Also $(Nu, u) \geq 0$ and N is an odd operator, so that all the results of Theorems 2, 4 and Corollary 5 apply here, enabling us to study the so-called large deflections of the plate.

(2) A Shell (i.e. nonzero initial curvature). In this case linearization is *not* locally valid (nontrivial solutions exist for $\lambda < \lambda_1$). Here the operator N is *not* odd. Theorem 3 applies to guarantee the bifurcation of solutions from every eigenvalue of the linearized problem.

We indicate the situation graphically by plotting the norms of solutions u versus λ.

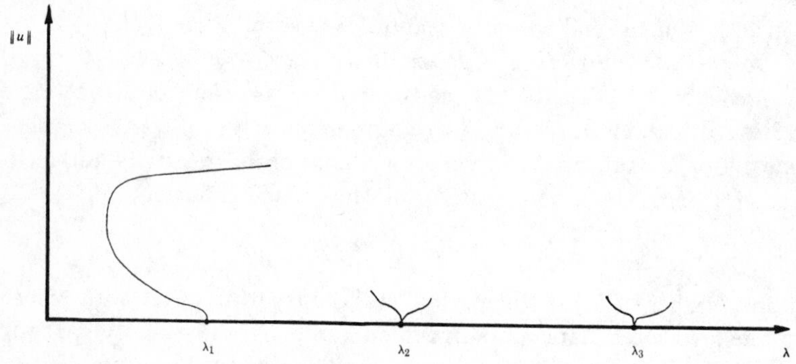

FIGURE 2. Illustrating Bifurcation Phenomena in Elastic Shells.

(v) *Physical instabilities due to the size of the domain.* In many physical problems governed by elliptic partial differential equations the number of solutions depends on the size of the domain. For example consider the system defined on a bounded domain G in $\boldsymbol{R}^2$ (or $\boldsymbol{R}^3$) with boundary ∂G

$$\Delta u + u - u^3 = 0 \tag{5}$$

$$u|_{\partial G} = 0.$$

THEOREM 7. *Let the diameter of the domain G be k. Then if k is sufficiently small the system* (5) *has only the solution $u \equiv 0$, but if k is sufficiently large the system* (5) *possesses other nontrivial solutions, and as $k \to \infty$ the number of distinct solutions of* (5) *also tends to become infinite.*

This result is obtained by making a change of scale $x = \lambda x'$ and noting that if $\lambda^2 \leq \lambda_1$, the equation

$$\begin{aligned} \Delta u + \lambda^2(u - u^3) &= 0 \\ u|_{\partial G} &= 0 \end{aligned} \tag{6}$$

has only the zero solution. On the other hand by the maximum principle for all solutions u, $|u| \leq 1$. Hence all solutions $(u_n(R), \lambda_n(R))$ bifurcating from λ_n for small R, by Theorem 4, can be extended to exist for large R with the property that $\lambda_n(R) \to \infty$ as $n \to \infty$. (See § H.2.) In particular in [**25**] it is shown that $u_1(R)$ exists for all R and as $\lambda_1(R) \to \infty$, $u_1(R) \to 1$ uniformly apart from a small boundary layer set concentrated near ∂G.

G. **Idea of the proofs of the results of §E.**

(a) *On the Proof of Theorem* 2. The basic idea of the proof is to study the nonlinear invariants of the problem

$$c_n(R) = \sup_{V_{n,R}} \inf_{V} b(u)$$

in accord with the basic idea set forth in §C. These numbers are to be thought of as a nonlinear generalization of minimax characterizations of the characteristic values of a compact selfadjoint linear operator L, namely

$$\lambda_n^{-1} = \sup_{E_n} \inf_{E} (Lu, u)$$

where E is an n-dimensional linear subspace of H with elements normalized to have length one, and E_n is the class of all such subspaces of H. The basic minimax principle of the Ljusternik-Schnirelmann theory [**12**], [**17**] insures that provided $V_{n,R}$ is not vacuous and the condition C of Palais-Smale is satisfied, the number $c_n(R)$ is attained and determines a critical value of $b(u)$ on the set ∂A_R, and consequently a critical point $u_n(R) \in \partial A_R$ which satisfies the equation

$$Au_n(R) = \lambda_n(R) Bu_n(R).$$

Since the hypotheses on the pair (A, B) following Theorem 2 imply that $V_{n,R}$ is not vacuous and the condition C of Palais-Smale for each $R > 0$, we note that there is a countably infinite sequence of solutions $(u_n(R), \lambda_n(R))$ $(n = 1, 2, \ldots)$ satisfying the equation $Au = \lambda Bu$ with $u_n(R) \in \partial A_R$. To show that the solutions $u_n(R)$ are distinct we observe that the numbers $c_n(R) \to 0$ as $n \to \infty$, for otherwise it would be possible to continuously deform a set $V \in V_{n,R}$ into a set W with cat $(W, \partial A_R/J) = n - 1$, a contradiction. Furthermore, as $c_n(R) \to 0$ when $n \to \infty$, a simple calculation shows $\lambda_n(R) \to \infty$. Actually another consequence of the Ljusternik-Schnirelmann theory [**17**], [**18**] insures that the numbers $c_n(R)$ are distinct, for otherwise the dimension of the critical set corresponding to the critical value $c_n(R)$ would be greater than one.

REMARK. The author's original paper on this subject contains an incorrect proof (Lemma 1.3.1) and several misprints. Furthermore, all operators considered there should be *explicitly* assumed to be locally Lipschitz continuous. A list of corrections will be found in [**2**(b)].

(b) *On the Proof of Theorem* 3. The case $n = 1$ is easy. Indeed, define

$$c_1(R) = \sup_{\partial A_R} b(u)$$

for R positive but sufficiently small. Then $c_1(R)$ is bounded above as ∂A_R is bounded and $b(u)$ is continuous with respect to weak convergence in H. Furthermore $c_1(R)$ is attained by some element $u \in \partial A_R$. Indeed otherwise there exists a sequence $u_n \to u$ weakly, such that $Au_n \to Au$ strongly, and $b(u_n) \to c_1(R)$; so that by hypothesis $u_n \to u$ strongly and so $u \in \partial A_R$ with $b(u) = c_1(R)$. Finally $c_1(R)$ is a critical value, for otherwise there exists a $\tilde{u} \in \partial A_R$ with $b(\tilde{u}) > c_1(R)$. This fact is proven by contradiction by considering the trajectory

$$du/dt = v - ((Au, v)/\|Au\|^2)Au$$
$$u(0) = u_0 \quad \text{where} \quad u_0 \in \partial A_R \cap b^{-1}(c_1(R))$$

and choosing v so that $dB(u(t))/dt = 1$ for sufficiently small t. Now as $R \to 0$, we show that

$$c_1(R) - \lambda_1^{-1}R = o(R),$$

since

$$|c_1(R) - \lambda_1^{-1}R| = \left| \sup_{\partial A_R} b(u) - \sup_{\partial S_R} \tfrac{1}{2}(Lu, u) \right|.$$

Consequently, an easy calculation shows that as $R \to 0$, $\lambda_1^{-1}(R) \to \lambda_1^{-1}$ so that $(u_1(R), \lambda_1(R)) \to (0, \lambda_1)$.

The case $n > 1$ is somewhat more complicated. First, if R is small, we may define a fixed point free involution i on ∂A_R by considering rays through the origin. Again we define, for sufficiently small $R > 0$,

$$c_n(R) = \sup_{W_{n,R}} \inf_{W} \tfrac{1}{2}[b(u) + b(iu)]$$

where W is a closed subset of ∂A_R and a member of the class of sets $W_{n,R}$. The class $W_{n,R}$ is defined by generalizing the construction mentioned in §E while preserving the properties:

(1) $W_{1,R} \supset W_{2,R} \supset W_{3,R} \cdots \supset W_{n,R} \supset \cdots$ is a strictly decreasing sequence and

(2) the class $W_{n,R}$ is invariant under continuous deformations.

For example, if $a(u)$ and $b(u)$ are even functionals of u then $W_{n,R}$ can be defined to be the class $V_{n,R}$ of §E. If $a(u)$ and $b(u)$ are not even functionals of u, the appropriate definition of $W_{n,R}$ can be obtained by using the generalization of the Ljusternik-Schnirelmann category due to Conner and Floyd [**8**]. The proof then proceeds by showing that the numbers $c_n(R)$ are "approximate" critical values of the functional $b(u)$ on the surface $a(u) = R$ in the sense that there is a critical value $\tilde{c}_n(R)$ "near" $c_n(R)$ such that $b^{-1}(\tilde{c}_n(R))$ contains a critical point $(u_n(R), \lambda_n(R))$ with the property that $(u_n(R), \lambda_n(R)) \to (0, \lambda_n)$ as $R \to 0$.

(c) *On the Proof of Theorem 4 and Corollary 5.* Since the functionals under consideration are even, we consider the one parameter family of manifolds $\partial A_R = \{u \mid a(u) = R\}$ and once again the invariants

$$c_n(R) = \sup_{V_{n,R}} \min_{V} b(u)$$

mentioned in the proof of Theorem 2. Once again $c_n(R)$ are critical values of the functional $b(u)$ on the hypersurface $a(u) = R$ and so determine critical points $u_n(R)$ which satisfy the operator equation

$$Au_n(R) = \lambda_n(R)Bu_n(R).$$

Then the proof is based on the following 3 lemmas.

LEMMA 1 (*The generalized minimax principle for quadratic functionals*).

$$R\lambda_n^{-1} = \sup_{V_{n,R}} \min_{V} \tfrac{1}{2}(Lu, u)$$

where λ_n is the nth eigenvalue of $u = \lambda Lu$ (ordered by magnitude and counted according to multiplicity) and $\partial A_R \equiv \partial S_R$.

LEMMA 2. $R\lambda_n^{-1} - c_n(R) = o(R)$.

LEMMA 3. $|\lambda_n^{-1} - \lambda_n^{-1}(R)| \to 0$ *as* $R \to 0$.

Thus the set $\{u_n(R), {}_n(R)\}$ defines a one parameter family of solutions of $Au = \lambda Bu$ for all $R \geq 0$ and can be considered as the continuation of a family of solutions bifurcating from λ_n at $u \equiv 0$, since as $R \to 0$ $(u_n(R), \lambda_n(R)) \to (0, \lambda_n)$. Corollary 5 then follows from Lemma 2, and the following property of Ljusternik-Schnirelmann category: if $c_i(R) = c_{i+1}(R) = \cdots = c_{i+p-1}(R) = c$, then the dimension of the critical set associated with the critical value c is at least $p - 1$, so that the families $(u_i(R), \lambda_i(R))$ $(i = 1, \ldots, p)$ bifurcating from $(0, \lambda_n)$ *are distinct.*

For simplicity we shall prove Lemmas 1–3 in case $M \equiv 0$ so that our equation becomes

$$(\dagger) \qquad u = \lambda(Lu + Nu).$$

PROOF OF LEMMA 1. Let S denote an n-dimensional subspace of H and $T_R = \{u \mid u \in S, \tfrac{1}{2}\|u\|^2 = R\}$, then we recall the following two facts:

(a) Let $P_R(n-1)$ be the set of elements obtained by identifying antipodal points of T_R and regarded as a subspace of $P_R(H)$. Then cat $(P_R(n-1), P_R^\infty(H)) = n$.

(b) The Courant-Fischer minimax principle can be rewritten

$$R\lambda_n^{-1} = \sup_{[T]_{n,R}} \min_{T_R} \tfrac{1}{2}(Lu, u)$$

where T_R is defined as above and $[T]_{n,R}$ is the class of all such sets for n fixed. Now we consider the numbers

$$\tilde{c}_n(R) = \sup_{[V]_{n,R}} \inf_{V} \tfrac{1}{2}(Lu, u).$$

By (a), $[T]_{n,R} \subset [A]_{n,R}$; so that $\tilde{c}_n(R) \geq R\lambda_n^{-1}$ for each n. Furthermore the numbers $\tilde{c}_n(R)$ are critical values of the function $\frac{1}{2}(Lu, u)$ on $P_R^\infty(H)$ and consequently on ∂S_R by the results of [**17**] so that $\tilde{c}_n(R) = R\lambda_{k(n)}^{-1}$ for some integer $k(n)$. To show that $\tilde{c}_n(R) = R\lambda_n^{-1}$ for each n, we proceed by induction. If $n = 1$, $\tilde{c}_n(R) = R\lambda_1^{-1}$ by definition. Suppose now that λ_1 is an eigenvalue of multiplicity exactly p, then $\lambda_1 \geq \lambda_{k(n)}$ for $n = 1, \ldots, p$. Hence $\lambda_1 = \lambda_{k(n)}$ $n = 1, \ldots, p$. Now we show $\tilde{c}_1(R) = \tilde{c}_2(R) = \cdots = \tilde{c}_p(R) \neq \tilde{c}_{p+1}(R)$. Indeed if $\tilde{c}_p(R) = \tilde{c}_{p+1}(R)$, then the critical set associated with the critical value would have dimension p on ∂S_R (see the remark after Corollary 4.9 of [**17**]) which contradicts the fact that λ_1 is an eigenvalue of multiplicity p. Hence as an induction hypothesis we assume that the distinct eigenvalues $\lambda_{(1)}, \lambda_{(2)}, \ldots, \lambda_{(n-1)}$ are consistent with the distinct numbers $\tilde{c}_{(1)}(R), \tilde{c}_{(2)}(R), \ldots, \tilde{c}_{(n-1)}(R)$, with multiplicities included, by means of the relation $\tilde{c}_{(p)}(R) = R\lambda_{(p)}^{-1}$ $p = 1, \ldots, n-1$. Now suppose $\lambda_{(n)}$ is an eigenvalue of multiplicity exactly t, then we show $\tilde{c}_{(n)}(R) = \tilde{c}_{(n)+1}(R) \cdots = \tilde{c}_{(n)+t}(R) = \lambda_{(n)}^{-1}R$. By our induction hypothesis clearly $\lambda_{n-1} < \lambda_{k(n+i)} \leq \lambda_n$ for $i = 1, 2, \ldots, t$. Thus $\lambda_{k(n+1)} = \lambda_n$ $i = 1, 2, \ldots, t$. Now suppose $\tilde{c}_{n+t+1}(R) = \lambda_n^{-1}R$, then the dimension of the critical set associated with the critical value of $\lambda_n^{-1}R$ exceeds $(t-1)$ on S_R by the above mentioned remark of [**17**], again contradicting the fact λ_n has multiplicity exactly t. Hence $\tilde{c}_{n+t+1}(R) \neq \lambda_n^{-1}R$ and hence the multiplicities of $\tilde{c}_{(n)}(R)$ and $\lambda_{(n)}$ agree. So the lemma is proven.

Proof of Lemma 2. First we note that as $\mathcal{N}(u) = \int_0^1 (u, N(su))\, ds$, for small R and $u \in \partial S_R$, $|\mathcal{N}(u)| \leq K(\|u\|)\,\|u\|^2$ where $K(\|u\|) \to 0$ as $\|u\| \to 0$. Hence $K_R = \sup_{S_R} |\mathcal{N}(u)| = o(R)$. Now

$$c_n(R) = \sup_{[V]_{n,R}} \inf_{V} \{\tfrac{1}{2}(Lu, u) + \mathcal{N}(u)\}$$

and by Lemma 2, $R\lambda_n^{-1} = \sup_{[V]_{n,R}} \inf_V \frac{1}{2}(Lu, u)$, so

$$\begin{aligned} |c_n(R) - R\lambda_n^{-1}| &\leq \left| \sup_{[V]_{n,R}} \inf_{V} \{\tfrac{1}{2}(Lu, u) + K_R\} - \sup_{[V]_{n,R}} \inf_{V} \tfrac{1}{2}(Lu, u) \right| \\ &\leq K_R = o(R). \end{aligned}$$

Proof of Lemma 3. Taking the inner product of (†) with $u_n(R)$ we obtain

$$\begin{aligned} R\lambda_n^{-1}(R) &= \tfrac{1}{2}(Lu_n(R), u_n(R)) + \tfrac{1}{2}(Nu_n(R), u_n(R)) \\ &= c_n(R) + \{\tfrac{1}{2}(Nu_n(R), u_n(R)) - \mathcal{N}(u_n(R))\} \\ &= c_n(R) + o(R), \quad \text{for small } R. \end{aligned}$$

Hence by Lemma 3,

$$R\lambda_n^{-1}(R) - R\lambda_n^{-1} = c_n(R) - \lambda_n^{-1}R + o(R).$$

So $|\lambda_n^{-1}(R) - \lambda_n^{-1}| = o(R)/R = o(1)$. Hence as $R \to 0$, $\lambda_n^{-1}(R) \to \lambda_n^{-1}$.

H. **Remarks on nonlinear elliptic partial differential equations.**

1. The results of §E apply to the study of solutions of quasi-linear elliptic partial differential equations (arising as Euler-Lagrange equations) of the form

$$\sum_{|\alpha|\leq m} D^\alpha(C_\alpha(x, u, Du, \ldots, D^m u)) = 0 \tag{7}$$

either defined on a compact n-dimensional Riemannian manifold $\mathcal{M}$ or defined on a bounded domain G (with boundary ∂G) in $\boldsymbol{R}^n$. In the latter case equation (7) is studied together with appropriate boundary conditions, for example, Dirichlet boundary conditions

$$D^\alpha u|_{\partial G} = 0 \quad \text{for } |\alpha| \leq m - 1. \tag{8}$$

In order to apply the results of §E, we must impose rather strong growth conditions on the function $C_\alpha(x, z, z^{(1)}, \ldots, z^{(m-1)}, z^{(m)})$ as a function of the vectors $z, z^{(i)}$ $(i = 1, \ldots, m)$ (see [**2**], [**7**]) as well as an ellipticity (positivity) condition on the expression

$$\sum_{|\alpha|=m} [C_\alpha(x, \eta, \xi) - C_\alpha(x, \eta, \xi')](\xi - \xi')$$

for $\xi \neq \xi'$. Here $\eta = (z^{(1)}, \ldots, z^{(m-1)})$ and $\xi = z^{(m)}$. The introduction of a real parameter λ into the problem as discussed in §D then transforms the problem into the form

$$\sum_{|\alpha|\leq m} D^\alpha \quad {}_\alpha(x, u, \ldots, D^m u) = \lambda \sum_{|\alpha|\leq m-1} D^\alpha B_\alpha(x, u, \ldots, D^{m-1}u). \tag{9}$$

2. *A priori estimates.* The scope of applications of the results of §E is considerably increased when taken together with a priori estimates for a given problem. Let us, for example, attempt to apply Theorem 2 to the system (6) mentioned in §F

$$\Delta u + \lambda^2(u - u^3) = 0$$
$$u|_{\partial G} = 0.$$

Here we must verify that the operator B defined by

$$(Bu, v) = \int_G (u - u^3)v$$

for all $u, v \in \dot{W}_{1,2}(G)$ must (i) define a completely continuous mapping of $\dot{W}_{1,2}(G)$ into itself and (ii) satisfy $(Bu, u) > 0$ for arbitrary $u \in \dot{W}_{1,2}(G)$ $(u \not\equiv 0)$. For general bounded domains G in $\boldsymbol{R}^n$, both (i) and (ii) are false. To overcome these difficulties in applying Theorem 2, we note the following a priori estimate:

If $u(x)$ is any smooth solution of (6), then $|u(x)| \leq 1$ in $\bar{G}$.

This result follows directly from the maximum principle. Thus the smooth solutions of (6) are identical with the smooth solutions of the "truncated" system

$$\Delta u + \lambda^2 f(u) = 0$$
$$u|_{\partial G} = 0$$

where $f(u) = \max(0, u - u^3)$. Note that $f(t)$ is a Lipschitz continuous function of t. Then, if we define

$$(Bu, v) = \int_G f(u)v \quad \text{for } u, v \in \mathring{W}_{1,2}(G),$$

B satisfies (i) and (ii) and so Theorem 2 is applicable to the system (6), provided R is not too large.

3. *Regularity of Solutions.* The results of §E yield only "weak" solutions $u(x) \in W_{m,p}(G)$ for systems of the form (8) and (9). To prove that such solutions are actually sufficiently smooth to satisfy (8), (9) in the classical pointwise sense, we apply the L_p regularity theory for linear elliptic partial differential equations [**1**] provided the top order part of the elliptic operator

$$I(u) = \sum_{|\alpha|=m} D^\alpha[A_\alpha(x, u, \ldots, D^m u) - A_\alpha(x, u, \ldots, 0)]$$

is a linear function of u. For example, applying these results to the system (6) and the a priori bound mentioned above, we can show that all "weak" solutions of (6) in $\mathring{W}_{1,2}(G)$ (after a possible redefinition on a set of measure zero) satisfy (6) in the classical pointwise sense in G and at all sufficiently smooth portions of ∂G. For a single second order equation even if $I(u)$ is not linear in u, the results of [**9**], [**15**] can also be applied to yield the required smoothness of "weak" solutions.

NOTE. Many of the proofs of the results discussed here will be found in the articles of the following bibliography and in the forthcoming volumes 14, 15 and 16 of Proceedings of Symposia in Pure Mathematics.

BIBLIOGRAPHY

1. S. Agmon, A. Douglis and L. Nirenberg, *Estimates near the boundary for solutions of elliptic differential equations.* I, Comm. Pure Appl. Math. **12** (1959), 623–727; II, ibid. **17** (1964), 35–92.

2. M. S. Berger, (a), *A Sturm-Liouville theorem for nonlinear elliptic partial differential equations*, Ann. Scuola Norm. Sup. Pisa (3) **20** (1966), 543–582. (b) Corrections, ibid, **22** (1968), 351–354.

3. ———, *Nonlinear perturbations of the eigenvalues of a compact self-adjoint operator*, Bull. Amer. Math. Soc. **73** (1967), 704–708.

4. ———, *Von Karman's equations and the buckling of a thin elastic plate.* I, Comm. Pure Appl. Math. **20** (1967), 687–719.

5. M. S. Berger and P. C. Fife, *Von Karman's equation.* II, Comm. Pure Appl. Math. **21** (1968), 227–241.

6. S. Bernstein, *Sur les équations du calcul des variations*, Ann. Sci. École Norm. Sup. **29** (1912), 431–485.

7. F. E. Browder, *Infinite dimensional manifolds and nonlinear eigenvalue problems*, Ann. of Math. (2) **82** (1965), 459–477.

8. P. E. Conner and E. E. Floyd, *Fixed point free involutions and equivariant maps*, Bull. Amer. Math. Soc. **66** (1960), 416–441.

9. E. Di Giorgi, *Sulla differenziabilita e l'analiticita delle estremali* . . . , Mem. Acca. Torino **3** (1957), 25–43.

10. M. A. Krasnoselskii, *Application of variational methods to the problem of bifurcation points*, Mat. Sb. **33** (1953), 199–214.

11. J. Leray and J. Schauder, *Topologie et equations functionelles*, Ann. Sci. École Norm. Sup. **51** (1934), 45–78.

12. L. Ljusternik, *The topology of the calculus of variations in the large*, Transl. Math. Monographs **16** (1966).

13. L. Ljusternik and L. G. Schnirelmann, *Topological methods in variational problems*, Moscow (1930) (in Russian).

14. M. Morse, *The calculus of variations in the large*, Amer. Math. Soc. Colloq. Publ., vol. 18, Providence, R.I., 1934.

15. J. Moser, *A new proof of Di Giorgi's theorem*, Comm. Pure Appl. Math. **13** (1960), 457–468.

16. R. Palais, *Morse theory on Hilbert manifolds*, Topology **2** (1963), 299–340.

17. ———, *Ljusternik-Schnirelmann theory on Banach manifolds*, Topology **5** (1966). 115–132.

18. J. Schwartz, *Generalizing the Ljusternik-Schnirelmann theory of critical points*, Comm. Pure Appl. Math. **17** (1964), 307–315.

19. S. Smale, *Morse theory and a nonlinear generalization of the Dirichlet problem*, Ann. of Math. **80** (1964), 382–396.

20. N. Trudinger, *Remarks concerning conformal deformation*, Ann. Scuola Norm. Sup. Pisa **22** (1968), 265–274.

21. H. Yamabe, *On the conformal deformations of Riemannian metrics*, Osaka J. Math. **12** (1966), 21–37.

22. M. S. Berger, *A generalization of Liapunov's theorem on periodic solutions of Hamiltonian systems* (to appear).

23. ———, *On the conformal equivalence of compact 2 dimensional manifolds*, J. Math. Mech. **18** (1969).

24. ———, *On bifurcation and continuability of solutions to nonlinear operator equations* (to appear).

25. M. S. Berger and L. E. Fraenkel, *On the asymptotic integration of a nonlinear Dirichlet problem*, J. Math. Mech. (to appear).

26. M. S. Berger and M. S. Berger, *Perspectives in nonlinearity*, Benjamin, New York, 1968.

27. M. S. Berger, "A bifurcation theory for nonlinear elliptic partial differential equations and related systems" in *Bifurcation theory*, Benjamin, New York, 1969, pp. 113–216.

UNIVERSITY OF MINNESOTA

BELFER GRADUATE SCHOOL OF SCIENCE, YESHIVA UNIVERSITY

ON SOME DEGENERATE NONLINEAR PARABOLIC EQUATIONS

Haim Brezis

Let V be a reflexive Banach space (over the reals) and V^* be the space of bounded linear functionals on V. Let Λ be an unbounded linear monotone operator defined on a subspace $D(\Lambda)$ of V, with values in V^*, and let A be a nonlinear monotone hemicontinuous bounded and coercive operator from V into V^*.

We consider the equation

$$\Lambda u + Au = f. \tag{1}$$

Equation (1) includes a number of parabolic and hyperbolic nonlinear partial differential equations.

It is known from a result of F. Browder [**5**] that equation (1) is solvable for any A satisfying these properties if and only if Λ is maximal monotone. Hence it is of interest to investigate the properties of linear maximal monotone operators.

In §I we prove a characterization of linear maximal monotone operators using the dual operator Λ^* of Λ. We mention briefly some applications of this result to nonlinear partial differential equations.

In §II we discuss the initial value problem for the equation

$$d(Eu)/dt + Au = f \tag{2}$$

where E is a linear selfadjoint monotone operator and A is the same as in (1). Equation (2) is of elliptic parabolic type which may be degenerate (for example if E equals zero). Consequently we must determine the sense we give to the initial condition "$u(0) = u_0$".

In §III we apply the result of §II to three examples:

(a) To a nonlinear elliptic parabolic degenerate equation.

(b) To a nonlinear gas diffusion equation.

(c) To a free boundary value problem including Stefan's equation.

I. **Characterization of linear maximal monotone operators.** Let V be a reflexive Banach space (over the reals) with norm $\| \;\|$ and V^* be its conjugate with norm $\| \;\|_*$. Let $D(\Lambda)$ be a linear subspace of V and Λ be a linear monotone (single-valued) operator from $D(\Lambda)$ into V^*.

THEOREM 1. *The following assertions are equivalent:*

(I.1) Λ *is maximal monotone.*

(I.2) Λ *is a densely defined closed linear operator such that its adjoint* Λ^* *is monotone.*

(I.3) Λ *is a densely defined closed linear operator such that* Λ^* *is maximal monotone.*

REMARKS. (i) Assertion (I.1) "Λ is maximal monotone" means that the graph of Λ in $V \times V^*$ is maximal among all monotone *sets* in $V \times V^*$. It is equivalent to assume that Λ is densely defined and that Λ is maximal among all linear densely defined monotone single-valued operators from V into V^*.

(ii) The fact that assertion (I.2) implies (I.1) is an answer to a question raised by F. Browder in [**5**] (and generalizes Proposition 3 of [**5**]). This result was already known in the Hilbert space case.

PROOF OF THEOREM 1. By a result of E. Asplund [**1**] there exists an equivalent norm for V such that both it and its dual norm are strictly convex. Hence we may assume that $\| \quad \|$ and $\| \quad \|_*$ are strictly convex norms.

Let J be the duality mapping from V into V^*. We recall that J (resp. J^{-1}) is monotone hemicontinuous and bijective from V onto V^* (resp. V^* onto V) and that

$$(Jv, v) = \|v\|^2, \qquad \|Jv\|_* = \|v\| \quad \text{for all } v \in V,$$

$$(g, J^{-1}g) = \|g\|_*^2, \qquad \|J^{-1}g\| = \|g\|_* \quad \text{for all } g \in V^*.$$

(I.2) *implies* (I.1). It suffices to show that $\Lambda + J$ is onto. We denote by X the space $D(\Lambda)$ with the graph norm, i.e. $\|v\|_X = \|v\| + \|\Lambda v\|_*$. It follows from (I.2) that X is a reflexive Banach space. Let f be given in V^* and $\varepsilon > 0$. We consider the mapping B_ε from X into X^* defined by $(B_\varepsilon u, v)_{X^*,X} = \varepsilon(\Lambda v, J^{-1}\Lambda u) + (\Lambda u, v) + (Ju, v) - (f, v)$. We remark that B_ε is monotone hemicontinuous and coercive from X into X^*. Hence there exists a solution u_ε in X of the equation $(B_\varepsilon u_\varepsilon, v)_{X^*,X} = 0$ for all $v \in X$ that is $u_\varepsilon \in D(\Lambda)$ and

$$\text{(I.4)} \qquad \varepsilon(\Lambda v, J^{-1}\Lambda u_\varepsilon) + (\Lambda u_\varepsilon, v) + (Ju_\varepsilon, v) = (f, v) \quad \text{for all } v \in D(\Lambda).$$

It follows from (I.4) that $\|u_\varepsilon\| \leqq \|f\|_*$. Since (I.4) holds for all v in $D(\Lambda)$ we have $J^{-1}\Lambda u_\varepsilon \in D(\Lambda^*)$ and

$$\text{(I.5)} \qquad \varepsilon\Lambda^* J^{-1}\Lambda u_\varepsilon + \Lambda u_\varepsilon + Ju_\varepsilon = f.$$

Next we take the scalar product of both sides of (I.5) with $J^{-1}\Lambda u_\varepsilon$. Using the hypothesis (I.2) we see that $\|\Lambda u_\varepsilon\|_* \leqq \|f\|_* + \|u_\varepsilon\| \leqq 2\|f\|_*$. We pass to the limit with respect to an ultrafilter $\mathscr{U}$ which converges to zero. Thus

$$\lim_{\mathscr{U}} u_\varepsilon = u \quad \text{weakly in } V, \qquad \lim_{\mathscr{U}} \Lambda u_\varepsilon = \Lambda u \quad \text{weakly in } V^*$$

(since the graph of Λ is weakly closed).

We replace v by $u_\varepsilon - u$ in (I.4) and we have

$$(\text{I.6}) \qquad (Ju_\varepsilon, u_\varepsilon - u) \leqq (f, u_\varepsilon - u) - (\Lambda u, u_\varepsilon - u) - \varepsilon(\Lambda u_\varepsilon - \Lambda u, J^{-1}\Lambda u).$$

Hence $\lim_{\mathscr{U}} \sup\, (Ju_\varepsilon, u_\varepsilon - u) \leqq 0$. It follows from a result of [**3**, Proposition 6] that $\lim_{\mathscr{U}} Ju_\varepsilon = Ju$ weakly in V^*. Passing to the limit in (I.4) we see that $(\Lambda u, v) + (Ju, v) = (f, v)$ for all $v \in D(\Lambda)$ and $\Lambda u + Ju = f$ since $D(\Lambda)$ is dense. It is obvious that (I.3) implies (I.2) and it suffices to prove that (I.1) *implies* (I.3).

$D(\Lambda)$ is a dense-subspace of V. Let f be an element of V^* such that $(f, v) = 0$ for all $v \in D(\Lambda)$. By the monotonicity of Λ we have (I.7) $(\Lambda v - f, v) \geqq 0$ for all $v \in D(\Lambda)$. Since Λ is maximal monotone, it follows from (I.7) that $f = \Lambda 0 = 0$.

Λ is a closed operator. Let u_n be a sequence of $D(\Lambda)$ such that $\lim_{n\to+\infty} u_n = u$ and $\lim_{n\to+\infty} \Lambda u_n = f$. Since Λ is monotone, we have $(\Lambda u_n - \Lambda v, u_n - v) \geqq 0$ for all $v \in D(\Lambda)$. Then, passing to the limit, we obtain

$$(f - \Lambda v, u - v) \geqq 0 \quad \text{for all } v \in D(\Lambda).$$

Hence u is in $D(\Lambda)$ and $\Lambda u = f$ since Λ is maximal monotone.

Λ^* is monotone. Let u be an element of $D(\Lambda^*)$. By a result of F. Browder [**5**, Theorem 1] there exists, for any $\varepsilon > 0$, u_ε in $D(\Lambda)$ such that

$$(\text{I.8}) \qquad \varepsilon\Lambda u_\varepsilon + J(u_\varepsilon - u) = 0.$$

Then $(J(u_\varepsilon - u), u) \leqq 0$ and $\|u_\varepsilon - u\| \leqq \|u\|$. Let v be an element of $D(\Lambda)$. We have

$$\begin{aligned}\|u_\varepsilon - u\|^2 = (J(u_\varepsilon - u), u_\varepsilon - u) &= (J(u_\varepsilon - u), u_\varepsilon - v) + (J(u_\varepsilon - u), v - u)\\ &\leqq \varepsilon(\Lambda v, v - u_\varepsilon) + (J(u_\varepsilon - u), v - u).\end{aligned}$$

Hence

$$\limsup_{\varepsilon\to 0} \|u_\varepsilon - u\|^2 \leqq \|u\|\, \|v - u\| \quad \text{for all } v \in D(\Lambda).$$

It follows that $\lim_{\varepsilon\to 0} u_\varepsilon = u$, since $D(\Lambda)$ is dense. In addition

$$(J(u_\varepsilon - u), u_\varepsilon) \leqq (J(u_\varepsilon - u), u_\varepsilon) = -\varepsilon(\Lambda u_\varepsilon, u_\varepsilon) \leqq 0.$$

Thus

$$(\text{I.9}) \qquad (\Lambda^* u, u) = (\Lambda u_\varepsilon, u) = -(\Lambda/\varepsilon)(J(u_\varepsilon - u), u) \geqq 0.$$

Passing to the limit in (I.9) as ε goes to zero we obtain $(\Lambda^* u, u) \geqq 0$ for all u in $D(\Lambda^*)$. Finally we apply the fact that (I.2) implies (I.1), Λ being replaced by Λ^* to prove that Λ is maximal monotone.

COROLLARY 1. *Let Λ be a densely defined closed linear operator such that Λ^* is monotone. Let A be a monotone hemicontinuous bounded and coercive mapping from V into V^*. Then $\Lambda + A$ is onto.*

Using the same technique as above, one can prove an existence theorem for "perturbed" equation of the form $\Lambda u + Au = f$ where Λ is linear maximal monotone and A satisfies modified monotonicity conditions. (To appear soon.)

These results are applicable to existence and uniqueness theorems for solutions of nonlinear parabolic boundary value problems (see F. Browder [**4**], T. Kato [**7**], J. L. Lions [**10**], J. L. Lions and W. Strauss [**11**], M. I. Višik [**15**], C. Bardos and H. Brezis [**2**]). Applications to nonlinear hyperbolic equations, first order symmetric systems are given in J. L. Lions [**9**], F. Browder [**4**], C. Bardos and H. Brezis [**2**], to wave equation and Schroedinger equation in J. L. Lions and W. Strauss [**11**], W. Strauss [**14**], C. Bardos and H. Brezis [**2**].

II. **Application to the initial value problem for the equation** $d(Eu)/dt + Au = f$. Let $\mathscr{V}$ be a reflexive Banach space with norm $\|\ \ \|$ and let $\mathfrak{H}$ be a Hilbert space with norm $|\ \ |$. $\mathscr{V}$ is contained in $\mathfrak{H}$ with continuous injection and $\mathscr{V}$ is dense in $\mathfrak{H}$. Making the usual identifications one has

$$\mathscr{V} \subset \mathfrak{H} \subset \mathscr{V}^*.$$

We set

$$V = L^p(0, T; \mathscr{V}), \quad H = L^2(0, T; \mathfrak{H}), \quad V^* = L^{p'}(0, T; \mathscr{V}^*)$$

with $p \geqq 2$, $1/p + 1/p' = 1$, $0 < T < +\infty$. We suppose that E is a bounded selfadjoint monotone linear operator from $\mathfrak{H}$ into $\mathfrak{H}$. We may thus introduce the square root $E^{1/2}$ of E. We denote by $C(0, T; \mathfrak{H})$ the space of continuous functions defined on $[0, T]$ with values in $\mathfrak{H}$.

THEOREM 2. *Let A be a monotone hemicontinuous bounded and coercive operator from V into V^*. For each f given in V^* and u_0 given in $\mathfrak{H}$, there exists u in V such that*

$$\text{(II.1)} \quad -\int_0^T (u, d(Ev)/dt)\, dt + \int_0^T (Au, v)\, dt = \int_0^T (f, v)\, dt + (E^{1/2}u_0, E^{1/2}v(0))$$

for all $v \in V$ such that $dv/dt \in H$ and $v(T) = 0$. Furthermore $E^{1/2}u \in C(0, T; \mathfrak{H})$ and $E^{1/2}u(0) = E^{1/2}u_0$. In addition equation (II.1) *is equivalent to the following*:

(II.2) *There exists a sequence u_n in V with $du_n/dt \in H$ such that u_n converges to u in V, dEu_n/dt converges to $-Au + f$ in V^* and $E^{1/2}u_n(0)$ converges to $E^{1/2}u_0$ in $\mathfrak{H}$.*

The solution of (II.1) *is unique if A is strictly monotone.*

PROOF OF THEOREM 2. We set $D(L_0) = \{u \in V$ such that $du/dt \in H\}$ and $L_0u = dEu/dt$. Obviously L_0 is preclosed. Let L be the closure of L_0 in $V \times V^*$. It follows easily from the definition of L that if u is in $D(L)$, then $E^{1/2}u$ is in $C(0, T; \mathfrak{H})$. We set $D(\Lambda) = \{u \in D(L)$ such that $E^{1/2}u(0) = 0\}$ and $\Lambda = L$.

The proof of Theorem 2 rests upon two lemmas.

LEMMA 1. *Let u be given in V and f in V^*. The following assertions are equivalent*:

(II.3) *u is in $D(\Lambda)$ and $\Lambda u = f$,*

$$\text{(II.4)} \qquad -\int_0^T (u, d(Ev)/dt)\, dt = \int_0^T (f, v)\, dt \quad \textit{for all } v \in V,$$

such that $dv/dt \in H$ and $v(T) = 0$. Consequently Λ is maximal monotone.

PROOF OF LEMMA 1. From the definition of Λ we see that (II.3) implies (II.4). Conversely suppose (II.4) is true. Let

$$u_n(t) = n\int_0^t e^{n(s-t)}u(s)\,ds$$

u_n is in $D(L_0)$ and $L_0u_n = dEu_n/dt = f_n$ where

$$f_n(t) = n\int_0^t e^{n(s-t)}f(s)\,ds.$$

Since u_n converges to f in V^* we have (II.3). Λ is a densely defined closed linear operator. By Theorem 1 it suffices to show that Λ^* is monotone. But Λ is equal to the adjoint of the operator $-dE/dt$ with domain $\{v \in V$ such that $dv/dt \in H$ and $v(T) = 0\}$.

Hence Λ^* is equal to the closure of the previous operator which is monotone. Actually one can deduce Theorem 2 from Corollary 1 in the special case where $u_0 = 0$.

LEMMA 2. *Let u be in V, f in V^* and u_0 in $\mathfrak{H}$. The following assertions are equivalent*:

(II.5) *u is in $D(L)$, $Lu = f$ and $E^{1/2}u(0) = E^{1/2}u_0$.*

$$\text{(II.6)}\qquad -\int_0^T (u, d(Ev)/dt)\,dt = \int_0^T (f, v)\,dt + (E^{1/2}u_0, E^{1/2}v(0)) \quad \textit{for all } v \in V$$

such that $dv/dt \in H$ and $v(T) = 0$.

PROOF OF LEMMA 2. It is obvious that the assertion (II.5) implies (II.6). Conversely suppose (II.6) is true. Let u_0^n be a sequence of $\mathscr{V}$ such that u_0^n converges to u_0 in $\mathfrak{H}$. Let J be the duality mapping from V into V^* (see §I). There exists w_n in $D(L)$ such that

$$Lw_n + Jw_n = 0, \qquad E^{1/2}w_n(0) = E^{1/2}u_0^n$$

(considering the new unknown function $w - u_0^n$ it suffices to apply Theorem 2 with $u_0 = 0$).

It is easy to show that w_n remains in a bounded set of V. Passing to the limit, w_n converges weakly to w and w is in $D(L)$ since the graph of L is weakly closed. Jw_n converges weakly to $-Lw$. Moreover $E^{1/2}w(0) = E^{1/2}u_0$. Hence we have

$$\text{(II.7)}\qquad -\int_0^T (w, dEv/dt)\,dt - \int_0^T (Lw, v)\,dt = (E^{1/2}u_0, E^{1/2}v(0)) \quad \text{for all } v \in V$$

such that $(dv/dt) \in H$ and $v(T) = 0$. Next we subtract (II.7) from (II.6) and we apply Lemma 1. It follows that $u - w$ is in $D(L)$ and that $L(u - w) = f - Lw$, $E^{1/2}(u - w)(0) = 0$. Hence u is in $D(L)$, $Lu = f$ and $E^{1/2}u(0) = E^{1/2}u_0$.

PROOF OF THEOREM 2 CONCLUDED. Using the same argument as in §I we see that for any $\varepsilon > 0$ there exists u_ε in $D(L)$ such that

$$\begin{aligned}\varepsilon\int_0^T (Lv, J^{-1}Lu_\varepsilon)\,dt + \int_0^T (Lu_\varepsilon, v)\,dt + \int_0^T (Au_\varepsilon, v)\,dt + (E^{1/2}u_\varepsilon(0), E^{1/2}v(0)) \\ = \int_0^T (f, v)\,dt + (Eu_0^{1/2}, Ev^{1/2}(0)) \quad \text{for all } v \in D(L).\end{aligned} \tag{II.8}$$

Then u_ε (resp. $E^{1/2}u_\varepsilon(0)$) remains in a bounded set of V (resp. $\mathfrak{H}$). It follows from (II.8) that $J^{-1}Lu_\varepsilon \in D(\Lambda^*)$ and

$$\varepsilon\Lambda^* J^{-1}Lu_\varepsilon + Lu_\varepsilon + Au_\varepsilon = f. \tag{II.9}$$

Since Λ^* is monotone, Lu_ε remains in a bounded set of V^*. Moreover

$$\begin{aligned}&\int_0^T (Au_\varepsilon, u_\varepsilon - u)\,dt \\ &\quad \leqq -\varepsilon\int_0^T (Lu_\varepsilon - Lu, J^{-1}Lu_\varepsilon)\,dt + \int_0^T (Lu, u - u_\varepsilon)\,dt + \int_0^T (f, u_\varepsilon - u)\,dt \\ &\qquad + (E^{1/2}u(0), E^{1/2}u(0) - E^{1/2}u_\varepsilon(0)) + (E^{1/2}u_0, E^{1/2}u_\varepsilon(0) - E^{1/2}u(0))\end{aligned}$$

Then passing to the limit with respect to an ultrafilter $\mathscr{U}$ which converges to zero we have

$$\lim_{\mathscr{U}} u_\varepsilon = u \quad \text{weakly in } V,$$

$$\lim_{\mathscr{U}} Lu_\varepsilon = Lu \quad \text{weakly in } V^*,$$

$$\lim_{\mathscr{U}} E^{1/2}u_\varepsilon(0) = E^{1/2}u(0) \quad \text{weakly in } \mathfrak{H}^*,$$

$$\lim_{\mathscr{U}} Au_\varepsilon = Au \quad \text{weakly in } V^*$$

(since $\lim_{\mathscr{U}} \sup (Au_\varepsilon, u_\varepsilon - u) \leqq 0$). Thus u is a solution of (II.1).

REMARKS. (i) Equation (II.1) was already discussed in [**2**] using slightly different methods.

(ii) Initial value problem can be replaced in Theorem 2 by the periodical problem $E^{1/2}u(0) = E^{1/2}u(T)$.

(iii) Under some additional hypothesis Theorem 2 can be extended to the case where E is a function of t.

III. **Examples.** Let Ω be a bounded domain in $\mathbf{R}^n$ with smooth boundary $\partial\Omega$. We set $Q = \Omega \times (0, T)$, $p \geqq 2$, $1/p + 1/p' = 1$. For any integer m let

$$W^{m,p}(\Omega) = \{u \in L^p(\Omega);\ \partial^\alpha u/\partial x_i^\alpha \in L^p(\Omega),\ |\alpha| \leqq m\}$$

$$W_0^{m,p}(\Omega) = \{u \in W^{m,p}(\Omega);\ \partial^\beta u/\partial x_i^\beta = 0 \text{ on } \partial\Omega,\ |\beta| \leqq m - 1\}.$$

(a) *An elliptic parabolic nonlinear degenerate equation.*

THEOREM 3. *Let $e(x)$ be an element of $L^\infty(\Omega)$ such that $e(x) \geqq 0$ a. e. in Ω. Let f be given in $L^{p'}(Q)$ and u_0 in $L^2(\Omega)$. Then there exists one and only one generalized solution u in $L^p(0, T; W_0^{1,p}(\Omega))$ of the equation*

$$\text{(III.1)} \quad \begin{aligned} &e(x)\frac{\partial u}{\partial t} - \sum_{i=1}^{n} \frac{\partial}{\partial x_i}\left(\left|\frac{\partial u}{\partial x_i}\right|^{p-2} \frac{\partial u}{\partial x_i}\right) = f \quad \text{in } Q \\ &u(x, t) = 0 \quad \text{for } x \in \partial\Omega, \quad t \in (0, T) \\ &u(x, 0) = u_0(x) \quad \text{for } x \in \Omega \text{ such that } e(x) \neq 0. \end{aligned}$$

In addition $\sqrt{e(x)}\, u(x, t) \in C(0, T; L^2(\Omega))$ *and* $\sqrt{e(x)}\, u(x, 0) = \sqrt{e(x)}\, u_0(x)$ *for* $x \in \Omega$.

Theorem 3 is a simple consequence of Theorem 2 applied with

$$\mathscr{V} = W_0^{1,p}(\Omega), \qquad H = L^2(\Omega) \qquad Eu(x) = e(x)u(x)$$

$$Au = -\sum_{i=1}^{n} \frac{\partial}{\partial x_i}\left(\left|\frac{\partial u}{\partial x_i}\right|^{p-2} \frac{\partial u}{\partial x_i}\right)$$

which is a strictly monotone hemicontinuous bounded and coercive operator. Note that equation (II.1) may be of parabolic type in $\Omega_1 \subset \Omega$ ($e(x) = 1$ for $x \in \Omega_1$) and of elliptic type in $\Omega_2 = \Omega - \Omega_1$ ($e(x) = 0$ for $x \in \Omega_2$). The transition condition on the surface which separates the two phases is then

$$\partial u/\partial n_1 + \partial u/\partial n_2 = 0.$$

(b) *A nonlinear gas diffusion equation.*

THEOREM 4. *Let f be given in $L^{p'}(Q)$ and u_0 in $L^2(\Omega)$. There exists one and only one function u such that*

$$\text{(III.2)} \qquad |u|^{(p-2)/2}u \in L^2(0, T; H_0^1(\Omega)),$$

$$\text{(III.3)} \qquad u \in L^\infty(0, T; L^2(\Omega))$$

and which satisfies in a generalized sense the following equations

$$\text{(III.4)} \qquad \frac{\partial u}{\partial t} - \sum_{i=1}^{n} \frac{\partial}{\partial x_i}\left(|u|^{p-2} \frac{\partial u}{\partial x_i}\right) = f \quad \text{in } Q,$$

$$\text{(III.5)} \qquad u(x, t) = 0 \quad \text{for } x \in \partial\Omega, \qquad t \in (0, T),$$

$$\text{(III.6)} \qquad u(x, 0) = u \quad \text{for } x \in \Omega.$$

PROOF OF THEOREM 4. First we remark that the operator

$$u \mapsto -\sum_{i=1}^{n} \frac{\partial}{\partial X_i}\left(|u|^{p-2} \frac{\partial u}{\partial X_i}\right)$$

is not monotone. Hence the theorems of [**4**], [**10**] or [**15**] do not apply. But we can write equations (III.4), (III.5) in the form

$$\partial u/\partial t - \Delta(|u|^{p-2}u)/(p-1) = f \quad \text{in } Q,$$

or

$$\partial Eu/\partial t + |u|^{p-2}u/(p-1) = Ef \tag{III.7}$$

where $E = -(\Delta)^{-1}$ is the inverse of Laplacian with zero boundary condition, and $E^{1/2}u(0) = E^{1/2}u_0$.

In this form it is clear we can apply Theorem 2 which proves that the equation (III.7) has one and only one solution u in $L^p(Q)$. This is a very weak solution of the equations (III.4), (III.5), (III.6). Nevertheless one can prove, using another approximation process, that $|u|^{(p-2)/2}u \in L^2(0, T; H_0^1(\Omega))$ thus $|u|^{p-2}u \in L^{p'}(0, T; W_0^{1,p'}(\Omega))$ and $u \in L^\infty(0, T; L^2(\Omega))$. In order to prove (III.2) we use the following lemma.

Lemma 3. *Let v be in $L^p(Q)$ and g in $L^{p'}(Q)$ with*

$$|v|^{p-2}v + (\Delta)^{-1}g = 0.$$

Then $|v|^{(p-2)/2}v$ is in $L^2(0, T; H_0^1(\Omega))$ and

$$\frac{4}{p^2}(p-1)\,\| |v|^{(p-2)/2}v\|_{L^2(0,T;H_0{}^1(\Omega))} \le \int_0^T (g, v)\,dt. \tag{III.8}$$

Proof of Lemma 3. By the monotonicity of $|w|^{p-2}w$ we have

$$\int_0^T (-(\Delta)^{-1}g - |w|^{p-2}w, v - w)\,dt \le 0 \quad \text{for all } w \in L^p(Q). \tag{III.9}$$

Let $v_\varepsilon \in L^p(0, T; W^{2,p}(\Omega)) \cap L^p(0, T; W_0^{1,p}(\Omega))$ be the solution of $-\varepsilon\Delta v_\varepsilon + v_\varepsilon = v$ in Q, $\varepsilon > 0$, $v_\varepsilon = 0$ on $\partial\Omega \times (0, T)$. It is known that v_ε converges to v in $L^p(Q)$. Setting $w = v_\varepsilon$ in (III.9) we have

$$\int_0^T (-(\Delta)^{-1}g - |v_\varepsilon|^{p-2}\,v_\varepsilon, -\Delta v_\varepsilon)\,dt \ge 0.$$

Thus

$$\frac{4(p-1)}{p^2}\,\| |v_\varepsilon|^{(p-2)/2}\,v_\varepsilon\|^2_{L^2(0,T;H_0{}^1(\Omega))} \le \int_0^T (g, v_\varepsilon)\,dt.$$

Passing to the limit we get the estimate (III.8).

Next we assume that $u_0 = 0$ and we approximate the equation (II.7) by

$$-(\Delta)^{-1}\frac{u_h - P(h)u_h}{h} + \frac{1}{p-1}|u_h|^{p-2}u_h = -(\Delta)^{-1}f$$

where

$$\begin{aligned} P(h)u(t) &= u(t-h) \quad \text{if } h < t < T \\ &= 0 \qquad \text{if } 0 < t < h. \end{aligned}$$

It is easy to prove that u_h converges to u in $L^p(Q)$ as h tends to zero and by the Lemma 3 we have the estimate

$$\frac{4}{p^2} \| |u_h|^{(p-2)/2} u_h \|_{L^2(0,T;H_0{}^1(\Omega))} \leqq \int_0^T (f, u_h)\, dt.$$

Hence (III.2).

The same argument applies also in the case where $u_0 \neq 0$. Equation (III.4) was discussed in a different framework by Dubinsky [**6**] (see also Raviart [**13**]).

(c) *A free boundary value problem.* Let $k(s)$ be a function defined for $s \in (-\infty, +\infty)$ continuous if $s \neq \sigma$ which has a jump at $s = \sigma$. Suppose that $0 < k_1 \leqq k(s) \leqq k_2 < +\infty$ for all s.

THEOREM 5. *Let f be given in $L^2(Q)$ and u_0 in $L^2(\Omega)$ such that $u_0(x) \neq \sigma$ a. e. in Ω. Then there exists one and only one weak solution*

$$u \in L^2(0, T; H_0^1(\Omega)) \cap L^\infty(0, T; L^2(\Omega))$$

of the equation

$$\frac{\partial u}{\partial t} - \sum_{i=1}^{n} \frac{\partial}{\partial x_i}\left(k(u)\frac{\partial u}{\partial x_i}\right) = f \quad \text{in } Q, \tag{III.10}$$

$$u(x, t) = 0 \qquad x \in \partial\Omega, \qquad t \in (0, T), \tag{III.11}$$

$$u(x, 0) = u_0(x) \qquad x \in \partial\Omega \tag{III.12}$$

and which satisfies the following transition condition on the "surface" S $u(x, t) = \sigma$ (i.e. on the surface which separates the two phases $u(x, t) < \sigma$ and $u(x, t) > \sigma$)

$$b \cos(n, t) - \left[\sum_{i=1}^{n} k(u) \frac{\partial u}{\partial x_i} \cos(n, x_i)\right]_{u=\sigma-0}^{u=\sigma+0} = 0 \tag{III.13}$$

b is a nonnegative given constant and n is the normal direction to S.

PROOF OF THEOREM 5. Following [**8**] and [**12**] we introduce the new unknown function

$$v(x, t) = \int_0^{u(x,t)} k(s)\, ds$$

i.e. $v = K(u)$ where $K(r) = \int_0^r k(s)\, ds$. Hence we have

$$\begin{aligned} &\partial\beta(v)/\partial t - \Delta(v) = f \quad \text{in } Q,\\ &v(x, t) = 0 \qquad x \in \partial\Omega, \quad t \in (0, T),\\ &v(x, 0) = v_0(x) \qquad x \in \Omega\\ &b \cos(n, t) - \left[\sum_{i=1}^{n} \frac{\partial v}{\partial x_i} \cos(n, x_i)\right]_{v=\tau-0}^{v=\tau+0} = 0 \quad \text{on } S \end{aligned} \tag{III.14}$$

where $v_0 = K(u_0)$, $\tau = K(\sigma)$ and β is a function such that $\beta'(t) = 1/k[K^{-1}(t)]$ for all $t \neq \tau$ with a jump at $t = \tau$

$$\beta(\tau + 0) - \beta(\tau - 0) = b.$$

Thus we may consider β as a multivalued function and we denote by $A = \beta^{-1}$ the inverse function which is singlevalued (by the hypothesis on k). A is monotone hemicontinuous and coercive from $L^2(Q)$ into itself. The function v is a weak solution of (III.14) in the sense that

$$\text{(III.15)} \quad -\int_0^T\int_\Omega \beta(v)\frac{\partial\varphi}{\partial t}\,dx\,dt + \int_0^T\int_\Omega \sum_{i=1}^n \frac{\partial v}{\partial x_i}\frac{\partial\varphi}{\partial x_i}\,dx\,dt = \int_0^T\int_\Omega f\cdot\varphi\,dx\,dt + \int_\Omega \beta(v_0)\cdot\varphi(x,0)\,dx$$

for all $\varphi \in L^2(0, T; H_0^1(\Omega))$ such that $(\partial\varphi/\partial t) \in L^2(Q)$ and $\varphi(x, T) = 0$.

Let $w = \beta(v)$ (thus $v = Aw$) and $\beta(v_0) = w_0$. From (III.15) it follows that

$$\text{(III.16)} \quad -\int_0^T\int_\Omega w\frac{\partial\varphi}{\partial t}\,dx\,dt + \int_0^T\int_\Omega \sum_{i=1}^n \frac{\partial}{\partial x_i}Aw\cdot\frac{\partial\varphi}{\partial x_i}\,dx\,dt = \int_0^T\int_\Omega f\cdot\varphi\,dx\,dt + \int_\Omega w_0\cdot\varphi(x,0)\,dx$$

for all $\varphi \in L^2(0, T; H_0^1(\Omega))$ such that $\partial\varphi/\partial t \in L^2(Q)$ and $\varphi(x, T) = 0$. Hence we have

$$\text{(III.17)} \qquad \partial w/\partial t - \Delta(Aw) = f \quad \text{in } Q$$

$$Aw(x,t) = 0 \qquad x \in \partial\Omega,\quad t \in (0, T) \qquad w(x, 0) = w_0(x),\quad x \in \Omega.$$

Using the same argument as in example (b) we deduce from Theorem 2 that there exists a solution w of (III.17) such that $Aw \in L^2(0, T; H_0^1(\Omega))$, $w \in L^\infty(0, T; L^2(\Omega))$. In addition one has $Aw^1 = Aw^2$ if w^1 and w^2 are two solutions of (III.17). Hence $v = Aw$ is the unique solution of (III.15), $v \in L^2(0, T; H_0^1(\Omega)) \cap L^\infty(0, T; L^2(\Omega))$, and $u = K^{-1}(v)$, $u \in L^2(0, T; H_0^1(\Omega)) \cap L^\infty(0, T; L^2(\Omega))$, u is the unique weak solution of (III.10)–(III.13). Let us finally refer to [8] and [**12**] for another approach to the Stefan problem.

ADDED IN PROOF. 1. Discussing Theorem 1 with Professor L. Nirenberg, the latter found a simple proof (which is also valid for multi-valued operators). (I.2) implies (I.1): assume that $f \in V^*$ and $u \in V$ satisfy $(\Lambda v - f, v - u) \geqq 0$ $\forall v \in D(\Lambda)$. The convex function $(\Delta v - f, v - u) + \frac{1}{2}\|v - u\|^2 + \frac{1}{2}\|\Lambda v - f\|^2$ achieves its minimum at $v_0 \in D(\Lambda)$ and it is not difficult to check that v_0 coincides with u. Λ maximal monotone implies that Λ^* is monotone since for $u \in D(\Lambda^*)$ we have $(\Lambda v + \Lambda^* u, v - u) \leqq -(\Lambda^* u, u)$ $\forall v \in D(\Lambda)$. But $\operatorname{Inf}_{v \in D(\Lambda)} (\Lambda v + \Lambda^* u, v - u) \leqq 0$ shows that $(\Lambda^* u, u) \geqq 0$.

2. In the proof of Theorem 5 it is useful to notice that $[A'(r)]^2 \leqq k\,A'(r)$ a.e. on $(-\infty, +\infty)$. In addition $u \in L^\infty(0, T, H_0^1(\Omega))$ and $\partial u/\partial t \in L^2(0, T, L^2(\Omega))$ in the case where $u_0 \in H_0^1(\Omega)$.

REFERENCES

1. E. Asplund, *Averaged norms*, Israel J. Math. **5** (1967), 227–233.

2. C. Bardos and H. Brezis, *Sur une classe de problemes d'evolution non linéaires*, J. Differential Equations **6** (1969).

3. H. Brezis, *Équations et inéquations non linéaires dans les espaces vectoriels en dualité*, Ann. Inst. Fourier **18** (1968), 115–175.

4. F. Browder, *Nonlinear initial value problems*, Ann. of Math. (2) **82** (1965), 51–87.

5. ———, *Non-linear maximal monotone operators in Banach spaces*, Math. Ann. **175** (1968), 89–113.

6. J. Dubinsky, *Weak convergence in nonlinear elliptic and parabolic equations*, Mat. Sb. **67** (**109**) (1965); 609–642; English transl., Amer. Math. Soc. Transl. (2) **67** (1968), 226–258.

7. T. Kato, *Nonlinear evolution equation in Banach spaces*, Proc. Sympos. Appl. Math., vol. 17, Amer. Math. Soc., Providence, R.I, 1965, pp. 50–67.

8. O. Ladyženskaja, V. Solonnikov and N. Ural′ceva, *Linear and quasilinear equations of parabolic type*, "Nauka", Moscow, 1967; English transl., Transl. Math. Monographs, vol. 23, Amer. Math. Soc., Providence, R.I., 1968.

9. J. L. Lions, *Sur certains systèmes hyperboliques non linéaires*, C. R. Acad. Sci. Paris. **267** (1963), 2057–2060.

10. ———, *Sur certaines équations paraboliques non linéaires*, Bull. Soc. Math. France **93** (1965), 155–175.

11. J. L. Lions and W. Strauss, *Some nonlinear evolution equations*, Bull. Soc. Math. France **93** (1965), 43–96.

12. O. Oleĭnik, *On Stefan-type free boundary problems for parabolic equations*, Seminari 1962–1963 Anal. Alg. Geom. e Topol., vol. I, Ist. Naz. Acta. Mat., p. 388–403. Ed. Cremonese.

13. D. Raviart, *Sur la résolution et l'approximation de certaines équations paraboliques nonlinéaires dégénérées*, Arch. Rational Mech. Anal. **25** (1967), 64–80.

14. W. Strauss, *Further applications of monotone methods to partial differential equations*, these Proceedings, pp. 282–288.

15. M. I. Višik, *Solvability of boundary problems for quasi-linear parabolic equations of higher order*, Mat. Sb. **59** (**101**) (1962), 289–325; English transl., Amer. Math. Soc. Transl. (2) **65** (1967), 1–40.

University of Paris, France

EXISTENCE THEOREMS FOR LAGRANGE PROBLEMS IN SOBOLEV SPACES[1]

Lamberto Cesari

We deal here with Lagrange problems of optimization in a fixed domain G of the t-space E_ν, $t = (t^1, \ldots, t^\nu)$, $\nu \geq 2$, G bounded or unbounded, with boundary conditions, partial differential equations, and possible unilateral constraints. As usual in such problems we deal with a set $z = (z^1, \ldots, z^n)$ of space, or state variables, and with a set $u = (u^1, \ldots, u^m)$ of control variables, all z^i and u^j being unknown functions of the independent variables $t = (t^1, \ldots, t^\nu) \in G$.

We shall be interested in the minimum of an integral

$$(1) \qquad I[z, u] = \int_G f_0(t, z(t), u(t))\, dt \qquad (dt = dt^1 \cdots dt^\nu),$$

in classes Ω of (admissible) pairs

$$z(t) = (z^1, \ldots, z^n), \qquad u(t) = (u^1, \ldots, u^m), \qquad t = (t^1, \ldots, t^\nu) \in G \subset E_\nu,$$

with each z^i belonging to a Sobolev class $W^1_{p_i}(G)$ of given exponent $p_i \geq 1$, $i = 1, \ldots, n$, and each u^s measurable in G, $s = 1, \ldots, m$, satisfying a given (canonic) system of νn partial differential equations

$$(2) \qquad D_j z^i = f_{ij}(t, z(t), u(t)), \qquad j = 1, \ldots, \nu, \quad i = 1, \ldots, n, \quad \text{a.e. in } G,$$

satisfying a system (B) of boundary conditions concerning the values of the functions z^i on the boundary ∂G of G, satisfying the constraints

$$(3) \qquad (t, z(t)) \in A \subset E_{\nu+n}, \qquad u(t) \in U(t, z(t)) \subset E_m, \quad \text{a.e. in } G,$$

and such that $f_0(t, z(t), u(t))$ is L-integrable in G.

If $m = \nu n$, $u = (u_{ij}, j = 1, \ldots, \nu, i = 1, \ldots, n)$, $U(t, z) = E_{\nu n}$, and $f_{ij} = u_{ij}$, the Lagrange problem above reduces to a free problem of calculus of variations.

In (2) $D_j z^i = \partial z^i / \partial t^j$, $j = 1, \ldots, \nu$, $i = 1, \ldots, n$, denote the generalized first order partial derivatives of the functions $z^i \in W^1_{p_i}(G)$ with respect to t^j in G.

If $z(t)$, $u(t)$, $t \in G$, is an admissible pair, as defined above, we say that $z(t)$ is an admissible state function, $u(t)$ is a control function relative to $z(t)$, and $u(t)$ generates

[1] Research partially supported by AF-OSR Grant 942-65 at the University of Michigan.

$z(t)$ (though different u may generate the same z, and a given u may correspond to different z).

We denote by cl S and co S the closure and the convex hull of a given set S. Concerning the constraints (3) we assume that for every $t \in \text{cl}\, G$ a nonempty subset $A(t)$ of E_n is given, and then $A = [(t, z) \mid t \in \text{cl}\, G,\ z \in A(t)] \subset E_{\nu+n}$. Analogously, we assume that for every $(t, z) \in A$ a nonempty subset $U(t, z)$ of E_m is given, and then we denote by M the set $M = [(t, z, u) \mid (t, z) \in A,\ u \in U(t, z)]$. Here $U(t, z)$ is said to be the control space relative to $(t, z) \in A$. We do not exclude the cases where $A(t) = E_n$ for every $t \in \text{cl}\, G$; hence $A = \text{cl}\, G \times E_n$, and $U(t, z) = E_m$ for every $(t, z) \in A$, hence $M = A \times E_m$, and in these cases constraints (3) are always satisfied. Constraints (3) are often denoted as unilateral constraints since in most cases the sets $A(t)$ and $U(t, z)$ are defined by inequalities.

We shall assume below that G has boundary ∂G satisfying the conditions required by Sobolev [**12**]; briefly we shall say that G is of class K.

As mentioned above, (B) denotes a given system of boundary conditions concerning the values of the functions z^i on the boundary ∂G of G. On such a system of boundary conditions (B) we shall require only the following closure property (P): If $z^i, z^i_k \in W^1_{p_i}(G)$, $i = 1, \ldots, n$, $k = 1, 2, \ldots$, if $z^k_i \to z^i$ strongly in $L_{p_i}(G)$, $D_j z^i_k \to D_j z^i$ weakly in $L_{p_i}(G)$ as $k \to \infty$, $j = 1, \ldots, \nu$, $i = 1, \ldots, n$, and if $z_k = (z^1_k, \ldots, z^n_k)$ satisfy boundary conditions (B), then $z = (z^1, \ldots, z^n)$ also satisfies (B).

A class Ω of admissible pairs z, u is said to be complete if for any sequence z_k, u_k, $k = 1, 2, \ldots$, of admissible pairs all in Ω and any other admissible pair z, u such that $z^i_k \to z^i$ strongly in $L_{p_i}(G)$, $D_j z^i_k \to D_j z^i$ weakly in $L_{p_i}(G)$ as $k \to \infty$, $j = 1, \ldots, \nu$, $i = 1, \ldots, n$, then the pair z, u belongs to Ω. The class of all admissible pairs is obviously complete.

If A is closed, then M is closed if and only if $U(t, z)$ satisfies Kuratowski's property of upper semicontinuity in A, that is,

$$U(t_0, z_0) = \bigcap_{\delta > 0} \text{cl} \bigcup_{(t,z) \in N_\delta(t_0, z_0)} U(t, z)$$

for all $(t_0, z_0) \in A$. Here $N_\delta(t_0, z_0)$ denotes the set of all $(t, z) \in A$ at a distance $\leq \delta$ from (t_0, z_0).

We shall consider below the sets $\tilde{Q}(t, z) = [\tilde{\xi} = (\xi^0, \xi) \mid \xi^0 \geq f_0(t, z, u)$, $\xi = f(t, z, u)$, $u \in U(t, z)] \subset E_{\nu n+1}$. We shall require that these sets are convex and satisfy the following property (Q) similar to Kuratowski's property:

$$\tilde{Q}(t_0, z_0) = \bigcap_{\delta > 0} \text{cl co} \bigcup_{(t,z) \in N_\delta(t_0, z_0)} \tilde{Q}(t, z)$$

for all $(t_0, z_0) \in A$.

We have been using property (Q) in one-dimensional problems [**1** (a)–(c)] and in multidimensional problems [**1** (d)–(h)] of optimal control and the calculus of variations. The same condition (Q) has been subsequently studied by C. Olech [**9**], by C. Olech and A. Lasota [**10**], and J. R. LaPalm [**5**].

The remarkable generality of canonic form (2) for systems of partial differential equations of any order has been mentioned by P. K. Rashevsky in his book on Pfaffian systems [**11**, p. 324], and recently the same canonic form has been used by K. A. Lurie [**7** (a)–(c)]. (In this paper we show the reduction of some second order partial differential equations to canonic systems (2) of first order equations.) For similar canonic forms see also T. Y. Thomas and E. W. Titt [**13**], and Hans Lewy [**6**].

If we denote the vector $D_j z^i$, $j = 1, \ldots, \nu$, $i = 1, \ldots, n$, as the Fréchet derivative dz/dt of $z \in E_n$ with respect to $t \in E_\nu$, then (2) takes the simple form $dz/dt = f(t, z(t), u(t))$, $t \in G \subset E_\nu$, where $f = (f_{ij}, \ j = 1, \ldots, \nu, \ i = 1, \ldots, n)$. The equation $dz/dt = F(t, z)$ with $t \in T$, $z \in Z$, T, Z Banach spaces, has been studied by J. Dieudonné [**2** (a),(b)] and by A. D. Michal and V. Elconin [**8**]. Recently the same equation $dz/dt = F(t, z)$ with both t and z in Montel spaces has been studied by E. Dubinsky [**3**].

Of the extensive work on the differential equation $\partial z/\partial t + A(z) = f$, ($t$ real, $z \in Z$, Z a Banach space, A an unbounded operator, f a given function), we merely mention S. Agmon and L. Nirenberg [Comm. Pure Appl. Math. **16** (1963), 121–139], T. Kato [Proc. Sympos. Appl. Math. **17** (1965), 50–67], H. Tanabe and M. Watanabe [Funkcial. Ekvac. **9** (1966), 163–170], and particularly J. L. Lions [*Optimisation pour certaines classes d'équations d'évolution non linéaires*, Ann. Mat. Pura Appl. **72** (1966), 275–293; *Équations différentielles opérationnelles*, Springer, 111, 1961].

EXISTENCE THEOREM 1. *Let G be bounded, open, and of class K, let A be compact, let $U(t, z)$ be nonempty and compact for every $(t, z) \in A$, and let us assume that the set M is compact. Let $\tilde{f}(t, z, u) = (f_0, f) = (f_0, f_{ij}, \ i = 1, \ldots, n, \ j = 1, \ldots, \nu)$ be continuous on M, and let us assume that the set $\tilde{Q}(t, z)$ of all $\tilde{\xi} = (\xi^0, \xi) = (\xi^0, \xi_{ij}) \in E_{n\nu+1}$ with $\xi^0 \geq f_0(t, z, u)$, $\xi = f(t, z, u)$, $u \in U(t, z)$, is a convex subset of $E_{n\nu+1}$ for every $(t, z) \in A$. Let $p_i \geq 1$, $i = 1, \ldots, n$, be given numbers, let* (B) *be a system of boundary conditions concerning the values of the n functions z^i on the boundary ∂G of G, and assume that* (B) *satisfies property* (P). *Let Ω be a nonempty complete class of admissible pairs $z(t) = (z^1, \ldots, z^n)$, $u(t) = (u^1, \ldots, u^m)$, $t \in G$, with $z^i \in W^1_{p_i}(G)$, $i = 1, \ldots, n$, u^s measurable, $s = 1, \ldots, m$, and satisfying given inequalities*

$$\int_G |z^i|^{p_i}\, dt \leq N_i \quad \text{for } i \in \{\beta\},$$

$$\int_G |D_j z^i|^{p_i}\, dt \leq N_{ij} \quad \text{for } i \in \{\beta\}_j,\ j = 1, \ldots, \nu,$$

for certain given constants N_i, N_{ij} and all i of certain systems $\{\beta\}$, $\{\beta\}_j$ of indices $1, 2, \ldots, n$, (which may be empty). Then the functional $I[z, u]$ possesses an absolute minimum in Ω.

EXISTENCE THEOREM 2. *Let G be bounded, open, and of class K, let A be closed, let $U(t, z)$ be nonempty and closed for every $(t, z) \in A$, and assume that the set*

M is closed. Let $\tilde{f}(t, z, u) = (f_0, f) = (f_0, f_{ij})$ be continuous on M, and let us assume that the set $\tilde{Q}(t, z)$ of all $\tilde{\xi} = (\xi^0, \xi) = (\xi^0, \xi_{ij}) \in E_{n\nu+1}$ with $\xi_0 \geq f_0(t, z, u)$, $\xi = f(t, z, u)$, $u \in U(t, z)$, is a convex closed subset of $E_{n\nu+1}$ for every $(t, z) \in A$, and that $\tilde{Q}(t, z)$ satisfies property (Q) in A. Let us assume that $f_0(t, z, u) \geq -\psi(t)$ for all $(t, z, u) \in M$ and some nonnegative function ψ L-integrable in G. Let $p_i > 1$, $i = 1, \ldots, n$, be given numbers, let (B) be a system of boundary conditions concerning the values of the functions z^i on the boundary ∂G of G, and assume that (B) satisfies property (P). Let Ω be a nonempty complete class of admissible pairs $z(t) = (z^1, \ldots, z^n)$, $u(t) = (u^1, \ldots, u^m)$, $t \in G$, with $z^i \in W^1_{p_i}(G)$, $i = 1, \ldots, n$, u^s measurable, $s = 1, \ldots, m$, and satisfying given inequalities

$$\int_G |z^i|^{p_i}\, dt \leq N_i \quad \text{for } i \in \{\beta\},$$

$$\int_G |D_j z^i|^{p_i}\, dt \leq N_{ij} \quad \text{for } i \in \{\beta\}_j,\ j = 1, \ldots, \nu,$$

for certain given constants N_i, N_{ij} and all i of certain systems $\{\beta\}$, $\{\beta\}_j$ of indices $1, 2, \ldots, n$, (which may be empty). Assume that $(z, u) \in \Omega$, $I[z, u] \leq L_0$ implies

$$\int_G |z^i|^{p_i}\, dt \leq L_i, \qquad \int_G |D_j z^i|^{p_i}\, dt \leq L_{ij}, \quad j = 1, \ldots, \nu,$$

for certain constants L_i, L_{ij} (which may depend on L_0, N_i, N_{ij}, G, (B), Ω) and for all $i = 1, \ldots, n$, which are not in $\{\beta\}$, $\{\beta\}_j$ respectively. Then the functional $I[z, u]$ possesses an absolute minimum in Ω.

EXISTENCE THEOREM 3. *Let G be bounded, open, and of class K, let A be closed, let $U(t, z)$ be nonempty and closed for every $(t, z) \in A$, and assume that the set M is closed. Let $\tilde{f}(t, z, u) = (f_0, f) = (f_0, f_{ij})$ be continuous on M, and let us assume that the set $\tilde{Q}(t, z)$ of all $\tilde{\xi} = (\xi^0, \xi) = (\xi^0, \xi_{ij}) \in E_{n\nu+1}$ with $\xi^0 \geq f_0(t, z, u)$, $\xi = f(t, z, u)$, $u \in U(t, z)$, is a convex closed subset of $E_{n\nu+1}$ for every $(t, z) \in A$, and that $\tilde{Q}(t, z)$ satisfies property (Q) in A. Let us assume that the following growth property (γ) holds: Given $\varepsilon > 0$ there is some nonnegative L-integrable function $\psi_\varepsilon(t)$, $t \in G$, such that $|f(t, z, u)| \leq \psi_\varepsilon(t) + \varepsilon f_0(t, z, u)$ for all $(t, z, u) \in M$. Let $p_i = 1$, $i = 1, \ldots, n$, let (B) be a system of boundary conditions concerning the values of the functions z^i on the boundary ∂G of G, and assume that (B) satisfies property (P). Let Ω be a nonempty complete class of admissible pairs $z(t) = (z^1, \ldots, z^n)$, $u(t) = (u^1, \ldots, u^m)$, $t \in G$, with $z^i \in W^1_1(G)$, $i = 1, \ldots, n$, u^s measurable, $s = 1, \ldots, m$, and satisfying given inequalities*

$$\int_G |z^i|\, dt \leq N_i \quad \text{for } i \in \{\beta\},$$

$$\int_G |D_j z^i|\, dt \leq N_{ij} \quad \text{for } i \in \{\beta\}_j,\ j = 1, \ldots, \nu,$$

for certain given constants N_i, N_{ij} and all i of certain systems $\{\beta\}$, $\{\beta\}_j$ of indices $1, 2, \ldots, n$, (which may be empty). Assume that $(z, u) \in \Omega$, $I[z, u] \leq L_0$ implies

$\int_G |z^i|\, dt \leq L_i$ *for certain constants* L_i *(which may depend on* L_0, N_i, N_{ij}, G, (B), Ω) *and for all* $i = 1, \ldots, n$ *which are not in* $\{\beta\}$. *Then the functional* $I[z, u]$ *possesses an absolute minimum in* Ω.

Theorems 1, 2, 3 have been proved in [**1** (d)—(g)] together with variants for the case of G unbounded. In these theorems it is sufficient to require that the boundary conditions (B) satisfy property (P) relatively to the complete class Ω under consideration and for pairs z, u of the class Ω for which $I[z, u] \leq L$.

Condition (Q) in Theorems 2 and 3 is certainly satisfied if, for instance, f, f_0 satisfy the following growth condition (ε): given $\varepsilon > 0$ there is a constant $\bar{u}_\varepsilon \geq 0$ such that $1 \leq \varepsilon f_0(t, z, u)$, $|f(t, z, u)| \leq \varepsilon f_0(t, z, u)$ for all $(t, z, u) \in M$ with $|u| \geq \bar{u}_\varepsilon$. This has been proved in [**1** (d)].

The existence theorems for Lagrange problems stated above are particular cases for analogous theorems concerning systems of partial differential equations in normal form $D^\alpha z^i = f_{i\alpha}(t, z, u)$, where $i = 1, \ldots, n$, and for every i the multi-index $\alpha = (\alpha_1, \ldots, \alpha_\nu)$ describes an arbitrary finite system of indices $\{\alpha\}_i$. Here $D^\alpha z^i = \partial^{|\alpha|} z^i / (\partial t^1)^{\alpha_1} \cdots (\partial t^{1\alpha}$ and $|\alpha| = \alpha_1 + \cdots + \alpha_\nu$ [**1** (d)–(f)].

We can only sketch here the proofs of the existence theorems. We start with Theorem 2 and we shall assume that G is bounded. It is not restrictive to assume that cl G lies in the interior of some hypercube $[0, a]$ of E_ν, $0 = (0, \ldots, 0)$, $a = (a_1, \ldots, a_\nu)$. If we introduce the auxiliary variable z^0 by taking

$$z^0(t) = \int_0^t \bar{f}_0(\tau, z(\tau), u(\tau))\, d\tau,$$

where $t = (t^1, \ldots, t^\nu) \in E_\nu$, the integration is performed in the interval $[0, t]$ of E_ν, and $\bar{f}_0 = f_0$ for $\tau \in G$ and $\bar{f}_0 = 0$ for $\tau \in E_\nu - G$, then $I[z, u] = z^0(a)$. Actually, z^0 satisfies the differential equation $D^\alpha z^0 = \bar{f}_0(t, z(t), u(t))$ a.e. in G, where $\alpha = (1, \ldots, 1)$, and the boundary conditions $z^0(t'_j, 0) = 0$, $j = 1, \ldots, \nu$. Here t'_j denotes the $(\nu - 1)$-vector obtained from $t = (t^1, \ldots, t^\nu)$ by suppressing t^j.

If i denotes the infimum of $I[z, u]$ in Ω, i finite, then there is a minimizing sequence $z_k(t), u_k(t), t \in G, k = 1, 2, \ldots$, such that $i \leq I[z_k, u_k] \leq i + k^{-1} \leq i + 1$, and if z_k^0 is the corresponding auxiliary variable, then $z_k^0(t'_j, 0) = 0$, $j = 1, \ldots, \nu$, $I[z_k, u_k] = z_k^0(a)$, $k = 1, 2, \ldots$.

In case of Theorem 2 the components z_k^i, $i = 1, \ldots, n$, belong to fixed balls in $W^1_{p_i}(G)$ which are weakly compact since $p_i > 1$. On the other hand, because of condition $f_0 \geq -\psi$, ψ L-integrable in G, and of the inequality $I[z_k, u_k] \leq i + 1$, it is possible to prove that the functions z_k^0 have uniformly bounded total variations, since

$$\int_0^a \bar{f}_0^-(t, z_k, (t), u_k(t))\, dt \leq \int_G \psi\, dt, \qquad \int_0^a \bar{f}_0^+(t, z_k(t), u_k(t))\, dt \leq |i| + 1 + \int_G \psi\, dt,$$

for all k. Thus, a Helly-type argument can be applied to the components z_k^0. As a consequence, there exists a subsequence, say still $[k]$, such that

$$z_k^0 \to z^0 \quad \text{pointwise,} \tag{4}$$

$$z_k^i \to z^i \text{ strongly in } L_{p_i}(G), \qquad D_s z_k^i \to D_s z^i \text{ weakly in } L_{p_i}(G), \tag{5}$$

as $k \to \infty$, $j = 1, \ldots, \nu$, $i = 1, \ldots, n$, where $z^i \in W^1_{p_i}(G)$, and z^0 admits of a Lebesgue decomposition $z^0 = Z + S$, Z AC and S singular, $Z(t'_j, 0) = S(t'_j, 0) = 0$, and, in addition, $S \geq 0$ and

$$Z(t) \leq z^0(t), \qquad t \in [0, a]. \tag{6}$$

In (4) pointwise convergence is assured in a suitable subset of $[0, a]$ containing 0 and a.

A closure theorem has been proved [**1** (e)], based on the convexity of the sets $\tilde{Q}(t, z)$, which shows that there exists some control function $u(t)$, $t \in G$, $u(t) \in U(t, z(t))$, such that

$$D_j z^i = f_{ij}(t, z(t), u(t)) \quad \text{a.e. in } G, \quad i = 1, \ldots, n, \quad j = 1, \ldots, \nu,$$
$$D^\alpha Z = \bar{f}_0(t, z(t), u(t)) \quad \text{a.e. in } [0, a], \quad \alpha = (1, \ldots, 1).$$

Then (z, u) is an admissible pair and even belongs to Ω [**1** (e)] so that $I[z, u] = Z(a) \geq i$. On the other hand

$$[Iz, u] = Z(a) \leq z^0(a) = \lim z^0_k(a) = \lim I[z_k, u_k] = i.$$

We conclude that $I[z, u] = Z(a) = i$ and the pair (z, u) is optimal.

The argument just sketched shows that the mentioned closure theorem is essentially a "lower semicontinuity" statement (cf. relation (6) above). The essence of this argument based on Helly's theorem has been used in [**1** (a)–(g) for one- and for multi-dimensional problems. Subsequently, it was used also by E. J. McShane [SIAM J. Control **5** (1967), 438–485], T. Nishiura [SIAM J. Control **5** (1967), 532–544], C. Olech [**9**], and J. R. LaPalm [**5**]. The same closure theorem mentioned above has been called a theorem of "lower closure" by P. Brunovsky [*On necessary conditions for lower closure of control problems*, SIAM J. Control **6** (1968), 174–185.]

For Theorem 3 the argument above remains the same but for a complement since now a ball in $W^1_1(G)$ is not weakly compact. Indeed, condition (γ) is now assumed, or $|f(t, z, u)| \leq \psi_\varepsilon(t) + \varepsilon f_0(t, z, u)$ for all $(t, z, u) \in M$ and every $\varepsilon > 0$, where $\psi_\varepsilon(t) \geq 0$ denotes an L-integrable function in G which may depend on ε. In particular, for $\varepsilon = 1$, we have $0 \leq \psi_1(t) + f_0(t, z, u)$, or $f_0 \geq -\psi_1$, where ψ_1 is L-integrable in G. Given $\varepsilon > 0$, we shall choose later another number $\sigma > 0$. For every measurable subset E of G and every admissible pair z, u with $I[z, u] \leq L_0$, we have

$$\begin{aligned}
\int_E |D_j z^i(t)|\, dt &= \int_E |f_{ij}(t, z(t), u(t))|\, dt \leq \int_E |f|\, dt \\
&\leq \int_E \psi_\sigma(t)\, dt + \sigma \int_E [f_0(t, z(t), u(t)) + \psi_1(t)]\, dt \\
&\leq \int_E \psi_\sigma(t)\, dt + \sigma \int_G f_0\, dt + \sigma \int_G \psi_1\, dt \\
&\leq \int_E \psi_\sigma(t)\, dt + \sigma L_0 + \sigma M_1,
\end{aligned}$$

where M_1 is a fixed constant. Thus, we first take $\sigma \leq \varepsilon 2^{-1}(M_1 + L_0)^{-1}$, and then from the L-integrability of ψ_σ in G we conclude that there is some $\delta > 0$ such that meas $E < \delta$ implies $\int_E \psi_\sigma \, dt < \varepsilon/2$, and finally

$$\int_E |D_j z^i(t)| \, dt \leq \varepsilon/2 + \varepsilon/2 = \varepsilon, \qquad i = 1, \ldots, n, j = 1, \ldots, \nu.$$

We see, therefore, that the growth condition (γ) of Theorem 3 guarantees the equiabsolute integrability of the functions $D_j z^i(t)$ in G, $j = 1, \ldots, \nu$, $i = 1, \ldots, n$, in the class of all z, u in Ω with $I[z, u] \leq L_0$. This implies a property of weak compactness of the Sobolev functions $z^i \in W_1^1(G)$ of the same class. In particular, the minimizing sequence z_k, u_k, $k = 1, 2, \ldots$, admits of a subsequence, say still $[k]$, for which (5) is valid with all $p_i = 1$, $i = 1, \ldots, n$. The same argument based on Helly's Theorem used for Theorem 2 guarantees the existence of a further subsequence also satisfying (4).

For Theorem 1, Ascoli's selection theorem can be used for all components z_k^i, $i = 0, 1, \ldots, n$.

Of the three existence theorems stated above, the first corresponds to Filippov's existence theorem for optimal control (the compact case), the second concerns the case where the unknown space functions z^i, $i = 1, \ldots, n$, are in given balls of Sobolev's spaces of exponents $p_i > 1$, the third concerns the case where all $p_i = 1$ and corresponds to the Tonelli-Nagumo existence theorem for unidimensional free problems of the calculus of variations. Nevertheless, the proofs are essentially new, based as they are on convexity properties, closure theorems, and Helly's selection process.

WEAK SOLUTIONS. In many cases the sets $\tilde{Q}$ are not convex, and examples show that an optimal solution may not exist. In these cases it has been proposed to replace the system of $n\nu$ partial differential equations (2) and the functional (1) by a new system of $n\nu$ first order partial differential equations and a new functional

$$(7) \qquad \begin{gathered} D_j z^i = g_{ij}(t, z, \lambda, v), \qquad j = 1, \ldots, \nu, i = 1, \ldots, n, \quad \text{a.e. in } G, \\ J[z, \lambda, v] = \int_G g_0(t, z, \lambda, v) \, dt, \end{gathered}$$

where

$$(8) \qquad \begin{gathered} g_{ij}(t, z, \lambda, v) = \Sigma_h \lambda_h f_{ij}(t, z, u^{(h)}), \\ g_0(t, z, \lambda, v) = \Sigma_h \lambda_h f_0(t, z, u^{(h)}), \\ \lambda = (\lambda_1, \ldots, \lambda_\mu), \qquad u^{(h)} = (u_h^1, \ldots, u_h^m), \\ \lambda \in \Gamma = [\lambda_h \geq 0, \lambda_1 + \cdots + \lambda_\mu = 1], \qquad u^{(h)} \in U(t, z), \quad h = 1, \ldots, \mu, \end{gathered}$$

where Σ ranges from 1 to some integer $\mu \geq 1$. We still have here n state variables, that is, $z = (z^1, \ldots, z^n)$, but the control variable, or m-vector u, is now replaced by a new control variable, or $(\mu m + \mu)$-vector $(v, \lambda) = (u^{(1)}, \ldots, u^{(\mu)}, \lambda_1, \ldots, \lambda_\mu)$

(Gamkrelidze's sliding regimes [4], or generalized or weak solutions). Each m-vector $u^{(h)} = (u_h^1, \ldots, u_h^m)$ is subject to the same constraint as before, or $u^{(h)} \in U(t, z)$, $h = 1, \ldots, \mu$, while the μ scalars λ_h are subject to the limitations $\lambda_h \geq 0$, $\lambda_1 + \cdots + \lambda_\mu = 1$. In other words, if Γ denotes the simplex in (8), we require $(v, \lambda) \in V(t, z) = U^\mu \times \Gamma$, where now $V(t, z)$ is the new control space. We shall require the n state functions z^i, $i = 1, \ldots, n$, to belong to the same Sobolev spaces $W^1_{p_i}(G)$, $p_i \geq 1$, as before, and to satisfy the same boundary conditions (B) on the boundary ∂G of G as before. We shall require all functions $u^{(h)}$, λ_h, $h = 1, \ldots, \mu$, to be measurable in G. Interpreting the λ_h as probability distributions, the new state variables z^i can be thought of as generated by a probability distribution of the μ controls $u^{(h)}$ (acting contemporaneously). The sets $\tilde{Q}(t, z)$ shall be replaced by analogous sets $\tilde{Q}^*(t, z)$. Here $\tilde{Q}^*(t, z)$ is the set of all points $\tilde{\xi} = (\xi^0, \xi) = (\xi^0, \xi_{ij}) \in E_{nv+1}$ with

$$\xi^0 \geq \Sigma_h \lambda_h f_0(t, z, u^{(h)}), \qquad \xi = \Sigma_h \lambda_h f(t, z, u^{(h)}),$$
$$\lambda \in \Gamma, \quad u^{(h)} \in U(t, z), \quad h = 1, \ldots, \mu.$$

Hence, each point of $\tilde{Q}^*$ can be thought of as the convex combination of μ points of the sets $\tilde{Q}$. Thus, for $\mu \geq 2n + 2$ we certainly have $\tilde{Q}^*(t, z) = \text{co}\, \tilde{Q}(t, z)$ and the sets $\tilde{Q}^*$ are convex, but it may well be that $\tilde{Q}^*(t, z) = \text{co}\, \tilde{Q}(t, z)$ for values of $\mu < 2n + 2$. Below we shall only require that μ is so chosen as to make the sets $\tilde{Q}^*$ all convex, namely $\tilde{Q}^*(t, z) = \text{co}\, \tilde{Q}(t, z)$. The remaining notations used below for weak solutions are obvious modifications of those introduced above.

EXISTENCE THEOREM 4 (FOR WEAK SOLUTIONS). *Let G be bounded, open, and of class K, let A be compact, let $U(t, z)$ be nonempty and compact for every $(t, z) \in A$, and let us assume that the set M be compact. Let $\tilde{f}(t, z, u) = (f_0, f) = (f_0, f_{ij})$ be continuous on M, and let us assume that $\tilde{Q}^*(t, z)$ is a convex closed subset of E_{nv+1} for every $(t, z) \in A$. Let $p_i \geq 1$, $i = 1, \ldots, n$, be given numbers, let* (B) *be a system of boundary conditions concerning the values of the n functions z^i on the boundary ∂G of G, and assume that* (B) *satisfies property* (P). *Let Ω^* be a nonempty complete class of admissible systems $z(t) = (z^1, \ldots, z^n)$, $\lambda(t) = (\lambda_1, \ldots, \lambda_\mu)$, $v(t) = (u^{(1)}, \ldots, u^{(\mu)})$, $t \in G$, with $z^i \in W^1_{p_i}(G)$, $i = 1, \ldots, n$, λ_h, $u^{(h)}$ measurable in G, $h = 1, \ldots, \mu$, and satisfying given inequalities*

$$\int_G |z^i|^{p_i}\, dt \leq N_i \quad \textit{for } i \in \{\beta\},$$

$$\int_G |D_j z^i|^{p_i}\, dt \leq N_{ij} \quad \textit{for } i \in \{\beta\}_j,\ j = 1, \ldots, \nu,$$

for certain given constants N_i, N_{ij} and all i of certain systems $\{\beta\}$, $\{\beta\}_j$ of indices $1, 2, \ldots, n$, (which may be empty). Then the functional $J[z, \lambda, v]$ possesses an absolute minimum in Ω.

EXISTENCE THEOREM 5 (FOR WEAK SOLUTIONS). *Let G be bounded, open, and of class K, let A be closed, let $U(t, z)$ be nonempty and closed for every $(t, z) \in A$,* *and assume that the set M is closed. Let $\tilde{f}(t, z, u) = (f_0, f) = (f_0, f_{ij})$ be continuous on M, and let us assume that $\tilde{Q}^*(t, z)$ is a closed convex subset of $E_{n\nu+1}$ for every $(t, z) \in A$, and that $\tilde{Q}^*(t, z)$ satisfies property* (Q) *in A. Let us assume that $f_0(t, z, u) \geq -\psi(t)$ for all $(t, z, u) \in M$ and some nonnegative L-integrable function ψ in G. Let $p_i > 1$, $i = 1, \ldots, n$, be given numbers, let* (B) *be a system of boundary conditions concerning the values of the n functions z^i on the boundary ∂G of G, and assume that* (B) *satisfies property* (P). *Let Ω^* be a nonempty complete class of admissible systems $z(t) = (z^1, \ldots, z^n)$, $\lambda(t) = (\lambda_1, \ldots, \lambda_\mu)$, $v(t) = (u^{(1)}, \ldots, u^{(\mu)})$, $t \in G$, with $z^i \in W^1_{p_i}(G)$, $i = 1, \ldots, n$, λ_h, $u^{(h)}$ measurable in G, $h = 1, \ldots, \mu$, and satisfying given inequalities*

$$\int_G |z^i|^{p_i}\, dt \leq N_i \quad \text{for } i \in \{\beta\},$$

$$\int_G |D_j z^i|^{p_i}\, dt \leq N_{ij} \quad \text{for } i \in \{\beta\}_j,\ j = 1, \ldots, \nu,$$

for certain given constants N_i, N_{ij} and all i of certain systems $\{\beta\}$, $\{\beta\}_j$ of indices $1, 2, \ldots, n$, (which may be empty). Assume that $(z, \lambda, v) \in \Omega^*$, $J(z, \lambda, v) \leq L_0$ implies

$$\int_G |z^i|^{p_i}\, dt \leq L_i,$$

$$\int_G |D_j z^i|^{p_i}\, dt \leq L_{ij}, \quad j = 1, \ldots, \nu,$$

for certain constants L_i, L_{ij} (which may depend on L_0, N_i, N_{ij}, G, (B), Ω) *and for all $i = 1, \ldots, n$, which are not in $\{\beta\}$, $\{\beta\}_j$ respectively. Then the functional $J[z, \lambda, v]$ possesses an absolute minimum in Ω^*.*

EXISTENCE THEOREM 6 (FOR WEAK SOLUTIONS). *Let G be bounded, open, and of class K, let A be closed, let $U(t, z)$ be nonempty and closed for every $(t, z) \in A$, and assume that the set M is closed. Let $\tilde{f}(t, z, u) = (f_0, f) = (f_0, f_{ij})$ be continuous on M, and let us assume that $\tilde{Q}^*(t, z)$ is a closed convex subset of $E_{n\nu+1}$ for every $(t, z) \in A$ and that $\tilde{Q}^*(t, z)$ satisfies property* (Q) *in A. Let us assume that the following growth property (γ) holds: Given $\varepsilon > 0$ there is some nonnegative L-integrable function $\psi_\varepsilon(t)$, $t \in G$, such that $|f(t, z, u)| \leq \psi_\varepsilon(t) + \varepsilon f_0(t, z, u)$ for all $(t, z, u) \in M$. Let $p_i = 1$, $i = 1, \ldots, n$, let* (B) *be a system of boundary conditions concerning the values of the n functions z^i on the boundary ∂G of G, and assume that* (B) *satisfies property* (P). *Let Ω^* be a nonempty complete class of admissible systems $z(t) = (z^1, \ldots, z^n)$, $\lambda(t) = (\lambda_1, \ldots, \lambda_\mu)$, $v(t) = (u^{(1)}, \ldots, u^{(\mu)})$, $t \in G$, with $z^i \in W^1_1(G)$, $i = 1, \ldots, n$, λ_h, $u^{(h)}$ measurable in G, $h = 1, \ldots, \mu$, and satisfying given inequalities*

$$\int_G |z^i|\, dt \leq N_i \quad \text{for } i \in \{\beta\}, \qquad \int_G |D_j z^i|\, dt \leq N_{ij} \quad \text{for } i \in \{\beta\}_j,\ j = 1, \ldots, \nu,$$

for certain given constants N_i, N_{ij} and all i of certain systems $\{\beta\}$, $\{\beta\}_j$ of indices $1, 2, \ldots, n$, (which may be empty). Assume that $(z, \lambda, v) \in \Omega^*$, $J[z, \lambda, v] \leq L_0$ implies $\int_G |z^i|\, dt \leq L_i$, for certain constants L_i (which may depend on L_0, N_i, N_{ij},

G, (B), Ω) and for all $i = 1, \ldots, n$, which are not in $\{\beta\}$. Then the functional $J[z, \lambda, v]$ possesses an absolute minimum in Ω^*.

In Theorems 4, 5, 6 it is sufficient to require that the boundary conditions (B) satisfy property (P) relatively to the complete class Ω^* under consideration and for systems z, λ, v of the class Ω^* for which $J[z, \lambda, v] \le L_0$.

In Theorem 6 we require growth condition (γ) to be verified for the original functions f_0, f. It is easily seen that condition (γ) implies the same property for the functions g_0, g.

Condition (Q) in both Theorems 5 and 6 is certainly satisfied if, for instance, the original functions f_0, f satisfy the following growth condition: (ε) given $\varepsilon > 0$ there is some constant $\bar{u}_\varepsilon \ge 0$ such that $1 \le \varepsilon f_0(t, z, u)$, $|f(t, z, u)| \le \varepsilon f_0(t, z, u)$ for all $(t, z, u) \in M$ with $|u| \ge \bar{u}_\varepsilon$. This was proved in [**1** (c)].

Let us remark that the typical equations in the xy-plane

$$z^1_{xx} \pm z^1_{yy} = f(x, y, z^1, z^1_x, z^1_y, u)$$

can be written in the canonic form

$$\begin{aligned} &z^1_x = z^2, \qquad z^2_x = w, \qquad z^3_x = v, \\ &z^1_y = z^3, \qquad z^2_y = v, \qquad z^3_y = \mp\, w \pm f(x, y, z^1, z^2, z^3, u), \end{aligned}$$

with $z = (z^1, z^2, z^3)$ state variables, (u, v, w) control variables. Analogously, the equation

$$z^1_{yy} = f(x, y, z^1, z^1_x, z^1_y, u)$$

can be written in the canonic form

$$\begin{aligned} &z^1_x = z^2, \qquad z^2_x = w, \qquad z^3_x = v, \\ &z^1_y = z^3, \qquad z^2_y = v, \qquad z^3_y = f(x, y, z^1, z^2, z^3, u). \end{aligned}$$

We have seen already at the beginning how free problems of the calculus of variations can be written as Lagrange problems with side conditions in canonic form. Now it is easy to formulate examples of Lagrange problems for which the existence of optimal solutions (usual or generalized) follows immediately from Theorems 1-6. In [**1** (f)–(g)] we discuss problems of minimum with unilateral constraints in an unbounded domain.

References

1. L. Cesari, (a) *Existence theorems for optimal solutions in Pontryagin and Lagrange problems*, SIAM J. Control **3** (1965), 475–498.—(b) *Existence theorems for weak and usual optimal solutions in Lagrange problems with unilateral constraints.* I, II, Trans. Amer. Math. Soc. **124** (1966), 369–412, 413–429.—(c) *Existence theorems for optimal control of the Mayer type*, SIAM J. Control **6** (1968), 517–552.—(d) "Existence theorems for multidimensional problems of optimal control" in *Differential equations and dynamical systems*, Academic Press, New York, 1967, pp. 115–132.—(e) *Existence theorems for multidimensional Lagrange problems*, J. Optimization Theory Appl. **1** (1967), 87–112.—(f) *Sobolev spaces and multidimensional Lagrange problems of optimization*, Ann. Scuola Norm. Sup. Pisa. **22** (1968), 193–227.—(g) *Multidimensional Lagrange problems of optimization in a fixed domain and an application to a problem of magnetohydrodynamics*,

Arch. Rational Mech. Anal. **29** (1968), 81–104.—(h) "Multidimensional Lagrange and Pontrjagin problems" in *Mathematical theory of control*, Academic Press, New York, 1967, pp. 272–284.

2. J. Dieudonné, (a) *Foundations of modern analysis*, Academic Press, New York, 1960. —(b) *Deux examples singuliers d'équations différentielles*, Acta Sci. Math. **12** (1950), 38–40.

3. E. Dubinsky, *Differential equations and differential calculus in Montel spaces*, Trans. Amer. Math. Soc. **110** (1964), 1–21.

4. R. V. Gamkrelidze, *On sliding optimal states*, Dokl. Akad. Nauk SSSR, **143** (1962), 1243–1246 = Soviet Math. Dokl. **3** (1962), 559–561.

5. J. R. LaPalm, *Existence theorems for problems of optimal control and the calculus of variations with exceptional sets*, Ph.D. thesis at the University of Michigan, November 1967.

6. H. Lewy, *Ueber das Anfangswertproblem einer hyperbolischen nichtlinearen partiellen Differentialgleichung zweiter Ordnung mit zwei unabhaengigen Veraenderlichen*, Math. Ann. **98** (1929), 179–191.

7. K. A. Lurie, (a) *The Mayer-Bolza problem for multiple integrals and the optimization of the performance of systems with distributed parameters*, Prikl. Mat. Meh. **27** (1963), 842–853; = J. Appl. Math. Mech. **27** (1964), 1284–1299.—(b) *Optimum control of conductivity of a fluid moving in a channel in a magnetic field*, Prikl. Mat. Meh. **28** (1964), 258–267; = J. Appl. Math. Mech. **28** (1964), 316–327.—(c) "The Mayer-Bolza problem for multiple integrals: some optimum problems for elliptic differential equations arising in magneto-hydrodynamics" in *Topics in optimization*, Academic Press, New York, 1967, 147–193.

8. A. D. Michal and V. Elconin, *Completely integrable differential equations in abstract spaces*, Acta Math. **68** (1937), 71–107.

9. C. Olech, *Existence theorems for optimal problems with vector-valued cost function*, Trans. Amer. Math. Soc. **136** (1969), 159–180.

10. C. Olech and A. Lasota, *On the closedness of the sets of trajectories of a control system* Bull. Acad. Polon. Sci. Sér. Sci. Math. Astonom. Phys. **14** (1966), 615–621.

11. P. K. Rashevsky, *Geometric theory of partial differential equations*, OGIZ, Moscow, 1947 (Russian)

12. S. L. Sobolev, *Applications of functional analysis in mathematical physics*, Transl. Math. Monographs., vol. 7, Amer. Math. Soc., Providence, R.I., 1963.

13. T. Y. Thomas and E. W. Titt, *Systems of partial differential equations and their characteristic surfaces*, Ann. of Math. (2) **33** (1932), 1–80.

UNIVERSITY OF MICHIGAN

UPPER AND LOWER BOUNDS FOR THE NUMBER OF SOLUTIONS OF NONLINEAR EQUATIONS[1]

Jane Cronin

1. **Introduction.** In a previous paper [**7**] the Leray-Schauder degree, used with the Smale-Sard Theorem [**11**], has been applied to obtain estimates on the number of solutions of certain nonlinear functional equations. Here we obtain further results in this direction. Our main interest is in functional equations involving polynomial operators and our results can be regarded as generalizations of aspects of the Fundamental Theorem of Algebra. The results are stated for polynomial operators although, as will be indicated, some of the results hold for larger classes of equations.

The operators studied are of the form $I + P$ where I is the identity operator and P is a completely continuous (i.e., compact) polynomial operator. One further hypothesis is imposed: in defining the Leray-Schauder degree of a map $I + F$, the usual procedure is to use the compactness of F to get a finite-dimensional approximation of $I + F$. More precisely, the degree of $I + F$ at 0 and relative to the closure $\overline{W}$ of a bounded open set W is defined to be the Brouwer degree of an approximation to $I + F$ in the finite-dimensional space which contains a certain ε-net in the compact set $F(\overline{W})$. The existence of the ε-net follows directly from the compactness, but the ε-net is not obtained constructively. Consequently, the finite-dimensional approximation of $I + F$ is not described explicitly and there is little hope, in most cases, of computing the degree. We impose the stronger hypothesis (see Assumptions 3 and 5 in §§2 and 3 respectively) that there exists a finite-dimensional linear map from which the finite-dimensional approximation can be obtained. This extra hypothesis is esthetically unattractive, but it is satisfied in applications to integral and elliptic differential equations and it makes possible the explicit computation of the Leray-Schauder degree in many cases.

In §2, we obtain an upper bound for the number of solutions of a functional equation in a complex Banach space. This result is a fairly strict analog of the statement that a polynomial of degree n has n roots. In §3, real solutions of a functional equation are studied. We require (Assumption 4) that an analog of the condition that the polynomial have real coefficients be satisfied and we obtain upper and lower bounds for the number of real solutions. These bounds contain

[1] This research was supported by the U.S. Army Research Office (Durham) (DA-ARO-D-31-124-G784).

generalizations of the statement that the number of real roots of a polynomial with real coefficients is less than or equal to the degree of the polynomial and is equal (mod 2) to the degree. In §4, we show how the results in §§2 and 3 can be applied to nonlinear elliptic differential equations.

2. The number of solutions of an equation in a complex Banach space. Let $\mathscr{B}$ be a Banach space over the complex numbers which is the complexification (see Bachman and Narici [**2**, p. 21 ff.]) of a real Banach space B and let $\mathscr{P}_m$ $(m = 1, \ldots, q)$ be a continuous homogeneous polynomial of degree m from $\mathscr{B}$ into $\mathscr{B}$ (see Hille and Phillips [**8**, pp. 760–770]). Let $\mathscr{P}_0$ denote the constant function defined by: for each $z \in B$, $\mathscr{P}_0(z) = z_0$ where z_0 is a fixed element of B.

ASSUMPTION 1. The operator $\mathscr{P} = \sum_{m=1}^{q} \mathscr{P}_m$ is compact, i.e., if $\mathscr{E}$ is a bounded set in $\mathscr{B}$, then $\mathscr{P}(\mathscr{E})$ is compact in the space (each infinite subset of $\mathscr{P}(\mathscr{E})$ has a limit point in $\mathscr{B}$).

ASSUMPTION 2. There exists a constant $r_0 > 0$ such that if $\|z\| = r_0$ then

$$\|(\mathscr{I} - \mathscr{P})z\| \neq 0,$$

where $\mathscr{I}$ is the identity map on $\mathscr{B}$.

REMARK. If Assumption 2 is not satisfied, then the equation

$$(\mathscr{I} - \mathscr{P})z = 0$$

is such that for each positive number r, the equation has a solution $z = z_1$, such that $\|z_1\| = r$.

LEMMA 1. *If $\mathscr{W}$ is a bounded open set in B such that $z \in \overline{\mathscr{W}} - \mathscr{W}$ implies $(\mathscr{I} - \mathscr{P})z \neq 0$, then there exists a constant $b > 0$ such that $z \in \overline{W} - W$ implies*

$$\|(\mathscr{I} - \mathscr{P})z\| > b. \tag{1}$$

PROOF. Since $\mathscr{P}$ is compact, then $(\mathscr{I} - \mathscr{P})$ is a closed map.

ASSUMPTION 3. Suppose $\mathscr{W}$ is a bounded open set in $\mathscr{B}$ such that $z \in \overline{\mathscr{W}} - \mathscr{W}$ implies $(\mathscr{I} - \mathscr{P})z \neq 0$. Let b be the constant given by Lemma 1. There exists a finite-dimensional (say of dimension n) subspace $\mathscr{N}$ of $\mathscr{B}$ and a linear (continuous) map $\mathscr{A} : \mathscr{B} \to \mathscr{B}$ such that *at* $z \in \overline{W}$ implies $\mathscr{A}\mathscr{P}(z) \in N$ and

$$\|\mathscr{A}\mathscr{P}(z) - \mathscr{P}(z)\| < b/2.$$

In order to discuss the topological degree of a map we must consider the corresponding real linear space $B \times B$. First we obtain, corresponding to each $\mathscr{P}_m$ $(m = 1, \ldots, q)$, a continuous homogeneous polynomial $\tilde{P}_m$ of degree m from $B \times B$ into $B \times B$. Let $\mathbf{P}(z_1, \ldots, z_m)$ denote the continuous symmetric m-linear form which is the polar form of $\mathscr{P}_m$ [**8**, p. 762]. If $z \in \mathscr{B}$ then z can be represented as (x, y) where $x, y \in B$ and we have

$$\mathscr{P}_m(z) = \mathbf{P}(x, y), \ldots, (x, y)).$$

If $(u, v) \in B \times B$, we define the projections:

$$\pi_1:(u, v) \to (u, 0) \qquad \pi_2:(u, v) \to (0, v).$$

Then it follows that the map

$$\tilde{\mathbf{P}}:((x_1, y_1), \ldots, (x_m, y_m)) \to \pi_1\mathbf{P}((x_1, y_1), \ldots, (x_m, y_m)) + \pi_2\mathbf{P}((x_1, y_1), \ldots, (x_m, y_m))$$

is a continuous symmetric m-linear form.

DEFINITION 1. Let $\tilde{P}_m$ be the homogeneous polynomial of degree m whose symmetric m-linear form is $\tilde{\mathbf{P}}$. Let $\tilde{P}_0$ be the constant map from $B \times B$ into (x_0, y_0) where $z_0 = (x_0, y_0)$.

Corresponding to map $\mathscr{A}$, we have a map $\tilde{A}:B \times B \to B \times B$ defined as follows:

DEFINITION 2. If $z = (x, y)$, then

$$\tilde{A}:(x, y) \to \pi_1\mathscr{A}_z + \pi_2\mathscr{A}_z.$$

Now let $\mathscr{S} = \{z \in B \mid \|z\| \leq r_0\}$ where r_0 is the constant in Assumption 2. Since the natural isomorphism between $\mathscr{B}$ and $B \times B$, i.e., $z \leftrightarrow (x, y)$ is norm-preserving, then from Assumption 2 and Assumption 3 with $\overline{\mathscr{W}} = \mathscr{S}$, it follows that the Leray-Schauder degree

$$d\left(I - \sum_{m=0}^{q} \tilde{P}_m, \tilde{S}, 0\right), \tag{3}$$

where I is the identity map on $B \times B$ and $\tilde{S}$ is the sphere in $B \times B$ with center 0 and radius r_0, is defined and is equal to

$$d\left(I - \tilde{A} \sum_{m=0}^{q} \tilde{P}_m, \tilde{S} \cap \tilde{N}, 0\right) \tag{4}$$

where $\tilde{N}$ is the $(2n)$-dimensional real Euclidean n-space underlying $\mathscr{N}$. Because it is more convenient to work with the maps in the complex spaces, we write (3) and (4) as $d(\mathscr{I} - \mathscr{P}, \mathscr{S}, 0)$ and $d(\mathscr{I} - \mathscr{A}\mathscr{P}, \mathscr{S} \cap \mathscr{N}, 0)$ respectively.

Now we compute the explicit form of

$$(\mathscr{I} - \mathscr{A}\mathscr{P})|\mathscr{S} \cap \mathscr{N}.$$

Let $z_1, \ldots, z_n$ be a fixed basis in $\mathscr{N}$. Then $z \in \mathscr{N}$ implies $z = a_1z_1 + \cdots + a_nz_n$. The constant $\mathscr{A}z_0$ is given, say, by $\mathscr{A}z_0 = a_1^0a_1 + \cdots + a_n^0z_n$. Hence

$$(\mathscr{I} - \mathscr{A}\mathscr{P})(a_1z_1 + \cdots + a_nz_n) = a_1z_1 + \cdots + a_nz_n - a_1^0z_1 - \cdots - a_n - \mathscr{A}\mathscr{P}_1(a_1z_1 + \cdots + a_nz_n) - \cdots - \mathscr{A}\mathscr{P}_q(a_1z_1 + \cdots + a_nz_n).$$

Since $\mathscr{P}_m$ is a homogeneous polynomial of degree m, then by [8, Theorem 26. 2.2],

$$\mathscr{P}_m(a_1z_1 + \cdots + a_nz_n) = \sum_{\substack{k_1+\cdots+k_n=m \\ k_i \geq 0}} (a_1^k \cdots a_n^k n)\mathscr{Q}^{(m)}_{k_1\cdots k_n}(z_1, \ldots, z_n) \tag{5}$$

where $\mathscr{Q}^{(m)}_{k_1\cdots k_n}(z_1, \ldots, z_n)$ depends on $z_1, \ldots, z_n$ only and is homogeneous of degree k_i in z_i $(i = 1, \ldots, n)$. Also $\mathscr{Q}^{(m)}_{k_1\cdots k_n}$ is continuous by [**8**, Theorem 26.2.2] and [**8**, Theorem 26.2.6].

Suppose

$$\mathscr{A}\mathscr{P}_1(z_1) = b_1^{(1)}z_1 + \cdots + b_n^{(1)}z_n$$
$$\cdots\cdots\cdots\cdots\cdots\cdots\cdots\cdots$$
$$\mathscr{A}\mathscr{P}_1(z_n) = b_1^{(n)}z_1 + \cdots + b_n^{(n)}z_n$$
$$\mathscr{A}\mathscr{Q}^{(m)}_{k_1\cdots k_n}(z_1, \ldots, z_n) = b_1^{(k_1\ldots k_n)m}z_1 + \cdots + b_n^{(k_1\ldots k_n)}.$$

Then

$$\begin{aligned}(\mathscr{I} - \mathscr{A}\mathscr{P})(z) = &-(a_1^0 z_1 + \cdots + a_n^0 z_n) + a_1 z_1 + \cdots + a_n z_n \\ &- a_1(b_1^{(1)}z_1 + \cdots + b_n^{(1)}z_n) \\ &- \cdots - a_n(b_1^{(n)}z_1 + \cdots + b_n^{(n)}z_n) \\ &- a_1^q[b_1^{(k_1 0\cdots 0)q}z_1 + \cdots + b_n^{(k_1 0\cdots 0)q}z_n] \\ &- \cdots.\end{aligned}$$

Thus if the map $(\mathscr{I} - \mathscr{A}\mathscr{P})/\mathscr{N}$ is described in terms of the coefficients of $z_1, \ldots, z_n$, it has the form:

$$\begin{aligned}a_i &= a_i - a_i^0 - a_1 b_1^{(i)} - \cdots - a_n b_n^{(i)} - \cdots - a_1^q b_i^{(k_1 0\cdots 0)q} - \cdots - \\ &= p_i(a_1, \ldots, a_n)\end{aligned}$$

where p_i is a polynomial of degree q in $a_1, \ldots, a_n$ $(i = 1, \ldots, n)$. Let $p_i^{(q)}$ denote the homogeneous polynomial which is the sum of the terms of degree q in p_i.

THEOREM 1. *If Assumptions* 1, 2, 3 *are satisfied, then* $d(\mathscr{I} - \mathscr{P}, \mathscr{S}, 0)$ *is defined and*

$$0 \leq d(I - P, S, 0) \leq q^n.$$

If the resultant of $p_1^{(q)}, \ldots, p_n^{(q)}$ *is nonzero and if the radius of* $\mathscr{S}$ *is sufficiently large, then*

$$d(\mathscr{I} - \mathscr{P}, \mathscr{S}, 0) = q^n.$$

PROOF. The proof of the first statement is by the same argument as for the proof of [**7**, Theorem 3]. It is only necessary to verify hypothesis (H-3), §3 of [**7**]. This is straightforward because $\mathscr{P}$ is differentiable. The proof of the second statement follows from [**4**, pp. 44–46].

REMARK. Notice that q is independent of the radius r_0 of $\mathscr{S}$ and of the number b. The number n depends on b in this sense: if b is very small, then the map $\mathscr{A}$ must, in general, be "fine," i.e., the dimension n of subspace $\mathscr{N}$ must be large.

THEOREM 2. *Let* $\mathscr{C}$ *be the open set which is the component (of connectedness) of* $\mathscr{B} - (\mathscr{I} - \mathscr{P})(\dot{\mathscr{S}})$ *(where* $\dot{\mathscr{S}}$ *is the point set boundary of* $\mathscr{S}$*) which contains* 0. *Then* *there is a first category set* $\mathscr{K} \subset \mathscr{C}$ *such that if* $v \in C - K$, *then the number of distinct solutions of*

$$(\mathscr{I} - \mathscr{P})u = v \tag{6}$$

in the interior of $\mathscr{S}$ *is equal to* $d(\mathscr{I} - \mathscr{P}, \mathscr{S}, 0)$. *If* $v \in \mathscr{K}$, *then the number of distinct solutions of* (6) *in the interior of* $\mathscr{S}$ *is infinite or is less than or equal to* $d(\mathscr{I} - \mathscr{P}, \mathscr{S}, 0)$.

PROOF. From the basic properties of topological degree, if $w \in \mathscr{C}$, then

$$d(\mathscr{I} - \mathscr{P}, \mathscr{S}, w) = d(\mathscr{I} - \mathscr{P}, \mathscr{S}, 0).$$

Let

$$\mathscr{K} = \{w \in \mathscr{C} \mid w \text{ is a singular value of } \mathscr{I} - \mathscr{P}\}.$$

Since $\mathscr{I} - \mathscr{P}$ is differentiable and $\mathscr{P}$ is compact, then $\mathscr{I} - \mathscr{P}$ is a Fredholm map of index zero and by the Smale-Sard Theorem (see Smale [**11**]), the set $\mathscr{K}$ is first category. If $v \in \mathscr{C} - \mathscr{K}$ and if $u_0 \in \mathscr{S} \cap (\mathscr{I} - \mathscr{P})^{-1}v$ then from the definition of regular value it follows that the differential of $\mathscr{I} - \mathscr{P}$ at u_0 is $1 - 1$. Hence [**7**, Lemma 5] the topological index of $\mathscr{I} - \mathscr{P}$ at u_0 is $+1$. From this fact and the fact that P is compact, it follows that $\mathscr{S} \cap (\mathscr{I} - \mathscr{P})^{-1}u_0$ is finite. Thus $d(\mathscr{I} - \mathscr{P}, \mathscr{S}, 0)$ is the sum of the topological indices of the points in $\mathscr{S} \cap (\mathscr{I} - \mathscr{P})^{-1}u_0$. Since each index is $+1$, this yields the first conclusion of the theorem.

Now suppose $v \in \mathscr{K}$ and suppose $\mathscr{S} \cap (\mathscr{I} - \mathscr{P})^{-1}v$ is a finite set with more than $d(\mathscr{I} - \mathscr{P}, \mathscr{S}, 0)$ elements. Call these elements $u_1, \ldots, u_t$ where $t > d(\mathscr{I} - \mathscr{P}, \mathscr{S}, 0)$. Since $\{u_1, \ldots, u_t\}$ is a finite set, then each u_i $(i = 1, \ldots, t)$ is an isolated v-point and hence the topological index of $\mathscr{I} - \mathscr{P}$ at u_i is defined and is positive (see Schwartz [**10**, p. 257]). The sum of the topological indices is greater than or equal to t and is equal to $d(\mathscr{I} - \mathscr{P}, \mathscr{S}, 0)$. Contradiction.

COROLLARY. *Let* $F: S \to B$ *be a completely continuous operator such that* $z \in \dot{\mathscr{S}}$ *implies*

$$\|(\mathscr{I} - \mathscr{F})z - (\mathscr{I} - \mathscr{P})z\| \leq \max\{\|(\mathscr{I} - \mathscr{P})z\|, \|(\mathscr{I} - \mathscr{F})z\|\}.$$

Then $d(\mathscr{I} - \mathscr{F}, \mathscr{S}, 0)$ *is defined and*

$$d(\mathscr{I} - \mathscr{F}, \mathscr{S}, 0) = d(\mathscr{I} - \mathscr{P}, \mathscr{S}, 0),$$

and if $\mathscr{F}$ *is differentiable, the conclusions of Theorem* 2 *hold for* $\mathscr{I} - \mathscr{F}$.

PROOF. That $d(\mathscr{I} - \mathscr{F}, \mathscr{S}, 0) = d(\mathscr{I} - \mathscr{P}, \mathscr{S}, 0)$ follows from the invariance under homotopy of the degree. The argument in the proof of Theorem 2 is applicable because $\mathscr{F}$ is differentiable.

3. The number of solutions of an equation in a real Banach space. Besides Assumptions 1, 2, 3, we want now to impose an assumption which corresponds to the condition that the coefficients of a polynomial are real. By [8, Theorem 26.2.2] we have for each positive integer n,

$$\mathscr{P}_m(a_1 z_1 + \cdots + a_n z_n) = \sum_{\substack{k_1+\cdots+k_n=m \\ k_i \geq 0 \quad (i=1,\ldots,n)}} (a_1^{k_1} \cdots a_n^{k_n}) \mathscr{Q}^{(m)}_{k_1\cdots k_n}(z_1, \ldots, z_n)$$

where $\mathscr{Q}^{(m)}_{k_1\cdots k_n}(z_1, \ldots, z_n)$ is a continuous homogeneous polynomial of degree k_i in z_i. If $z \in B$, then z can be represented as (x, y) where $x, y \in B$. Let z^* denote $(x, -y)$.

ASSUMPTION 4. For $m = 1, \ldots, q$ and for all $k_1, \ldots, k_n$ such that $k_1 + \cdots + k_n = m$ and for all $z_1, \ldots, z_n \in B$,

$$\mathscr{Q}^{(m)}_{k_1\cdots k_n}(z_1^*, \ldots, z_n^*) = [\mathscr{Q}^{(m)}_{k_1\cdots k_n}(z_1, \ldots, z_n)]^* \tag{7}$$

and for all $z \in B$ $[P_0(z)]^* = P_0(z)$, i.e., $z_0^* = z_0$.

Assumption 4 implies that $\mathbf{P}$, the polar form of P_m, is such that

$$\mathbf{P}((x, 0), \ldots, (x, 0)) \in B \times \{0\}$$

because

$$\begin{aligned}[\mathbf{P}(x, 0), \ldots, (x, 0))]^* &= [\mathscr{P}_m((x, 0))]^* \\ &= \mathscr{P}_m([(x, 0)]^*) \\ &= \mathscr{P}_m((x, 0)) \\ &= \mathbf{P}((x, 0), \ldots, (x, 0)).\end{aligned}$$

Then if j is the standard map from $B \times \{0\}$ into B, the function defined by $j\mathbf{P}((x, 0), \ldots, (x, 0))$ is a symmetric m-linear form on B.

DEFINITION 3. Let P_m denote the continuous homogeneous polynomial of degree m defined by: if $x \in B$, then

$$P_m(x) = j\mathbf{P}((x, 0), \ldots, (x, 0)).$$

Let $S = j[\mathscr{S} \cap (B \times \{0\})]$ and let $P = \sum_{m=0}^{q} P_m$. Since we will be concerned with $d(I - P, S, 0)$, we need an approximation map in the space B.

ASSUMPTION 5. There is a finite-dimensional subspace N of B and a linear continuous map $A: B \to B$ such that if $x \in S$, then $AP(x) \in N$ and

$$\|AP(x) - P(x)\| < b/2.$$

Also $\mathscr{N}$ (the linear subspace of $\mathscr{B}$ described in Assumption 3 which corresponds to $\bar{\mathscr{W}} = \mathscr{S}$) is the complexification of N and if $z \in \mathscr{S}$ and $z = (x, y)$, then

$$\mathscr{A}(z) = \mathscr{A}(x, y) = (Ax, Ay).$$

THEOREM 3. *If Assumptions* 1, 2, 3, 4, 5 *hold and if* $C = j[\mathscr{C} \cap (B \times \{0\})]$ (*where* $\mathscr{C}$ *is the open set in* B *which is the component of connectedness of* $\mathscr{B} - (\mathscr{I} - \mathscr{P})(\bar{\mathscr{S}})$ *that contains* 0) *then there exists a first category set* $K \subset C$ *such that the number* t *of distinct solutions of*

$$(\mathscr{I} - \mathscr{P})x = y, \tag{8}$$

where $y \in C - K$, *in the interior of* S *is such that*

$$t \equiv d(\mathscr{I} - \mathscr{P}, \mathscr{S}, 0) \pmod 2 \tag{9}$$

and

$$|d(I - P, S, 0)| \leq t \leq d(\mathscr{I} - \mathscr{P}, \mathscr{S}, 0). \tag{10}$$

If $y \in K$, *the set of solutions in the interior of* $\mathscr{S}$ *of the equation*

$$(\mathscr{I} - \mathscr{P})(u, v) = (y, 0)$$

is infinite or has q *elements where* $q \leq d(\mathscr{I} - \mathscr{P}, \mathscr{S}, 0)$.

PROOF. Since $I - P$ is a Fredholm map of index zero and the Leray-Schauder degree $d(I - P, S, 0)$ is defined (this latter follows from Assumption 2 and the definition of P) then by the Smale-Sard Theorem and the invariance under homotopy of the degree, it follows that there is a first category set $K \subset C$ such that if $u \in C - K$, then

(i) $d(I - P, S, 0) = d(I - P, S, u)$;

(ii) $d(\mathscr{I} - \mathscr{P}, \mathscr{S}, 0) = d(\mathscr{I} - \mathscr{P}, \mathscr{S}, (u, 0))$;

(iii) the set $[(I - P)^{-1}u]\ S$ is finite and the differential of $I - P$ at each point of this set has an inverse.

(From now on, we work entirely within S or $\mathscr{S}$. So we omit "$\cap S$" or "$\cap \mathscr{S}$" in the remainder of this discussion.) Let $w_1, \ldots, w_t = [(I - P)^{-1}u]$ and denote the differential of $(I - P)$ at w by $d(I - P)_w$. Since $d(I - P)_{w_i}$ is 1-1 ($i = 1, \ldots, t$), then if $d(\mathscr{I} - \mathscr{P})_{(w_i, 0)}$ denotes the differential of $(\mathscr{I} - \mathscr{P})$ at $(w_i, 0)$, the Leray-Schauder index of $d(\mathscr{I} - \mathscr{P})_{(w_i, 0)}$ is $+1$ by [7, Lemma 5].

LEMMA 2. *There exists an open set* $\mathscr{U} \subset \mathscr{S}$ *such that if* $U = j[\mathscr{U} \cap (B \times \{0\})]$ *then*

$$j\{(\bar{U} \times \{0\}) \cap [(\mathscr{I} - \mathscr{P})^{-1}(u, 0)] = \{w_1, \ldots, w_t\}.$$

PROOF. Since $d(\mathscr{I} - \mathscr{P})_{(w_i, 0)}$ is 1-1 in a neighborhood of $(w_i, 0)$, there is an open sphere σ_i in $\mathscr{B}$ with center $(w_i, 0)$ such that

$$\bar{\sigma}_i \cap [(\mathscr{I} - \mathscr{P})^{-1}(u, 0)] = (w_i, 0)$$

and

$$d(\mathscr{I} - \mathscr{P}, \bar{\sigma}_i, (u, 0)) = +1. \tag{11}$$

Also the σ_i can be chosen so that $\bar{\sigma}_1, \ldots, \bar{\sigma}_t$ are pairwise disjoint. Let $\mathscr{U} = \bigcup_{i=1}^t \sigma_i$. Now let $\mathscr{V}$ denote the open set Int $\mathscr{S} - \bar{\mathscr{U}}$. Then $S = \bar{\mathscr{V}} \cap \bar{\mathscr{U}}$, and $\bar{\mathscr{V}} \cap \bar{\mathscr{U}}$ contains only boundary points of $\mathscr{U}$ and $\mathscr{V}$. Therefore

$$\begin{aligned} d(\mathscr{I} - \mathscr{P}, \mathscr{S}, 0) &= d(\mathscr{I} - \mathscr{P}, \mathscr{S}, (u, 0)) \\ &= d(\mathscr{I} - \mathscr{P}, \bar{\mathscr{V}}, (u, 0)) + d(\mathscr{I} - \mathscr{P}, \bar{\mathscr{U}}, (u, 0)). \end{aligned} \tag{12}$$

From the definition of $\mathscr{U}$ and (11), we have

$$d(\mathscr{I} - \mathscr{P}, \bar{\mathscr{U}}, (u, 0)) = t. \tag{13}$$

LEMMA 3. *$d(\mathscr{I} - \mathscr{P}, \bar{\mathscr{V}}, 0)$ is a nonnegative even number.*

PROOF. It follows from Assumption 3 and the Reduction Theorem [**4**, p. 51] that there is a map $\mathscr{I} - \mathscr{A}\mathscr{P}$ from a complex finite-dimensional Euclidean space $\mathscr{N}$ into itself such that $(u, 0) \in \mathscr{N}$ and

$$d(\mathscr{I} - \mathscr{P}, \bar{\mathscr{V}}, 0) = d(\mathscr{I} - \mathscr{A}\mathscr{P}, \bar{\mathscr{V}} \cap \mathscr{N}, 0) = d(\mathscr{I} - \mathscr{A}\mathscr{P}, \bar{\mathscr{V}} \cap \mathscr{N}, (u, 0)).$$

By Assumption 5, $\mathscr{N}$ is the complexification of a finite-dimensional subspace N of B and $u \in N$. Let $x_1, \ldots, x_n$ be a basis for N. Since $\mathscr{N}$ is the complexification of N, then $(x_1, 0), \ldots, (x_n, 0)$ is a basis for N. Let $z_i = (x_i, 0)$. Then $z_i = z_i^*$ and by Assumption 4,

$$\begin{aligned} \mathscr{Q}^{(m)}_{k_1 \cdots k_n}(z_1, \ldots, z_n) &= \mathscr{Q}^{(m)}_{k_1 \cdots k_n}(z_1^*, \ldots, z_n^*) \\ &= [\mathscr{Q}^{(m)}_{k_1 \cdots k_n}(z_1, \ldots, z_n)]^*. \end{aligned}$$

Thus $\mathscr{Q}^{(m)}_{k_1 \cdots k_n}(z_1, \ldots, z_n)$ has the form $(v^{(m)}_{k_i}, 0)$ where $v^{(m)}_{k_i}$ denotes an element of B which depends on $m, k_1, \ldots, k_n$. Since $v^{(m)}_{k_i}$ can be expressed uniquely as

$$v^{(m)}_{k_i} = \gamma_1^{(m)} x_1 + \cdots + \gamma_n^{(m)} x_n,$$

where $\gamma_1^{(m)}, \ldots, \gamma_n^{(m)}$ are real, then equation (5) becomes

$$\mathscr{P}_m(a_1 z_1 + \cdots + a_n z_n) = \sum_{\substack{k_1 + \cdots + k_n = m \\ k_i \geq 0}} (a_1^{k_1} \cdots a_n^{k_n})(\gamma_1^{(m)} x_1 + \cdots + \gamma_n^{(m)} x_n, 0).$$

Hence the map

$$a_i \to p_i(a_1, \ldots, a_n)$$

is such that each polynomial p_i has real coefficients. Since $(u, 0) \in \mathscr{N}$, then there exist unique real numbers $\eta_1, \ldots, \eta_n$ such that $u = \eta_1 x_1 + \cdots + \eta_n x_n$. Thus $d(\mathscr{I} - \mathscr{A}\mathscr{P}, \bar{\mathscr{V}} \cap \mathscr{N}, (u, 0))$ is the degree at 0 and relative to $\bar{\mathscr{V}} \cap \mathscr{N}$ of the map defined by

$$a_i \to p_i(a_1, \ldots, a_n) - \eta_i \quad (i = 1, \ldots, n)$$

which is a map described by polynomials in $a_1, \ldots, a_n$ with real coefficients. From the definition of $\bar{\mathscr{V}}$, it follows that the system of equations

$$p_i(a_1, \ldots, a_n) - \eta_i = 0 \quad (i = 1, \ldots, n) \tag{14}$$

has no real solutions (i.e., n-tuples $(a_1, \ldots, a_n)$ such that $a_1, \ldots, a_n$ are real) in $\bar{\mathscr{V}} \cap \mathscr{N}$. We use this fact to prove that $d(\mathscr{I} - \mathscr{A}\mathscr{P}, \bar{\mathscr{V}} \cap \mathscr{N}, (u, 0))$ or $d(\mathscr{I} - \mathscr{A}\mathscr{P} - (u, 0), \mathscr{V} \cap \mathscr{N}, (u, 0))$ is an even number. (The underlying idea of the proof is the same as the idea in the proof of the Index Lemma [**6**, pp. 342–344].) Let F denote the map $(\mathscr{I} - \mathscr{A}\mathscr{P} - (u, 0)/\bar{\mathscr{V}} \cap \mathscr{N}$. Let

$$E = \{p \mid p \in \operatorname{Int}\{\bar{\mathscr{V}} \cap \mathscr{N}\} \text{ and } F(p) = 0\}.$$

Suppose

$E_{j_1}, \ldots, j_h = \{(\xi_1, \ldots, \xi_n) \in E \mid$ imaginary parts of $\xi_{j_1}, \ldots, \xi_{j_h}$ are nonnegative and the imaginary parts of the other coordinates ξ_j are negative$\}$.
Since the coefficients in the polynomials $p_i(a_1, \ldots, a_n) - \eta_i$ are real, the set

$$(E_{j_1}, \ldots j_h)^* = \{(\xi_1, \ldots, \xi_n) \, / \, (\xi_1^*, \ldots, \xi_n^*) \in E_{j_1}, \ldots j_h\},$$

where ξ_i^* denotes the conjugate of ξ_i, is contained in E. Hence we may write

$$E = \bigcup_{i=1}^{M} (B_i \cup B_i^*)$$

where $B_1, \ldots, B_M$ are sets of the form $E_{j_1 \cdots j_h}$ and

$$B_i^* = \{(\xi_1, \ldots, \xi_n) \mid (\xi_1^*, \ldots, \xi_n^*) \in B_i\}$$

and $B_1, \ldots, B_M, B_1^*, \ldots, B_M^*$ are pairwise disjoint. Since (14) has no real solutions in $\overline{\mathscr{V}} \cap \mathscr{N}$, there exist pairwise disjoint open sets $U_1, \ldots, U_M$, $U_1^*, \ldots, U_M^*$ in $\overline{\mathscr{V}} \cap \mathscr{N}$ such that

$$U_i^* = \{(\xi_1, \ldots, \xi_n) \mid (\xi_1^*, \ldots, \xi_n^*) \in U_i\}$$

and such that $B_i \subset U_i$, $B_i^* \subset U_i^*$. By a basic property of topological degree, we have

$$d(\mathscr{I} - \mathscr{A}\mathscr{P}) - (u, 0), \overline{\mathscr{V}} \cap \mathscr{N}, (0, 0)) = \sum_{i=1}^{M} [d(\overline{U}_i) + d(\overline{U}_i^*)]$$

where $d(\overline{U}_i) = d(F, \overline{U}_i, 0)$ and $d(\overline{U}_i^*) = d(F, \overline{U}_i^*, 0)$. To complete the proof of Lemma 3, it is sufficient to prove that $d(\overline{U}_i) = d(\overline{U}_i^*)$.

To prove this observe first that map F, regarded as a map from real Euclidean $(2n)$-space into itself, can be approximated on $\bigcup_{i=1}^{M} \overline{U}_i$ by a map f such that the 0-points of f are regular (see [**4**, Theorem (5.4), p. 28 and Lemma (6.1), p. 30]) and such that $d(F, \overline{U}_i, 0) = d(f, \overline{U}_i, 0)$. Next extend map f to $\bigcup_{i=1}^{M} \overline{U}_i^*$ in this way: if $p^* \subset \overline{U}_i^*$, let $f(p^*) = [f(p)]^*$. That $d(f, \overline{U}_i, 0) = d(f, \overline{U}_i^*, 0)$ is a consequence of the following remarks.

Suppose $p_0, p_1, \ldots, p_{2n}$ is a set of $(2n + 1)$ points in complex Euclidean n-space. Then p_j can be written as

$$p_j = \bar{p}_j + i\bar{q}_j$$

where $\bar{p}_j$, $\bar{q}_j$ are points in real Euclidean n-space and if $\bar{p}_j = (x_1^{(j)}, \ldots, x_n^{(j)})$ and $\bar{q}_j = (y_1^{(j)}, \ldots, y_n^{(j)})$ then p_j may be made to correspond to the point

$$P_j = (x_1^{(j)}, y_1^{(j)}, \ldots, x_n^{(j)}, y_n^{(j)})$$

in real Euclidean $(2n)$-space. Assume that $p_0, \ldots, p_{2n}$ are such that $P_0, P_1, \ldots, P_{2n}$ are linearly independent and let σ denote the $(2n)$-simplex determined by $P_0, P_1, \ldots, P_{2n}$. Let σ^* denote the $(2n)$-simplex determined by the points

$$\overline{P}_j = (x_1^{(j)}, -y_1^{(j)}, \ldots, x_n^{(j)}, -y_n^{(j)}) \quad (j = 0, 1, \ldots, 2n).$$

(Since $P_0, \ldots, P_{2n}$ are linearly independent, it follows at once that $\bar{P}_0, \ldots, \bar{P}_{2n}$ are linearly independent.) If n is odd, σ and σ^* have opposite orientations. From the definition of f, it follows that $f(\sigma^*) = [f(\sigma)]^*$. Hence $f(\sigma)$ and $f(\sigma^*)$ also have opposite orientations. If n is even, σ and σ^* have the same orientation, and $f(\sigma)$ and $f(\sigma^*)$ also have the same orientation. This completes the proof of Lemma 3.

Equation (9) and the second half of inequality (10) follow from equations (12) and (13) and Lemma 3. The first half of inequality (10) follows from the facts that the index of $\mathscr{d}(I - P)_{w_i}$ is $+1$ or -1 and that

$$d(I - P, S, 0) = \sum_{i=1}^{t} (\text{index of } \mathscr{d}(I - P)_{w_i}).$$

The proof of the last statement in Theorem 3 is practically the same as the proof of the corresponding statement in Theorem 2.

COROLLARY. *If $d(\mathscr{I} - \mathscr{P}, \mathscr{S}, 0)$ is odd, then the number of solutions of* (7) *is nonzero.*

REMARK. If $\mathscr{I} - \mathscr{F}$ is such that

$$\mathscr{I} - \mathscr{F} : \mathscr{B} \to \mathscr{B}$$

and $\mathscr{F}$ is completely continuous and differentiable and if appropriate forms of Assumptions 4 and 5 are imposed (e.g., since $\mathscr{I} - \mathscr{F}$ can be represented by a power series [8, Chapter III] we can require that each term of the power series, which is a homogeneous polynomial $\mathscr{P}_m$, be completely continuous and satisfy Assumption 4 and we can require that $F = \mathscr{F}/B$ [obtained by applying Assumption 4 to each $\mathscr{P}_m$, just as P was obtained] satisfy the condition that P satisfies in Assumption 5.) then Theorem 3 applies to $\mathscr{I} - \mathscr{F}$.

THEOREM 4. *If $\mathscr{P} = \sum_{1 \le m \le q; m \text{ odd}} \mathscr{P}_m$ and if $d(I - P, S, 0)$ is defined where $P = \sum P_m$ and S is a sphere in B with center 0, then $d(I - P, S, 0)$ is odd. If $d(\mathscr{I} - \mathscr{P}, \mathscr{S}, 0)$ is defined (and a fortiori $d(I - P, S, 0)$ is defined where $S = j\{\mathscr{S} \cap [B \times \{0\}]\}$) there is a set K of first category such that the conclusions of Theorem 3 holds.*

PROOF. From the hypothesis, it follows that the map $(I - AP)/N$ contains, in this case, only polynomial terms of odd degree, i.e., $(I - AP)/N$ is defined by:

$$I - AP : (x_1, \ldots, x_n) - (x'_i, \ldots, x'_n)$$

where $x_i = T_i(x_1, \ldots, x_n)$ and T_i is a polynomial in $x_1, \ldots, x_n$ with real coefficients which contains only terms of odd degree and which has no constant term. But this shows that $I - AP$ is an odd map and hence its degree is odd (Krasnosel'skii [9, Theorem 2.2, p. 97]).

THEOREM 5. *Suppose that P is a completely continuous operator which maps a real Banach space B into itself and is such that*

$$P = \sum_{1 \le m \le q; \, m \text{ odd}} P_m$$

where each P_m is a continuous homogeneous polynomial from B into B and P satisfies the first statement in Assumption 5. Suppose there exist positive numbers M, b, r such that: if $\|x\| \geq M$, then

$$\|(I - P)x\| \geq r(\|x\|)^b.$$

Then for each $y \in B$, there exists $x \in B$ such that

$$(I - P)x = y. \tag{15}$$

PROOF. Let $y \in B$ and let $\|y\| = R_1$. Then there exists $R_2 > 0$ such that if $\|x\| > R_2$, then

$$r\,\|x\|^q > R_1. \tag{16}$$

From (16) it follows that $d(I - P, S, y)$ is defined. From (16) and invariance under homotopy, it follows that

$$d(I - P, S, y) = d(I - P, S, 0).$$

But $d(I - P, S, 0)$ is odd and therefore nonzero.

4. **Application to elliptic equations.** We show that the Dirichlet problem for a class of nonlinear elliptic equations can be formulated in terms of the functional equations studied in §§2 and 3. We use the notation and terminology of [**4**, pp. 157–162, pp. 171–176]. For simplicity, we describe the case of two independent variables, but more general cases can be considered.

We study the equation

$$a(x, y)z_{xx} + b(x, y)z_{xy} + c(x, y)z_{yy} = p(x, y, z, z_x, z_y) + \psi(x, y) \tag{17}$$

where a, b, c satisfy the hypotheses of [**4**, Theorem (6.2), p. 161], $\psi \in C_\mu(\overline{\mathscr{D}})$ and $p(x, y, \xi, \eta, \zeta)$ is a polynomial in ξ, η, ζ with coefficients which are μ-Hölder continuous functions from R^2 into the reals and $p(x, y, 0, 0, 0) = 0$. Let ϕ be a fixed element of $C_{2+\mu}(\overline{\mathscr{D}}')$, $\rho(x, y) \in C_\mu(\overline{\mathscr{D}})$, and let $w(\rho, \phi)$ be the solution in $C_{2+\mu}(\overline{\mathscr{D}})$, of the Dirichlet problem for

$$a(x, y)z_{xx} + b(x, y)z_{xy} + c(x, y)z_{yy} = \rho(x, y).$$

Then (17) may be rewritten as:

$$\rho(x, y) - p\left[x, y, w(\rho, \phi), \frac{\partial}{\partial x} w(\rho, \phi), \frac{\partial}{\partial y} w(\rho, \phi)\right] = \psi(x, y) \tag{18}$$

and by the Schauder Theorem [**4**, p. 1612], the map P_ϕ described by

$$P_\phi\colon \rho(x, y) \to p\left[x, y, w(\rho, \phi), \frac{\partial}{\partial x} w(\rho, \phi), \frac{\partial}{\partial y} w(\rho, \phi)\right]$$

maps bounded sets in $C_\mu(\overline{\mathscr{D}})$ into bounded sets in $C_{1+\mu}(\overline{\mathscr{D}})$ and thus is a completely continuous map from $C_\mu(\overline{\mathscr{D}})$ into $C_\mu(\overline{\mathscr{D}})$.

Thus the Dirichlet problem becomes the problem of solving the functional equation $(I - P_\phi)\rho = \psi$ in $C_\mu(\overline{\mathscr{D}})$. Clearly this equation can be extended to an equation in the complexification of $C_\mu(\overline{\mathscr{D}})$.

Suppose that Assumption 2 is satisfied and that b is the positive constant given in Lemma 1 for the set $\overline{\mathscr{W}}$. It remains to show that Assumption 5 is satisfied. To show this, let $R = [a_1, b_1] \times [a_2, b_2]$ be a closed rectangle which contains $\overline{\mathscr{D}}$. Assume that $\overline{\mathscr{D}}$ is nice enough so that for each $f \in C_\mu(\overline{\mathscr{D}})$ there exists $F \in C_\mu(R)$ such that $F/\overline{\mathscr{D}} = f$ and such that $\|F\|_\mu \leq M \|f\|_\mu$ where M is a positive constant that is independent of f and that a similar condition holds for $f \in C_{1+\mu}(\overline{\mathscr{D}})$. Function F can be approximated uniformly on R by a polynomial $U(x, y)$ (See Courant and Hilbert [**3**, pp. 65–68]). The map A is defined by $A: f \to u = U/\overline{\mathscr{D}}$. From the definition of U [**3**, p. 68], it follows that A is additive. Also the elements of $P_\phi(\overline{W})$ are a bounded set of equicontinuous functions. Hence [**3**, p. 67] there is an integer n such that for all $f \in P_\phi(\overline{\mathscr{W}})$, $A(f)$ is a polynomial of degree not exceeding n, i.e., $A[P_\phi(\overline{\mathscr{W}})]$ is contained in a finite-dimensional linear subspace. Finally it must be shown that A is bounded and hence continuous. This follows from the fact that the first derivatives of f are uniformly approximated by the first derivatives of U and the uniform approximation is independent of f for $f \in P_\phi(\overline{\mathscr{W}})$ because $P_\phi(\overline{\mathscr{W}})$ is contained in a bounded set in $C_{1+\mu}(\overline{\mathscr{D}})$.

BIBLIOGRAPHY

1. P. Alexandroff and H. Hopf, *Topologie*. I, Berlin, 1935 (Reprinted by Edwards Brothers, Ann Arbor, Michigan).
2. G. Bachman and L. Narici, *Functional analysis*, Academic Press, New York, 1966.
3. R. Courant and D. Hilbert, *Methods of mathematical physics*, Vol. 1, Interscience, New York, 1956.
4. Jane Cronin, *Fixed points and topological degree in nonlinear analysis*, Amer. Math. Soc., Providence, R.I., 1964.
5. ———, *Topological degree of some mappings*, Proc. Amer. Math. Soc. **5** (1954), 175–178.
6. ———, *The Dirichlet problem for nonlinear elliptic equations*, Pacific J. Math. **5** (1955). 335–344.
7. ———, *Using Leray-Schauder degree*, J. Math. Anal. Appl. **25** (1969), 414–424.
8. Einar Hille and Ralph S. Phillips, *Functional analysis and semi-groups*, Amer. Math. Soc., Colloq. Publ. **31**, rev. ed. Amer. Math. Soc. Providence, R.I., 1957.
9. M. A. Krasnosel'skii, *Topological methods in the theory of nonlinear integral equations*, (translated by A. H. Armstrong) Macmillan, New York, 1964.
10. J. Schwartz, *Compact analytical mappings of B-spaces and a theorem of Jane Cronin*, Comm. Pure Appl. Math. **16** (1963), 253–260.
11. S. Smale, *An infinite dimensional version of Sard's theorem*, Amer. J. Math., **87** (1965), 861–866.

RUTGERS, THE STATE UNIVERSITY

FREDHOLM STRUCTURES

James Eells, Jr.[1]

1. **Introduction.** *Fredholm structures* appeared in the early work of A. D. Michal [**43**], [**44**, II], in his study of affine connections in the local differential geometry of certain Banach spaces. The Fredholm group was prominent there—in a role which in present-day formulation is played by reductions of the structural groups of fibre bundles. The idea of a *Fredholm map* was introduced by Smale [**59**] as the natural invariant and suitably general form for a class of maps of Banach spaces, studied intensively by Schauder, Leray–Schauder, and Rothe. (See [**21**] and the references therein.) These two concepts are closely related; in their linear form that is seen in the diagram in §2B below. A fundamental theorem of the present theory establishes their relationship in the context of manifolds; see §§6, 7.

A Fredholm structure on an infinite dimensional differentiable manifold provides natural extensions—as we shall see—of many of the concepts and tools of finite dimensional differential and algebraic topology: (1) characteristic classes of Fredholm bundles and manifolds; (2) transversality theorems; (3) various forms of Poincaré and Alexander–Pontrjagin duality; (4) Brouwer–Jordan separation theorems; (5) Brouwer degree theory; (6) a Lefschetz coincidence theorem.

Concrete examples of Fredholm structures on differentiable manifolds of maps (or, more generally, on manifolds of sections of fibre bundles) arise from elliptic problems—as we shall find in Example 3 of §6C. More examples and more concrete applications of the topological results to elliptic problems (in the spirit of Leray–Schauder [**40**]) are urgently needed.

This report is an attempt to give a coherent picture of recent developments in the theory, most of which arose in eight doctoral dissertations of the past year: those of J. Callahan (New York University), K. D. Elworthy (Oxford University), U. Koschorke (Brandeis and Bonn Universities), N. Krikorian (Cornell University), J. Morava (Rice University), K. K. Mukherjea (Cornell University), F. Quinn (Princeton University), and A. J. Tromba (Princeton University). In the preparation of this paper I have had the benefit of their generous cooperation; I am happy to express here my gratitude.

---

[1] Research partially supported by NSF Grant GP-4216.

I have maintained the terminology and notation of [**21**] and think of the present article as a supplement to that.

2. **Linear theory.**

(A) If E and F are Banach spaces over the field $\boldsymbol{K}$ ($= \boldsymbol{R}$ or $\boldsymbol{C}$), then we let $L(E, F)$ denote the Banach space of all continuous linear maps $u\colon E \to F$, with norm

$$\|u\| = \sup \{|u(x)|_F/|x|_E : 0 \neq x \in E\}.$$

$C(E, F)$ denotes the closed linear subspace consisting of those u which map bounded sets of E into compact sets of F. Set $L(E) = L(E, E)$. We will think of $L(E)$ as a Banach (associative) algebra with unit I, and also as a Banach Lie algebra with bracket $[u, v] = u \circ v - v \circ u$. Then $C(E)$ is a closed bilateral ideal (without unit if $\dim E = \infty$) in $L(E)$; and a Lie ideal. If E is a Hilbert space, then the quotient algebra $L(E)/C(E)$ can be represented $*$-isometrically as a uniformly closed self-adjoint Banach algebra of operators on some Hilbert space [**11**].

A *Fredholm operator* from E to F is a map $u \in L(E, F)$ with finite dimensional kernel $\operatorname{Ker} u$ and cokernel $\operatorname{Coker} u = F/u(E)$; it follows that $u(E)$ is a closed linear subspace of F. Its *index*

$$\operatorname{ind}(u) = \dim \operatorname{Ker} u - \dim \operatorname{Coker} u.$$

We let $\Phi_n(E, F)$ denote the totality of such Fredholm operators of index n and $\Phi(E, F) = (\Phi_n(E, F))_{n \in \boldsymbol{Z}}$. Then $\Phi(E, F)$ and $\Phi_n(E, F)$ are open subsets of $L(E, F)$ and the index function is constant on each component. If E and F are Hilbert spaces, then two Fredholm operators with the same index are in the same component.

(B) Let $GL(E)$ be the group of units of the Banach algebra $L(E)$; i.e., the group (under composition) of invertible automorphisms of E. Then $GL(E)$ is an open subset of $L(E)$, and a Banach Lie group whose Lie algebra is $L(E)$. We let $GC(E) = \{I + u \in GL(E) : u \in C(E)\}$; then $GC(E)$ is also a Banach Lie group, whose Lie algebra is $C(E)$. We follow tradition [**43**], [**15**, p. 25] in calling $GC(E)$ the *Fredholm group of E*.

If $p\colon L(E) \to L(E)/C(E)$ is the coset map and $G = G[L(E)/C(E)]$ is the Lie group of units of the indicated quotient algebra, then $\Phi(E) = \Phi(E, E) = p^{-1}G$. Set $p\Phi_0(E) = G_0$. This is a closed subgroup of G; if $GL(E)$ is connected, G_0 is the identity component of G. Also $GL(E)/GC(E) = G_0$. The restriction map $p \mid \Phi(E) \to G$ is a fibre bundle with contractible fibres and therefore p is a homotopy equivalence. $p \mid \Phi_0(E) \to G_0$ is also a homotopy equivalence. We have the diagram

$$\begin{array}{ccccccc} GL(E) & \longrightarrow & \Phi_0(E) & \longrightarrow & \Phi(E) & \longrightarrow & L(E) \\ \downarrow & & \downarrow & & \downarrow & & \downarrow p \\ GL(E)/GC(E) & \xrightarrow{=} & G_0 & \longrightarrow & G & \longrightarrow & L(E)/C(E) \end{array}$$

where the horizontal arrows represent inclusion maps and the vertical arrows are fibrations. The homogeneous space $GL(E)/GC(E)$ has a natural Lie group structure, with Banach Lie algebra $L(E)/C(E)$; we shall call it *the space of Fredholm structures on* E, following the terminology of the general theory of G-structures. The above diagram displays the basic relation between Fredholm operators and Fredholm structures—which (in later sections) we shall develop in its nonlinear context.

$GL(E)$ may not be connected; for instance, it may have infinitely many components [**16**]. However, *it is contractible for a large class of Banach spaces* E: Infinite dimensional Hilbert spaces [**38**] and certain classes of sequence spaces [**48**], [**49**]. If $GL(E)$ is contractible, then *the left vertical arrow defines a universal* $GC(E)$*-bundle. Also,* $\Phi_0(E)$ *is a classifying space for* $GC(E)$ [**31**].

(C) Now let E be a separable Hilbert space. For any real number $1 \leq p < \infty$ and $u \in C(E)$ we define

$$\|u\|_p = [\mathrm{Tr}\,(u^*u)^{p/2}]^{1/p}$$

where Tr denotes the trace of the nonnegative self-adjoint compact operator u^*u. Let $L^pC(E)$ denote the space of all $u \in C(E)$ for which the norm $\|u\|_p < \infty$; then $L^pC(E)$ is a Banach algebra [**18**, XI, §9], [**50**]. If $1 \leq p \leq r < \infty$, then $\|u\|_r \leq \|u\|_p$, so that we have the continuous injections

$$L^pC(E) \rightarrowtail L^rC(E) \rightarrowtail C(E);$$

and each $L^pC(E)$ is an ideal (algebraically speaking) in $C(E)$, but of course (if $\dim E = \infty$) not closed in the topology of $C(E)$. The elements of $L^2C(E)$ are called the *Hilbert–Schmidt operators of* E. These algebras play an important role in the classification of Hilbert Lie groups; see [**15**], [**55**], [**56**]. The elements of $L^1C(E)$ are said to be of *trace class*. $L^1C(E)$ can be interpreted as the conjugate space of $C(E)$, as a Banach space [**54**, Chapter IV].

Using techniques as in §8A below, Palais [**50**] has shown that, relative to an orthonormal base in E, *the natural inclusion map*

$$\lim_{n\to\infty} GL(\boldsymbol{R}^n) \to GL^pC(E)$$

is a homotopy equivalence ($1 \leq p < \infty$) *where* $GL^pC(E)$ *is the group* $\{I + u \in GL(E): u \in L^pC(E)\}$, topologized by the condition that $I + u \to u$ is a bicontinuous map of $GL^pC(E) \to L^pC(E)$.

(D) A sequence $\mathscr{C} = (C_i, d_i)_{i\in \boldsymbol{Z}}$ of Banach spaces C_i and continuous linear maps $d_i: C_i \to C_{i+1}$ is a *Fredholm complex* if

(1) $d_{i+1} \circ d_i = 0$ for all i;

(2) $d_i = 0$ except for finitely many i;

(3) the cohomology spaces

$$H^i = \mathrm{Ker}\,\{d_i: C_i \to C_{i+1}\}/\mathrm{Im}\,\{d_{i-1}: C_{i-1} \to C_i\}$$

are all finite dimensional. The *index* of the complex is

$$\mathrm{ind}\,(\mathscr{C}) = \sum (-1)^i \dim H^i.$$

Thus ind ($\mathscr{C}$) is the Euler characteristic of $\mathscr{C}$. Of course, any Fredholm operator $d_0: C_0 \to C_1$ determines a Fredholm complex (with all $C_i = 0$ $(i \neq 0, 1)$) whose index is ind (d_0).

EXAMPLE [**5**]. Let M be a compact C^∞-manifold without boundary and $(E_i)_{0 \leq i \leq n}$ a sequence of C^∞-vector bundles over M. Let $d_i: C^\infty(E_i) \to C^\infty(E_{i+1})$ be linear differential operators of degree m_i on the indicated space of C^∞-sections such that (1) $d_{i+1} \circ d_i = 0$ for all i, and (2) for any nonzero covector $\eta \in T_x^*(M)$ at $x \in M$, the symbol sequence

$$0 \longrightarrow E_{0,x} \xrightarrow{\sigma(d_0,\eta)} E_{1,x} \xrightarrow{\sigma(d_1,\eta)} \cdots \longrightarrow E_{n,x} \longrightarrow 0$$

of vector spaces is exact. Such a system is called an *elliptic complex*. Let $C_i = C^{m_i + \cdots + m_n}(E_i)$ the Banach space (relative to some Riemannian structure on E_i) of the sections of E_i of differentiability class $C^{m_i + \cdots + m_n}$; then d_i has a natural extension to a continuous linear map $d_i: C_i \to C_{i+1}$. It is a fundamental property of linear elliptic operators that the cohomology spaces of elliptic complexes are always finite dimensional (over the compact base manifold M). In particular, (C_i, d_i) is a Fredholm complex.

Fredholm complexes provide the natural context (1) for the Atiyah–Bott fixed point theorem; see [**5**], [**29**] for examples; (2) for the analytic deformation theory of Spencer [**60**], [**61**].

3. **Topology of certain spaces of operators.**

(A) If E is an infinite dimensional Banach space, we define a *flag* in E as a sequence of (topological) direct sum decompositions $E = E_n \oplus E^n$ $(n \geq 1)$, where $E_n \subset E_{n+1}$, $E^{n+1} \subset E^n$, and $\dim E_n = n$; set $E_\infty = \operatorname{inj} \lim E_n$ (with direct limit topology). Flags always exist.

A flag determines a direct system of Lie subgroups $(GL(E_n))_{n \geq 1}$ of $GC(E)$, under inclusion; let $GL(E_\infty)$ denote their direct limit (with direct limit topology). The following result was proved by Palais [**50**] and Švarc [**63**] for a wide class of Banach spaces and by Elworthy [**25**] and Gęba [**26**] in general: *For any Banach space E and any flag, the inclusion map induces a homotopy equivalence*

$$GL(E_\infty) = \operatorname{inj} \lim GL(E_n) \to GC(E). \tag{1}$$

In particular, *$GC(E)$ has precisely two components if E is real* and these are distinguished by the Leray–Schauder index; we let $GC^+(E)$ denote the identity component. *$GC(E)$ is connected if E is complex.*

The map (1) induces a homotopy equivalence of the classifying spaces $BGL(E_\infty) \to BGC(E)$, which is independent of the flag used in its construction [**34**]. If $GL(E)$ is contractible, then as in §2B we have the homotopy equivalence

$$BGL(E_\infty) = \operatorname{inj} \lim BGL(E_n) \to \Phi_0(E).$$

(B) These results provide explicit descriptions of the homotopy and cohomology of the Fredholm group and its classifying space. We appeal to Bott [**8**] for the periodicity theorem for the stable homotopy of the classical groups, and to Borel [7] for the cohomology of those groups and their classifying spaces.

1) If E is real, the homotopy groups

$$\pi_i(GC(E)) = \pi_{i+8}(GC(E)) \qquad (i \geq 0),$$

and the first eight groups are $\boldsymbol{Z}_2$, $\boldsymbol{Z}_2$, 0, $\boldsymbol{Z}$, 0, 0, 0, $\boldsymbol{Z}$. The cohomology

$H^*(GC^+(E); \boldsymbol{Z}_2) = \Lambda_{\boldsymbol{Z}_2}(x_1, x_2, \ldots)$, an exterior algebra over $\boldsymbol{Z}_2$ with generators x_i of degree $x_i = i - 1$.

$H^*(GC^+(E); \boldsymbol{R}) = \Lambda_{\boldsymbol{R}}(y_1, y_2, \ldots)$, an exterior algebra over $\boldsymbol{R}$ with generators y_i of degree $y_i = 4i - 1$.

2) If E is real and $GL(E)$ is contractible, then

$H^*(\Phi_0(E); \boldsymbol{Z}_2) = \boldsymbol{Z}_2[w_1, w_2, \ldots]$, a polynomial algebra over $\boldsymbol{Z}_2$ with generators w_i of degree $w_i = i$.

$H^*(\Phi_0(E); \boldsymbol{R}) = \boldsymbol{R}[p_1, p_2, \ldots]$, a polynomial algebra over $\boldsymbol{R}$ with generators p_i of degree $p_i = 4i$.

3) If E is complex, the homotopy groups

$$\pi_i(GC(E)) = \pi_{i+2}(GC(E)) \qquad (i \geq 0),$$

and the first two groups are 0, $\boldsymbol{Z}$. The cohomology

$H^*(GC(E); \boldsymbol{Z}) = \Lambda_{\boldsymbol{Z}}(z_1, z_2, \ldots)$, an exterior algebra over $\boldsymbol{Z}$ with generators z_i of degree $z_i = 2i - 1$.

4) If E is complex and $GL(E)$ is contractible, then

$H^*(\Phi_0(E); \boldsymbol{Z}) = \boldsymbol{Z}[c_1, c_2, \ldots]$, a polynomial algebra over $\boldsymbol{Z}$ with generators c_i of degree $c_i = 2i$.

Relative to a flag in a real Banach space E with contractible $GL(E)$, let $G_rC(E) = \{v \in GC(E) : v \mid E^r = \text{identity}\}$. Then $V^r(E) = GC(E)/G_rC(E)$ is the *Stiefel manifold of r-coframes of C(E)-vectors in E*. $V^r(E)$ is $(r - 1)$-connected. We can choose the generators (x_r), (w_r) so that w_r is an obstruction class, and the pairs (w_r, x_r) are transgressive: Relative to the diagram (here q is the homogeneous fibration with group $GC(E)$ and fibre $V^{r-1}(E)$, and i is the inclusion map onto a fibre)

$$\begin{array}{ccccccc} GL(E) & \longrightarrow & GL(E)/G_{r-1}C(E) & \xleftarrow{\ i\ } & V^{r-1}(E) & \xleftarrow{\ \pi\ } & GC(E) \\ & \searrow^{p} & \downarrow{\scriptstyle q} & & & & \\ & & GL(E)/GC(E) & & & & \end{array}$$

we have the cohomology relation $q^*w_r = d\xi$ and $\pi^* \circ i^*\xi = x_r$. If the (w_r, x_r) are so related, we call the w_r *the universal Stiefel–Whitney classes*. Similarly for (c_r, z_r) in the case of complex E, where $V^r(E)$ is now $2r$-connected; we call c_r *the universal Chern classes*. There is a related description of $(p_r\, y_r)$; we call p_r *the universal Pontrjagin classes*.

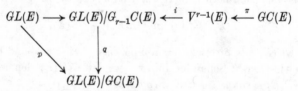

(C) Let E be the separable infinite dimensional real Hilbert space. Then $\Phi(E)$ is an H-space [**31**], whence there is a Pontrjagin multiplication in its homology giving a ring structure to $H_*(\Phi(E); Z_2)$. Then [**34**] *there is a natural embedding of the unoriented cobordism ring* $\mathfrak{N}_*$ *into* $H_*(\Phi(E); Z_2)$.

(D) For Banach spaces E, F (over the same field $\boldsymbol{K} = \boldsymbol{R}$ or $\boldsymbol{C}$; let $\dim_{\boldsymbol{R}} \boldsymbol{K} = d$) we introduce the open dense subsets $V_{p,q} = V_{p,q}(E, F) = \{u \in \Phi_{p-q}(E, F)$: $\dim \operatorname{Ker} u \leq p\}$ of $\Phi_{p-q}(E, F)$; here $p, q \geq 0$. Then

$$\cdots V_{p-1,q-1} \subset V_{p,q} \subset \cdots \subset \bigcup \{V_{p+j,q+j}: j \geq 0\} = \Phi_{p-q}(E, F).$$

Let $A_{p,q} = V_{p,q} - V_{p-1,q-1}$. $A_{p,q}$ *is a closed analytic submanifold of* $V_{p,q}$; Koschorke [**33**] has proved that $A_{p,q}$ *has transverse bundle* Hom (Ker (p, q), Coker $(p, q)) \to A_{p,q}$. *In particular*, codim $(\Phi_{p-q}, A_{p,q}) = dpq < \infty$. Here Ker (p, q) is the subbundle of $A_{p,q} \times E$ whose fibre at $u \in A_{p,q}$ is Ker u; Coker (p, q) is the obvious quotient bundle of $A_{p,q} \times F$.

Alexander-Pontrjagin duality determines a canonical isomorphism

$$\psi: H^i(V_{p,q}, V_{p-1,q-1}) \to H^{i-dpq}(A_{p,q}),$$

where now and henceforth the coefficients are dictated by the field $\boldsymbol{K}$: The coefficient ring is

$$\Lambda(\boldsymbol{K}) = \boldsymbol{Z}_2 \quad \text{if } \boldsymbol{K} = \boldsymbol{R}$$
$$= \boldsymbol{Z} \quad \text{if } \boldsymbol{K} = \boldsymbol{C}.$$

Applying this successively to pairs $(V_{p+i,q+i}, V_{p-1+i,q-1+i})$ produces in particular an isomorphism $\bar{\psi}: H^{dpq}(\Phi_{p-q}, V_{p-1,q-1}) \to H^0(A_{p,q})$. We define the *canonical class*

$$a_{p,q}(E, F) = \bar{\psi}^{-1}(1) \in H^{dpq}(\Phi_{p-q}, V_{p-1,q-1}).$$

Now specialize so that $E = F$, an infinite dimensional Hilbert space. If $G_p(E)$ denotes the Grassmann manifold of p-planes in E and γ_p the classifying bundle over $G_p(E)$, we have natural maps

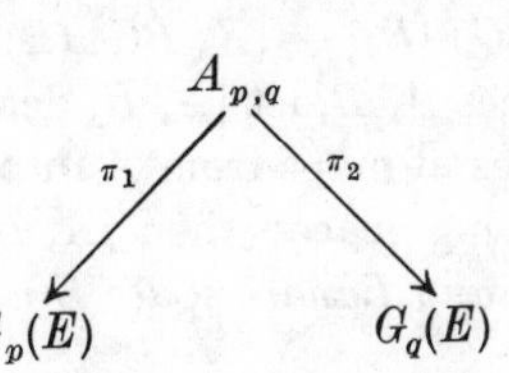

where π_1 (resp., π_2) assigns to each element of $A_{p,q}$ its kernel (resp., the orthogonal complement of its range). *These maps induce a homotopy equivalence* $\pi: A_{p,q} \to G_p(E) \times G_q(E)$; *and* Ker $(p, q) = \pi^{-1}\gamma_p$, Coker $(p, q) = \pi^{-1}(\gamma_q)$ [**33**]. In particular, we have an isomorphism

$$H^*(A_{p,q}) = \Lambda(\boldsymbol{K})[v_1', \ldots, v_p'; v_1'', \ldots, v_q''],$$

a polynomial algebra over the coefficient ring $\Lambda(\boldsymbol{K})$ whose generators v_i', v_j'' are the indicated Stiefel–Whitney or Chern classes. As an application, *the normal bundle of* $A_{p,q}$ *in* $V_{p,q}$ *is orientable if and only if* $p \equiv q \equiv 0 \bmod 2$. Furthermore, Deligne, using methods of Koschorke [**33**], [**34**] (see also §4D below) has shown (to confirm a conjecture of A. Douady) that if $j^*: H^{dpq}(\Phi_{p-q}, V_{p-1,q-1}) \to H^{dpq}(\Phi_{p-q})$ is the natural homomorphism induced by the inclusion maps of pairs, then

$$
{}^*a_{p,q} = (-1)^{pq} \begin{vmatrix} v_p & v_{p+1} & \cdots & v_{p+q-1} \\ v_{p-1} & v_p & \cdots & v_{p+q-2} \\ \cdot & & & \cdot \\ \cdot & & & \cdot \\ \cdot & & & \cdot \\ v_{p-q+1} & & \cdots & v_p \end{vmatrix}
$$

where $(-1)^i v_i = w_i$ or c_i depending on whether $d = 1$ or 2. This is the qth Hankel determinant of the sequence (v_k).

4. **Fredholm bundles.**

(A) Fix a Banach space E and a topological subgroup G of $GL(E)$. We are interested in G-reductions of principal $GL(E)$-bundles over a paracompact space X. *If G is closed in $GL(E)$, then the G-reductions of a given $GL(E)$-bundle ξ are in natural bijective correspondence with the continuous sections of the associated $(GL(E), GL(E)/G)$-bundle of ξ* [**62**]. In the special case of $G = GC(E)$ we have a normal subgroup, so that the associated bundle is itself principal; therefore, it has sections if and only if it is trivial. A $GC(E)$-reduction of a bundle ξ is called a *Fredholm reduction* of ξ (or a *Fredholm bundle*). Examples show that not all $GL(E)$-bundles admit a Fredholm reduction [**25**]; however, they do in case $GL(E)$ is contractible. In particular, if $GL(E)$ is contractible then every $GL(E)$-bundle over X is trivial.

A tool of primary importance in the study of Fredholm reductions is the possibility of approximation by maps with locally finite dimensional ranges (a technique used by Leray–Schauder [**40**]). In the present context we have the following result [**25**]: *Let $GF(E) = \{I + u \in GC(E) : u(E)$ is finite dimensional$\}$. Then the natural map $[X, BGF(E)] \to [X, BGC(E)]$ is bijective.*

(B) For topological spaces A, B, let $[A, B]$ denote the set (with distinguished element) of homotopy classes of maps from A into B. For a topological group G we let BG denote a classifying space; then $[A, BG]$ is independent of choice of representative for BG. *For any Banach space E we have* [**25**] *the exact sequence*

$$
\begin{aligned}
[X, GC(E)] \to [X, GL(E)] \to [X, \Phi_0(E)] \to [X, BGC(E)] \\
\to [X, BGL(E)] \to [X, B(GL(E)/GC(E))],
\end{aligned}
$$

where the first two arrows are induced from inclusion maps. This is the sequence of Atiyah–Jänich [**4**], [**31**] if X is compact and E is a Hilbert space. As applications:

(1) Given a $GL(E)$-bundle ξ over X, a choice of one Fredholm reduction identifies all others with the continuous maps $X \to GL(E)/GC(E)$, as in §4A.

(2) Suppose that $GL(E)$ is contractible. Then as in §2B, we have natural bijections

$$
[X, \Phi_0(E)] \to [X, BGC(E)] \to [X, GL(E)/GC(E)].
$$

(3) *A choice of flag determines a bijection*

$$\mathrm{ind}:[X, \Phi_0(E)] \to \tilde{K}_K(X),$$

where

$$\tilde{K}_K(X) = [X, BGL(E_\infty)].$$

The map ind is a bundle-theoretic generalization of the Fredholm index [**4**], [**31**].

(C) A choice of flag in E determines the homotopy equivalence (1), and therefore characteristic classes for Fredholm bundles over X. In particular, we have the notion of *orientable Fredholm bundle* ξ: one for which the first Stiefel–Whitney class $w_1(\xi) = 0$. An *orientation of* ξ is an isomorphism class of a reduction to the identity component $GC^+(E)$ of $GC(E)$.

(D) Let $\xi: V \to X$, $\eta: W \to X$ be Banach vector bundles over the paracompact space X, and let $\Phi_n(V, W) \to X$ denote the bundle of Φ_n-bundle maps of $V \to W$. Similarly for the bundles $V_{p,q}(V, W) \to X$. Fix p, $q \geq 0$, and take a Fredholm bundle map $f: V \to W$ which induces the identity map on X; thus for each $x \in X$, f defines a Fredholm operator $f_x: V_x \to W_x$ on the fibres over x. There are finite dimensional trivial bundles 1_V, 1_W over X such that $f: V \oplus 1_V \to W \oplus 1_W$ is a Φ_{p-q}-bundle map. Thus f defines a continuous section s_f of $\Phi_{p-q}(V \oplus 1_V, W \oplus 1_W) \to X$. Again we let $j: \Phi_{p-q} \to (\Phi_{p-q}, V_{p,q})$ be the inclusion map of pairs. The following construction is due to Koschorke [**33**]:

There is a unique cohomology class

$$a_{p,q}(V, W) \in H^{dpq}(\Phi_{p-q}(V, W), V_{p-1,q-1}(V, W))$$

which restricts to the class $a_{p,q}(V_x, W_x)$ *for each* $x \in X$. *The class* $x_{p,q}(f) = (js_f)^*(a_{p,q}(V, W)) \in H^{dpq}(X)$ *does not depend on the stabilization. In fact,*

$$x_{p,q}(f) = (-1)^{pq} \begin{vmatrix} v_p & v_{p+1} & \cdots & v_{p+q-1} \\ v_{p-1} & v_p & \cdots & v_{p+q-2} \\ \cdot & & & \cdot \\ \cdot & & & \cdot \\ \cdot & & & \cdot \\ v_{p-q+1} & & \cdots & v_p \end{vmatrix}$$

where $(-1)^i v_i = x_{i,1}(f)$.

Alternatively in the case that ξ *and* η *have common fibre an infinite dimensional Hilbert space* E, f *determines a unique element of* $[X, \Phi(E)]$; *and for any* p, $q \geq 0$, $x_{p,q}(f) = f^*j^*(a_{p,q})$. Agree that $x_{0,1}(f) = 1$. $a_{p,q}(V, W)$ and $x_{p,q}(f)$ play the roles of the Thom and Euler classes of finite dimensional vector bundles. (If V, W have finite fibre dimensions p, q, then any bundle map $f: V \to W$ is Fredholm, and $x_{p,q}(f)$ is the Euler class of the vector bundle $L(V, W) \to X$ of linear maps of the fibres.) Note that we have the analogue of both the obstruction and the universal definition of the characteristic class $x_{p,q}(f)$.

$x_{p,q}(f)$ depends only on the stable isomorphism classes of ξ, η. Also, if f_1 is homotopic to f in $\Phi(V, W)$, then $x_{p,q}(f_1) = x_{p,q}(f)$. If $f: V_1 \to V_2$ and $g: V_2 \to V_3$ are Fredholm bundle maps and $v(f) = \sum_{i \geq 0} (-1)^i x_{i,1}(f)$, then

$$v(g \circ f) = v(g) \cup v(f).$$

$v(f)$ is called *the Stiefel–Whitney or Chern class of the Fredholm bundle map f.*

In the case of real bundles, if we choose an orientation of the normal bundle of $A_{2p,2q}$, we can define *oriented characteristic classes* $a^{or}_{p,q} \in H^{4pq}(\Phi_{2(p-q)}, V_{2p-1,2q-1})$, $x^{or}_{p,q}(f) \in H^{4pq}(X)$ and *Pontrjagin class* $v^{or}(f) = \sum_{i \geq 0} (-1)^i x^{or}_{i,1}(f)$, with similar properties.

5. **Compact maps.** Abstraction of classical integral equations leads to the theory of compact and Fredholm maps. (Roughly speaking, the integral equations $f(s) = \int K(s, t, h(t))\, dt$ of first kind give rise to compact maps, and those $f(s) = h(s) + \int K(s, t, h(t))\, dt$ of second kind to Fredholm maps.) A compact map is a purely topological concept. The definition of a Fredholm map is more special, requiring manifold structure. We could formulate a theory of Fredholm C^0-maps of C^0-manifolds using the local representation (§6A) as guide; however, for the most part we will treat them only in the differentiable context.

(A) Let X and Y be Hausdorff topological spaces with Y completely regular (equivalently, uniform); and $C(X, Y)$ the totality of continuous maps $\varphi: X \to Y$. There are several standard topologies [**9**], [**32**, Chapter 7] that can be put on $C(X, Y)$; among them, the topology of uniform convergence on the compact subsets of X (i.e., the *compact-open topology*). In it $C(X, Y)$ is completely regular, and is complete in the induced uniform structure of a complete uniform structure on Y. If Z is a third space, then the natural map

$$\eta: C(X \times Z, Y) \to C(Z, C(X, Y))$$

defined by $(\eta(\theta)z)x = \theta(x, z)$ is 1–1, open and continuous; i.e., η is a homeomorphism onto its image. η is surjective if and only if the evaluation map $C(X, Z) \times X \to Z$ is continuous; for instance, this is so if X is locally compact [**9**]. The compact-open topology on $C(X, Y)$ is particularly convenient if X is locally compact; but for our purposes that topology is too small.

Suppose now that Y is metrizable, and that ρ is an admissible metric. The *fine topology on* $C(X, Y)$ is that generated by the following subbase: For each continuous function $\epsilon: X \to \boldsymbol{R}$ (>0) and $\varphi \in C(X, Y)$, let $U(\varphi, \epsilon) = \{\psi \in C(X, Y): \rho(\varphi(x), \psi(x)) < \epsilon(x)$ for all $x \in X\}$. That topology is given by a uniform structure with écarts (= semi-metrics, possibly taking on the value ∞)

$$\rho_\epsilon(\varphi, \psi) = \sup \{\rho(\varphi(x), \psi(x))/\epsilon(x): x \in X\}.$$

We let $C_{\text{fine}}(X, Y)$ denote the resulting completely regular space. The following properties are established in Krikorian [**36**]:

(1) *If X is paracompact, then the topology of $C_{\text{fine}}(X, Y)$ is independent of the choice of admissible metric on Y.*

(2) *If Y is complete in the metric ρ, then* $C_{\text{fine}}(X, Y)$ *is complete as a uniform space.*

(3) *If X is metrizable and not compact, and if Y is a* C^0*-manifold modeled on a Fréchet space, then* $C_{\text{fine}}(X, Y)$ *is not metrizable.*

(4) *Suppose X has the topology generated by its compact subsets* (i.e., a subset of $U \subset X$ is open if $U \cap C$ is open in C for all compact subsets $C \subset X$. For instance, that is insured if X satisfies the first axiom of countability [**32**, p. 231]). *If* (Y, ρ) *is complete, then* $C_{\text{fine}}(X, Y)$ *is a Baire space.*

(5) *If X is metrizable and not locally compact at any point, then* $C_{\text{fine}}(X, Y)$ *is totally disconnected.* (E.g., we could take for X a C^0-manifold modeled on a Fréchet space of infinite dimension.) $C_{\text{fine}}(\boldsymbol{R}, \boldsymbol{R})$ *is not locally connected at any point.*

(B) For any space X and metric space (Y, ρ_Y) let $BC(X, Y)$ denote the set of those continuous maps $\varphi: X \to Y$ such that the image $\varphi(X)$ is bounded. Then $BC(X, Y)$ has the natural metric

$$\rho(\varphi, \psi) = \sup \{\rho_Y(\varphi(x), \psi(x)) : x \in X\}.$$

A continuous map $\varphi: X \to Y$ is said to be *compact* if $\varphi(X)$ is relatively compact; i.e., has compact closure in Y. Thus a compact linear map is compact on bounded subsets of its domain. The totality $KC(X, Y)$ of compact maps is a closed subset of $BC(X, Y)$, provided that (Y, ρ_Y) is complete [**28**]. The evaluation map ev: $KC(X, Y) \times X \to Y$ defined by ev $(\varphi, x) = \varphi(x)$ is continuous.

We have the following version of Ascoli's theorem [**18**, p. 260], [**66**]: *Suppose that* (Y, ρ_Y) *is a complete metric space. A subset A of* $KC(X, Y)$ *is relatively compact if and only if* (1) *A has equal variation (i.e., for every* $\epsilon > 0$ *there is a finite covering of X such that* $\rho_Y(\varphi(x), \varphi(x')) < \epsilon$ *for all* $\varphi \in A$, *whenever x, x′ belong to the same element of the covering); and* (2) *for each* $x \in X$ *the set* $A(x)$ *of all* $\varphi(x)$ *for which* $\varphi \in A$ *is relatively compact in Y.*

If Y is a Banach space V with norm $| \ \ |_V$, then $BC(X, V)$ has the structure of a Banach space (algebraic operations defined pointwise) with norm

$$|\varphi|_{BC(X,V)} = \sup \{|\varphi(x)|_V : x \in X\};$$

and $KC(X, V)$ is a closed linear subspace of $BC(X, V)$. If $g: V \to V_1$ is a continuous map of Banach spaces, then g induces a continuous map $\bar{g}: KC(X, V) \to KC(X, V_1)$ by $(\bar{g}\varphi)x = g(\varphi(x))$ for all $x \in X$. If $f: X \to X_1$ is continuous, then f induces a continuous linear map $\bar{f}: KC(X_1, V) \to KC(X, V)$ by $(\bar{f}\psi)x = \psi(f(x))$ for all $x \in X$. In particular, if A is a closed subset of X, then the map $KC(X, V) \to KC(A, V)$ induced from the restriction map is surjective [**17**]. If U is an open subset of V, then $KC(X, U) = \{\varphi \in KC(X, V) : \overline{\varphi(X)} \subset U\}$ is open in $KC(X, V)$.

Example. Let Y be a complete Riemannian manifold of finite dimension. The Hopf–Rinow theorem asserts that the compact subsets of Y are those which are closed and bounded in the induced metric on Y. In this case we have $KC(X, Y) = BC(X, Y)$. For instance, $KC(X, \boldsymbol{R}) = BC(X, \boldsymbol{R})$; in particular, $KC(X, Y)$ is often nonseparable.

(C) Suppose that U is an open subset of a Banach space E. For any Banach space F the totality $BC^r(U, F)$ of C^r-maps $\varphi: U \to F$, all of whose differentials $d^k\varphi$ $(0 \leq k \leq r < \infty)$ are bounded on U, is a Banach space (as in [**21**, §6]) with norm

$$|\varphi|_{BC^r(U,F)} = \sup \left\{ \sum_{k=0}^{r} |d^k\varphi(x)| : x \in U \right\}.$$

Let $KC^r(U, F)$ denote the closed subspace consisting of those $\varphi \in BC^r(U, F)$ such that each $d^k\varphi: U \to SL^k(E, F)$ is compact $(0 \leq k \leq r)$.

If $\varphi \in KC^r(U, F)$, then each $d^k\varphi(x) \in SC^k(E, F)$, the space of symmetric compact (i.e., carrying bounded sets of $E \times \cdots \times E$ into compact sets of F) k-linear maps $E \to F$. In particular, if $\varphi \in KC(U, F) = KC^0(U, F)$ and if $\varphi \in C^1$, then every $d\varphi(x) \in C(E, F)$. The following example of R. Bonic [**6**] (other examples have been produced by A. Tromba [**64**]) shows that a C^1 map can have a compact differential at every point without being compact.

EXAMPLE. Let E be the Banach space l^p $(1 \leq p < \infty)$ with its standard base $(e_i)_{i \geq 1}$; we can express each $x \in E$ in the form $x = \sum x_i e_i$. Then the map $\varphi: E \to E$ defined by $\varphi(x) = \sum x_i^2 e_i$ is not compact, but its differentials $v \to d\varphi(x)v = \sum 2x_i v_i$ are. As contrast, if $\varphi: U \to F$ has every $\varphi_*(x) \in C(E, F)$ and if $\varphi_*: U \to C(E, F)$ is a compact map, then φ is compact [**65**, p. 51].

Let V be open in F, and G a Banach space. If $g \in C^{r+s}(V, G)$, then the induced map $\bar{g}: KC^r(U, V) \to KC^r(U, G)$ is of class C^s for all $0 \leq s \leq r$. Again, $KC^r(U, V)$ is open in $KC^r(U, F)$.

(D) The next result is the point of departure for the thesis of N. Krikorian [**37**]. Its proof follows the method of [**19**], with account taken of the theory of sprays [**1**]:

Let X be a space, and Y a separable C^{s+2}-manifold with C^{s+2}-smooth model F. Then $KC(X, Y)$ is a C^s-manifold modeled on the Banach space $KC(X, F)$. Similarly, if X is a C^r-manifold, then $KC^r(X, Y)$ is a C^s-manifold modeled on $KC^r(X, F)$.

(E) EXAMPLE [**18**, p. 518]. Let S be a σ-finite measure space with positive measure μ. Take real numbers $1 < p, p' < \infty$ with $1/p + 1/p' = 1$. Let $K: S \times S \to C$ be a measurable function with finite norm

$$\|K\| = \left[\int_S \left\{ \int_S |K(s, t)|^p \, d\mu(s) \right\}^{p'/p} d\mu(t) \right]^{1/p'}.$$

Then the map $u: L^p(S, \mu) \to L^p(S, \mu)$ defined by

$$(u(x))s = \int_S K(s, t)x(t) \, d\mu(t)$$

is a compact linear operator, and the restriction of u to any bounded subset of $L^p(S, \mu)$ defines a compact map; that is a consequence of Ascoli's theorem. Those kernels K (sometimes assumed to be Hermitian symmetric $K(s, t) = \overline{K(t, s)}$) with $\|K\| < \infty$, computed for $p = 2 = p'$, are called *Hilbert-Schmidt kernels*.

EXAMPLE [**35**, §3]. Let S be a compact Hausdorff space with positive Radon measure μ, and $k: S \times S \times \mathbf{C} \to \mathbf{C}$ a continuous function. Then the map

$\varphi: C(S) \to C(S) = C(S; \boldsymbol{C})$ defined by

$$(\varphi(x))s = \int_S k(s, t, x(t))\, d\mu(t)$$

is compact on bounded subsets of $C(S)$.

An important special case is that of the *Hammerstein maps*, of the form

$$(\varphi(x))s = \int_S K(s, t) f(t, x(t))\, d\mu(t)$$

for a suitable function $f: S \times \boldsymbol{C} \to \boldsymbol{C}$. Thus φ factors through a compact linear operator, and the function f carries all the nonlinearity:

$$\begin{array}{ccc} C(S) & \xrightarrow{\ \varphi\ } & C(S) \\ & {\scriptstyle \bar{f}}\searrow \qquad \nearrow{\scriptstyle \psi} & \\ & C(S) & \end{array}$$

where $(\bar{f}(x))t = f(t, x(t))$ and $\psi(y)t = \int_S K(s, t) y(t)\, d\mu(t)$.

Many variations of these examples are possible; e.g., (1) we could consider functions of various Sobolev classes L_r^p, in case S is a (compact) C^r-manifold of dimension n, taking into account compact embeddings of the form

$$L_r^p \subset L_s^q \quad (1 \le p, q < \infty; 0 \le s < r \text{ and } 1/q > 1/p - (r - s)/n \ge 0);$$

$$L_r^p \subset C^k \quad (r > n/p + k); \qquad C^{k+\beta} \subset C^{k+\alpha} \quad (0 < \alpha < \beta < 1);$$

(2) we could form similar integral operators on spaces of sections of vector bundles.

6. Fredholm maps.

(A) If X, Y are C^r-manifolds ($r \ge 1$) modeled on Banach spaces E, F, and $\varphi: X \to Y$ is a C^r-map, then *φ is a Fredholm map* if every differential $\varphi_*(x) = d\varphi(x)$ is a Fredholm operator; i.e., $\varphi_*(x) \in \Phi(X(x), Y(\varphi(x)))$. We will say that such a φ belongs to ΦC^r. If X is connected, then its index $\operatorname{ind}(\varphi) = \operatorname{ind}(\varphi_*(x))$ is independent of the choice of the point $x \in X$. The inverse function theorem insures that locally φ has the form $\varphi(x_1, x_2) = (\eta(x_1, x_2), x_2)$ as a map $U_1 \times U_2 \to \boldsymbol{R}^q \times E_2$, where $U_1 \subset \boldsymbol{R}^p$ and $U_2 \subset E_2$ are open balls of a splitting $E = \boldsymbol{R}^p \times E_2$ and $F = \boldsymbol{R}^q \times E_2$ [**2**, p. 42].

An important tool in the study of Fredholm maps is Smale's theorem [**59**]: *If X, Y are C^r-manifolds which are connected and separable and $\varphi: X \to Y$ is a $\Phi_k C^r$-map with $r > \max(0, k)$, then the regular values of φ form a residual subset of Y.* Taking a suitable generalization of this as starting point, F. Quinn [**53**] has derived the following transversality theorem (in the spirit of Abraham's approach [**2**]).

Say that a Banach space E is *uniformly C^∞-smooth* if there is a C^∞-function $\alpha: E \to \boldsymbol{R}$ with bounded nonvoid support, each of whose differentials $d^k\alpha$ is uniformly bounded on E. Thus all separable Hilbert spaces, as well as the Banach spaces L^p (for p an even integer), are uniformly C^∞-smooth.

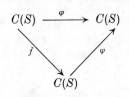

Let X be a separable C^∞-manifold modeled on a uniformly C^∞-smooth Banach space, and $\psi: B \to Y$ a C^∞-map which is left Fredholm and σ-proper (i.e., B can be expressed as a countable union of subsets on each of which ψ is proper). Assume that either X or B is finite dimensional. Then $\{\varphi \in C^\infty(X, Y): \varphi$ is transversal to $\psi\}$ is residual in the C^∞_{fine}-topology on the subset of $C^\infty(X, Y)$ consisting of left Fredholm maps.

J. Callahan [**12**] has derived this (for purposes described in §8F below; a somewhat stronger result can be obtained from the theorem of Quinn):

Let Y be a separable C^r-manifold $(r \geq 2)$ with C^r-smooth model, and B a closed n-codimensional submanifold of Y. Suppose that X is a compact C^r-manifold (possibly with boundary) of dimension n or $n + 1$, and $\varphi: X \to Y$ a C^2-map. Then the set $\{\psi \in \Phi C^2(Y, Y): \psi \circ \varphi$ is transversal to $B\}$ is dense in $\Phi C^2(Y, Y)$ with its compact-open topology (on all differentials of orders ≤ 2).

(B) In general, the differential properties of a Fredholm map show that such maps can occur only in highly restrictive situations. On the other hand, the following three fundamental theorems of K. D. Elworthy [**25**] do provide a sufficiently rich theory:

Let X, Y be C^r-manifolds $(r \geq 3)$ with the same C^r-smooth separable model E, and suppose their principal bundles admit a Fredholm reduction. Then

(1) *There is a proper $\Phi_0 C^r$-map $X \to E$.*

(2) *The subset $\Phi_0 C^r(X, Y)$ of Φ_0-maps are a dense subset of $C^0_{\text{fine}}(X, Y)$.*

(3) *Given $\varphi \in C^0_{\text{fine}}(X, Y)$ and a metric on Y, there is a continuous function $\epsilon: X \to R\ (>0)$ such that any two ϵ-approximations to φ are homotopic through Φ_0-maps.*

As an application of (1) Elworthy has given the following embedding theorem, proved in the Hilbert space case by J. H. McAlpin [**21**, §4]:

Let X be a separable C^r-manifold $(r \geq 3)$ modeled on a C^r-smooth model E. There is a split C^r-embedding of X onto a closed submanifold of $l^2(E)$ (= the Hilbert direct sum of countably many copies of E).

(C) EXAMPLE 1. If $\varphi: X \to Y$ is a locally proper C^1-foliation map, then φ is Fredholm of index ≥ 0. If $\varphi: X \to Y$ is a C^1-immersion onto a finite codimensional image, then φ is Fredholm of index ≤ 0.

EXAMPLE 2 [**2**, Chapter 4]. Let S be a compact C^r-manifold $(r < \infty)$ of dimension n, M a separable C^{2r+2}-manifold modeled on a C^{2r+2}-smooth Banach space. Then the space $C^r(S, M)$ is a separable C^r-manifold [**21**, §6]; and the evaluation map $\text{ev}: C^r(S, M) \times S \to M$ is C^r. Suppose that M_0 is a locally closed submanifold of M of finite codimension q, and assume that ev is transversal over M_0; then $\text{ev}^{-1}(M_0)$ is a q-codimensional locally closed submanifold of $C^r(S, M) \times S$. *If $r > \max(0, n - q)$, then the map $\varphi: \text{ev}^{-1}(M_0) \to C^r(S, M)$ given by $\varphi(x, s) = x$ is a Fredholm map of index $n - q$.* This fact is established (in a somewhat more general context) in [**2**] as part of Abraham's treatment of Thom's transversality theory.

EXAMPLE 3. Elliptic differential (or pseudo-differential) operators are a rich source of Fredholm maps:

Let S be a compact C^∞-manifold without boundary, and $\xi: V \to S$, $\eta: W \to S$ finite dimensional C^∞-fibre bundles over S. We denote by $J^r(\xi): J^r(V) \to S$ the r-jet bundle of sections of ξ; let us assume (for simplicity of exposition [**52**, §17]) that each element of $J^r(V)$ is the r-jet of a global section. A *C^∞-differential operator from ξ to η of order r* is a C^∞-map (of the indicated Banach C^∞-manifolds of sections)

$$\begin{array}{ccc} A: C^{s+r}(\xi) & \longrightarrow & C^s(\eta) \\ \Big\downarrow j^r & \nearrow F & \\ C^s(J^r(\xi)) & & \end{array}$$

which factors through the rth prolongation j^r by a map induced from a C^∞-bundle map $F: J^r(\xi) \to \eta$ $(0 \le s < \infty)$. Say that *A is elliptic* if the differential $A_*(\varphi): T_\varphi C^r(\xi) \to T_{A(\varphi)} C^0(\eta)$, which is a linear differential operator from the tangent space $T_\varphi C^r(\xi)$ of $C^r(\xi)$ at φ to that of $C^0(\eta)$ at $A(\varphi)$, is elliptic for all $\varphi \in C^r(\xi)$. *In that case $A: C^r(\xi) \to C^0(\eta)$ is a Fredholm map.* The index of A is defined [**52**, §18] as an element of $K_{\boldsymbol{K}}(C^\infty(\xi)) = [C^\infty(\xi), \Phi(E)]$, where E is an infinite dimensional Hilbert space over $\boldsymbol{K}$.

Examples of Fredholm bundle maps derived from families of differential operators are given in [**10**], [**58**].

7. Fredholm manifolds.

(A) Let X be a C^r-manifold $(r \ge 1)$ modeled on a Banach space E. *A Fredholm structure on X* is a maximal atlas $(\theta_i, U_i)_{i \in I}$ (the elements of which are called Fredholm charts) for the differential structure of X, such that all differentials

$$(\theta_j \circ \theta_i^{-1})_*(\theta_i(x)) \in GC(E) \quad \text{for } x \in U_i \cap U_j.$$

Such a structure determines a unique $GC(E)$-reduction of the principal bundle of X, and our definition requires that the reduction be integrable. A *Fredholm manifold* X is a C^r-manifold with a specific Fredholm structure. Similarly, we have the notion of integrable $GL^pC(E)$-structures $(1 \le p < \infty)$ on X (in the notation of §2C).

There are C^∞-manifolds X modeled on separable Banach spaces E whose principal $GL(E)$-bundles do not have any Fredholm reduction [**25**]. Therefore, such an X admits no Fredholm structure. The existence theorem of §6B together with §7C below insure the existence of Fredholm structures in case X is separable and parallelizable (i.e., has trivial principal bundle); that is the case, for instance, if $GL(E)$ is contractible.

The *characteristic classes* of a Fredholm manifold are those of its principal $GC(E)$-bundle. In particular, we have the notion of *orientable* Fredholm manifold [**64**]: One whose principal bundle has vanishing first Stiefel–Whitney class. That is equivalent to saying that the principal bundle has a reduction to the identity component $GC^+(E)$. An *oriented* Fredholm manifold has a definite choice of such a reduction. An application of §7C below shows [**34**], [**64**]:

A C^r-manifold X whose model has contractible general linear group is orientable in all its Fredholm structures if and only if $H^1(X; \mathbf{Z}_2) = 0$. The real projective space $P(\boldsymbol{R})$ of infinite dimensional separable Hilbert space admits both an orientable and a nonorientable Fredholm structure.

A map $\varphi: X \to Y$ of Fredholm manifolds (with the same model E) is said to *respect the structures* if for any Fredholm charts (α, U), (β, V) with $\varphi(U) \subset V$ we have $(\beta \circ \varphi \circ \alpha^{-1})_*(\alpha(x)) \in C(E)$ for all $x \in U$. Similarly for maps of oriented Fredholm manifolds.

(B) EXAMPLES 1. If X is an open subset of a Banach space E, then E induces an obvious Fredholm structure on X. The Cartesian product of two Fredholm manifolds has the Fredholm product structure. A covering space of a Fredholm manifold has a naturally induced Fredholm structure; as does the quotient of a Fredholm manifold by a properly discontinuous group of automorphisms of its structure. The tangent vector bundle of a Fredholm manifold has a compatible Fredholm structure.

EXAMPLE 2 [**25**], [**64**]. The concept of Fredholm submanifold is clearly defined. Any finite dimensional submanifold of a Fredholm manifold is a Fredholm submanifold. If A is a closed finite codimensional submanifold of the Fredholm manifold X, then the structure of X induces a Fredholm structure on A as a Fredholm submanifold.

Let X, Y be Fredholm C^r-manifolds $(r \geq 1)$, and B a locally closed Fredholm submanifold of Y; assume that codim $(Y, B) = q < \infty$. *If $f: X \to Y$ is a C^r-map transversal over B, then $A = f^{-1}(B)$ is a Fredholm submanifold of X with* codim $(X, A) = q$. *Furthermore, if X, Y, B are oriented, then A has a natural orientation.*

EXAMPLE 3. If A is a closed Fredholm C^r-submanifold $(r \geq 3)$ of the separable Fredholm C^r-manifold X and if the model of X is C^r-smooth, then A has a Fredholm tubular neighborhood C^{r-2}-manifold in X.

(C) The next result was established independently by Elworthy [**25**] and Tromba [**64**]:

Let X and Y be C^r-manifolds $(r \geq 1)$ modeled on E. Suppose that Y has a Fredholm structure. Then every Φ_k-map $\varphi: X \to Y$ induces a Fredholm structure on X; in terms of Fredholm charts, φ has a local representation whose differential $\varphi_*(x) = A(x) + \Phi(x)$, where $\Phi(x) \in C(E)$ and A is a linear injection, the identity, or linear projection, depending on whether $k < 0$, $k = 0$, $k > 0$.

If E is C^r-smooth and X is separable, a Fredholm structure on X determines a Φ_0-map $X \to E$, unique to within perturbations by maps with compact differentials.

(D) Let $\Phi_k C^r[X, Y]$ denote the set of homotopy classes of Φ_k-maps $X \to Y$ of class C^r. Elworthy [**25**] has established the following classification theorem:

Let X and Y be separable C^r-manifolds $(r \geq 3)$ with the same C^r-smooth infinite dimensional model E. Suppose that X is parallelizable and that Y admits a Fredholm reduction. Then there is a bijection

$$\theta: \Phi_0 C^r[X, Y] \to [X, \Phi_0(E)] \times [X, Y].$$

The map θ can be constructed as follows: Choose (1) a C^{p-1}-trivialization

$$\tau: X \times E \to TX$$

of the tangent vector bundle of X, and (2) a $\Phi_0 C^r$-map $\psi: Y \to E$. Then for any $\varphi: X \to Y$ representing the class $[\varphi]_0 \in \Phi_0 C^r[X, Y]$ and for each $x \in X$ we have the composition $(\psi \circ \varphi)_*(x) \circ \tau_x: E \to E$, which is a Φ_0-operator; let

$$t(\psi \circ \varphi): X \to \Phi_0(E)$$

be defined by $x \to (\psi \circ \varphi)_*(x) \circ \tau_x$. Then set $\theta[\varphi]_0 = [t(\psi \circ \varphi)] \times [\varphi]$.

In the special case (which is proved first) that $Y = E$ *we have the bijection* $\theta: \Phi_0 C^r[X, E] \to [X, \Phi_0(E)]$. Furthermore, from §4B we see that *if* $GL(E)$ *is contractible then a choice of flag determines a bijection*

$$\text{ind} \circ \theta: \Phi_0 C^r[X, E] \to \tilde{K}_K(X),$$

in the notation of §4B.

Taken in conjunction with the results in §6B, we have the following integrability theorem [**25**]:

Let X be a separable C^r-manifold ($r \geq 3$) *with C^r-smooth model E, and suppose its principal bundle has a Fredholm reduction. Then X admits a compatible Fredholm structure.*

In this section and in §6B we have often assumed that X is a separable manifold with a C^r-smooth model E. In many cases it would have sufficed to require merely that X admit C^r-partitions of unity (subordinate to given open covers).

8. **Topology of Fredholm manifolds.**

(A) A *manifold filtration* of a C^r-manifold X is a nested sequence $\cdots X_n \subset X_{n+1} \subset \cdots$ of finite dimensional closed submanifolds of X such that if $X_\infty = \bigcup \{X_n : n \in \mathbf{Z}\}$ is given its direct limit topology, then the natural map $i: X_\infty \to X$ is a homotopy equivalence. The limit space X_∞ has the homotopy type of a CW-complex. Therefore, in order to affirm that i is a homotopy equivalence, it suffices [**67**, Theorem 1] to show that i induces an isomorphism of the homotopy groups $i_*: \pi_k(X_\infty) = \lim_{n\to\infty} \pi_k(X_n) \to \pi_k(X)$ for all k.

EXAMPLE [**50**], [**63**]. Let $E_1 \subset E_2 \subset \cdots$ be a flag in the Banach space E; suppose that the union $\bigcup \{E_n : n \geq 1\}$ is dense in E. If U is open in E and $U_n = U \cap E_n$, then $(U_n)_{n\geq 1}$ is a manifold filtration of U.

The following theorem is due to K. K. Mukherjea [**46**]:

Let X be a separable Fredholm C^∞-manifold modeled on a C^∞-smooth Banach space E. Then for some k, X admits a filtration $(X_n)_{n\geq k}$ with $\dim X_n = n$, *and with X_∞ dense in X.* In fact, if $(E_n)_{n\geq 1}$ is a flag of E, there is a $\Phi_0 C^r$-map $f: X \to E$ associated with the Fredholm structure as in §7C which is transversal to every E_n; then $X_n = f^{-1}E_n$; we call (X_n) a *Fredholm filtration* of X. In this situation we have $w_1(X_n) = i^*_{n,n+1} w_1(X_{n+1})$, *where $i_{n,n+1}: X_n \to X_{n+1}$ is the inclusion map. The inverse system $(w_1(X_n))$ determines the* 1*st Stiefel-Whitney class $w_1(X)$ of the Fredholm structure.*

In case X is modeled on a Hilbert space E, *each* X_n *has a tubular neighborhood* U_n *in* X *such that* $U_n \subset U_{n+1}$ *and* $X = \bigcup \{U_n : n \geq k\}$. As an application [**22**], *there is a* C^∞*-embedding of* X *onto an open subset of* E.

(B) Let X be a Fredholm manifold and $(X_n)_{n\geq 1}$ a Fredholm filtration. For each n let $\Omega_n \to X_n$ be the orientation sheaf. For any coefficient group or ring Λ we have the cohomology groups $H^i_{\mathscr{C}}(X_n : \Omega_n \otimes \Lambda)$ of X_n, compact supports and twisted coefficients. Then the Poincaré duality isomorphisms $\mathscr{D}_n$ [**13**] together with the inclusion maps $i_{n,n+1} : X_n \to X_{n+1}$ determine the homomorphisms η_n:

$$\begin{array}{ccc} H^{n-i}_{\mathscr{C}}(X_n; \Omega_n \otimes \Lambda) & \xrightarrow{\eta_n} & H^{n+1-i}_{\mathscr{C}}(X_{n+1}; \Omega_{n+1} \otimes \Lambda) \\ \downarrow{\scriptstyle \mathscr{D}_n} & & \downarrow{\scriptstyle \mathscr{D}_{n+1}} \\ H_i(X_n; \Lambda) & \xrightarrow{(i_{n,n+1})_*} & H_i(X_{n+1}; \Lambda) \end{array}$$

Following Mukherjea [**46**] we define the *finite codimensional cohomology group* $H^{\infty - i}_{\mathscr{C}}(X; \Lambda)$ as the direct limit of the system $(H^{n-i}_{\mathscr{C}}(X_n; \Omega_n \otimes \Lambda), \eta_n)$. Then we have the *canonical duality isomorphism*

$$\mathscr{D} = \mathscr{D}_X : H^{\infty - i}_{\mathscr{C}}(X; \Lambda) \to H_i(X; \Lambda).$$

In particular, the group $H^{\infty - i}_{\mathscr{C}}(X; \Lambda)$ is independent of the Fredholm structure and the manifold filtration of X. For a coefficient ring Λ, the cup product induces an $H^*(X; \Lambda)$-module structure in $H^{\infty - *}_{\mathscr{C}}(X; \Lambda)$.

A similar finite codimensional cohomology theory was introduced by Gęba–Granas [**27**] for closed bounded subsets of Banach spaces. See also [**21**, §12] for a related construction in locally finite homology.

(C) Next, suppose that $A \subset X$ is a closed subset; set $A_n = A \cap X_n$. For each i the Alexander–Pontrjagin duality theorem [**13**] asserts that there is an isomorphism

$$\alpha_n : \check{H}^{n-i}_{\mathscr{C}}(A_n; \Omega_n \otimes \Lambda) \to H_i(X_n; X_n - A_n; \Lambda),$$

where $\check{H}_{\mathscr{C}}$ denotes Čech cohomology with compact supports. Then the composition ξ_n:

$$\begin{array}{ccc} \check{H}^{n-1}_{\mathscr{C}}(A_n; \Omega_n \otimes \Lambda) & \xrightarrow{\xi_n} & \check{H}^{n+1-i}_{\mathscr{C}}(A_{n+1}; \Omega_{n+1} \otimes \Lambda) \\ \downarrow{\scriptstyle \alpha_n} & & \downarrow{\scriptstyle \alpha_{n+1}} \\ H_i(X_n; X_n - A_n; \Lambda) & \longrightarrow & H_i(X_{n+1}; X_{n+1} - A_{n+1}; \Lambda) \end{array}$$

determines a direct system $(\check{H}^{n-i}_{\mathscr{C}}(A_n; \Omega_n \otimes \Lambda), \xi_n)$; we define the group $\check{H}^{\infty - i}_{\mathscr{C}}(A \subset X; \Lambda)$ as its limit. That group may depend on the position of A in X:

EXAMPLE. Let A be a p-codimensional ($p < \infty$) closed submanifold of X. Then combining α (below) with the Gysin-Thom isomorphism gives the isomorphism

$$\check{H}^{\infty - i}_{\mathscr{C}}(A \subset X; \Lambda) \to H_{i-p}(A; \mathscr{T} \otimes \Lambda),$$

where $\mathscr{T} \to A$ is the orientation sheaf of the transverse bundle of A in X.

Finally, we set $H^{\infty - i}_{\mathscr{C}}(X, A; \Lambda) = H^{\infty - i}_{\mathscr{C}}(X - A; \Lambda)$, where the open manifold $X - A$ is given the Fredholm structure induced from that of X. The following diagram is a commutative exact ladder [**47**]:

$$\begin{array}{ccccccccc} \cdots \to H_{\mathscr{C}}^{\infty-i}(X, A; \Lambda) & \to & H_{\mathscr{C}}^{\infty-i}(X; \Lambda) & \to & \check{H}_{\mathscr{C}}^{\infty-i}(A \subset X; \Lambda) & \to & H_{\mathscr{C}}^{\infty-i+1}(X, A; \Lambda) \to \cdots \\ \downarrow \mathscr{D}_{X-A} & & \downarrow \mathscr{D}_X & & \downarrow \alpha & & \downarrow \mathscr{D}_{X-A} \\ \cdots \to H_i(X - A; \Lambda) & \to & H_i(X; \Lambda) & \to & H_i(X, X - A; \Lambda) & \to & H_{i-1}(X - A; \Lambda) \to \cdots \end{array}$$

The isomorphism α is called *Alexander–Pontrjagin duality.*

As an application, we have the following version of the Brouwer–Jordan separation theorem:

Let X be a connected separable Fredholm manifold with C^∞-smooth model. Assume that $H_1(X; \mathbf{Z}_2) = 0$. If A is a closed subset of X with $\check{H}_{\mathscr{C}}^{\infty-1}(A \subset X; \mathbf{Z}_2) \neq 0$, then A disconnects X. In the special case that A is a 1-codimensional submanifold this was established in [**20**]. Another special case is in [**27**]. On the other hand, it has been shown recently by Anderson–Henderson–West [**3**] that *if X is a C^0-manifold modeled on a separable infinite dimensional Fréchet space and A is a closed subset of X with the following property, then there is a homeomorphism of $X - A$ onto X*: Each point of A has an open neighborhood V in X such that for each contractible, nonvoid, open subset $U \subset V$, we have $U - A \cap V$ contractible and nonvoid. For instance, that property is present if A is closed and locally compact. Furthermore, *if A has that property, then the inclusion map $X - A \to X$ is a homotopy equivalence* [**24**], *so that $\check{H}_{\mathscr{C}}^{\infty-i}(A \subset X; \Lambda) = 0$ for all i.*

(D) If X and Y are Fredholm manifolds and $\varphi: X \to Y$ is a map which induces the Fredholm structure of X from that of Y, we say that φ *is oriented* if $w_1(X) = \varphi^* w_1(Y)$. Then [**47**]:

If $\varphi: X \to Y$ is an oriented proper $\Phi_k C^\infty$-map of Fredholm manifolds ($k \geq 0$), then φ induces cohomology-module homomorphisms

$$\varphi^*: H_{\mathscr{C}}^{\infty-i}(Y; \Lambda) \to H_{\mathscr{C}}^{\infty-(i+k)}(X; \Lambda);$$

i.e., $\varphi^* y \cup \varphi^* \eta = \varphi^*(y \cup \eta)$ for all $y \in H^*(Y; \Lambda)$, $\eta \in H_{\mathscr{C}}^{\infty-i+k}(Y; \Lambda)$. Note, however, that in finite codimensional cohomology φ^* may depend on the filtration used in its construction.

Dually, we have the Gysin homomorphism in homology

$$\varphi_\natural: H_i(Y; \Lambda) \to H_{i+k}(X; \Lambda)$$

given by $\mathscr{D}_X \circ \varphi^* \circ \mathscr{D}_Y^{-1} = \varphi_\natural$, and satisfying $\varphi_\natural(y \cap b) = \varphi^* y \cap \varphi_\natural b$ for all $y \in H^*(Y; \Lambda)$, $b \in H(Y; \Lambda)$.

(E) Let X, Y be connected C^∞-manifolds over separable Banach spaces. Smale's theorem [**59**, Theorem 3.3] indicates that a proper Φ_k-map $\varphi: X \to Y$ induces a homomorphism $\varphi_!: \Omega_i^O(Y) \to \Omega_{i+k}^O(X)$ of the unoriented bordism groups (here $O = \lim O(\mathbf{R}^n)$). In his thesis, J. Morava [**45**] showed that $\varphi_!$ is a module homomorphism (so that it plays the role of a Gysin homomorphism), in the sense that $\varphi_!(\beta \cap b) = \varphi^* \beta \cap \varphi_! b$ for $\beta \in \Omega_O^*(Y)$, the unoriented cobordism group of Y, and $b \in \Omega^O(Y)$.

The construction proceeds as follows: Any element of $\Omega_p^O(Y)$ can be represented by a compact p-dimensional submanifold B of Y and we can suppose that φ is transversal over B [**59**]. Then $\varphi^{-1}(B)$ is a compact $(p + k)$-dimensional submanifold A of X, whose bordism class $[A] = \varphi_![B]$. Furthermore, we have the commutative diagram

$$\begin{array}{ccccc} & & \Omega^{O}_{i+k}(A) & & \\ & \swarrow{\scriptstyle \alpha_*} & & \nwarrow{\scriptstyle \bar{\varphi}_!} & \\ \Omega^{O}_{i+k}(X) & & & & \Omega^{O}_{i}(B) \\ & \nwarrow{\scriptstyle \varphi_!} & & \swarrow{\scriptstyle \beta_*} & \\ & & \Omega^{O}_{i}(Y) & & \end{array}$$

where $\alpha: A \to X$ and $\beta: B \to Y$ are the embeddings. $\bar{\varphi}_!$ is the finite dimensional Gysin homomorphism of $\varphi \mid A \to B$.

We have a commutative diagram

$$\begin{array}{ccc} \Omega^{O}_{i}(Y) & \xrightarrow[\varphi_!]{} & \Omega^{O}_{i+k}(X) \\ \downarrow & & \downarrow \\ H_i(Y; \boldsymbol{Z}_2) & \xrightarrow[\varphi_\natural]{} & H_{i+k}(X; \boldsymbol{Z}_2). \end{array}$$

Morava extends this situation in several directions; e.g., he obtains analogous conditions for $\boldsymbol{G}$-bordism, for certain topological subgroups $\boldsymbol{G} \subset \boldsymbol{O}$. For instance, in the case $\boldsymbol{G} = \boldsymbol{SO} = \lim SO(\boldsymbol{R}^n)$ and the map φ is oriented.

(F) In this subsection I will hazard the following *conjectures*—which certainly can be made precise and are correct for a large (and I believe representative) class of examples.

There is a Poincaré duality relating singular cohomology and finite codimensional homology as follows: Fix a Fredholm C^r-manifold X of infinite dimension, a coefficient ring Λ and an integer $i \geq 0$. Consider oriented Fredholm C^r-manifolds A with 1-codimensional C^r-boundaries ∂A, given the induced orientation. A *singular manifold of codimension i of X* is a pair (φ, A), where φ is a proper Φ_{-i} C^r-map of A into X. Its boundary $\partial(\varphi, A) = (\varphi, \partial A)$ is a singular manifold of codimension $i + 1$. The locally finite chains of singular manifolds of X, with coefficients in the ring Λ twisted by the orientation sheaf of X, determine a complex whose ith homology group we denote by $\mathscr{H}_{\infty-i}(X; \Lambda)$. (This group is not the same as that introduced (temporarily) in [**21**, §12D] via a projective limit; they are isomorphic, however, if Λ is a field.)

Methods of sheaf theory [**13**] establish a canonical isomorphism

$$\mathscr{D} = \mathscr{D}_X : H^i(X; \Lambda) \to \mathscr{H}_{\infty-i}(X; \Lambda).$$

If X is oriented (so that we can avoid twisted coefficients), then intersection theory defines an associative pairing

$$\circ : \mathscr{H}_{\infty-i}(X; \Lambda) \times \mathscr{H}_{\infty-j}(X; \Lambda) \to \mathscr{H}_{\infty-(i+j)}(X; \Lambda),$$

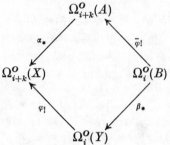

with

$$a_1 \circ a_2 = \mathscr{D}[\mathscr{D}^{-1} a_1 \cup \mathscr{D}^{-1} a_2].$$

Furthermore, $\mathscr{H}_{\infty-*}(X;\Lambda)$ is an $H^*(X)$-module.

We have a commutative exact ladder as in §8C producing an Alexander–Pontrjagin duality isomorphism of the form

$$\alpha: \check{H}^i(A;\Lambda) \to \mathscr{H}_{\infty-i}(X, X - A;\Lambda),$$

where A is any closed subset of X.

From its very definition a proper Φ_k-map $\psi: X \to Y$ induces a homomorphism $\psi_\#: \mathscr{H}_{\infty-i}(X;\Lambda) \to \mathscr{H}_{\infty-i+k}(Y;\Lambda)$, which in turn determines a Gysin homomorphism $\psi^\#$ through the diagram

$$\begin{array}{ccc} H^i(X;\Lambda) & \xrightarrow{\psi^\#} & H^{i-k}(Y;\Lambda) \\ \downarrow{\scriptstyle \mathscr{D}_X} & & \downarrow{\scriptstyle \mathscr{D}_Y} \\ \mathscr{H}_{\infty-i}(X;\Lambda) & \xrightarrow{\psi_\#} & \mathscr{H}_{\infty-i+k}(Y;\Lambda) \end{array}$$

Again, $\psi^\#(\psi^* y \cup x) = y \cup \psi^\#(x)$; $\psi_\#(\psi^* y \cap \alpha) = y \cap \psi_\#(\alpha)$. In particular, if ψ is an embedding of an oriented $(-k)$-codimensional submanifold of Y, then $\psi^\#(1) \in H^{-k}(Y)$ is the fundamental class of X as constructed in [**20**]. In his thesis, J. Callahan [**12**] used singular manifolds (in an open set of a Banach space) to define a class $\psi^\#(1)$; he went on to make intersection computations in the homology of certain loop spaces. See also Lusternik [**41**], [**42**], where such intersections are used to give lower bounds on Lusternik–Schnirelmann category.

We also have compatibility with the finite codimensional bordism groups introduced in [**23**]:

$$\begin{array}{ccc} \Lambda^{SO}_{\infty-i}(X) & \longrightarrow & \Lambda^{SO}_{\infty-i+k}(Y) \\ \downarrow & & \downarrow \\ \mathscr{H}_{\infty-i}(X) & \xrightarrow[\psi_\#]{} & \mathscr{H}_{\infty-i+k}(Y) \end{array}$$

9. **Degrees and indices of maps.**

(A) From the theorem in §8D we see that if X is a connected Fredholm manifold, then we have an isomorphism $H^{\infty-0}_{\mathscr{C}}(X;\mathbf{Z}) \to \mathbf{Z}$. Consequently, for any oriented proper $\Phi_0 C^\infty$-map $\varphi: X \to Y$ of connected Fredholm manifolds we can define [**47**] its *degree* $d_\varphi \in \mathbf{Z}$ by $\varphi^*(\eta) = d_\varphi \xi$, where $\eta = \mathscr{D}_Y^{-1}(y_0)$ and $y_0 \in H_0(Y;\mathbf{Z})$ is the canonical generator; similarly, $\xi = \mathscr{D}_X^{-1}(x_0)$. The absolute value $|d_\varphi|$ is independent of the Fredholm filtration used in the definition of φ^*. Most of the expected properties of degree remain valid in this context; however, Example 3 below shows that this degree is not an invariant of proper Φ_0-homotopy classes (although its absolute value is). Furthermore, in the special case that X is a bounded open subset of a Banach space $E = Y$ and $\varphi: X \to E$ has the form $\varphi(x) = x + \Phi(x)$ for a compact map Φ, then *this degree determines the Leray-Schauder degree* $d[\varphi, X, b]$ *for* $b \in E - \varphi(\text{bdy } X)$ [**40**].

If X and Y are both oriented Fredholm manifolds and $\varphi: X \to Y$ is a proper $\Phi_0 C^r$-map which respects the Fredholm structures, then we can compute its degree as follows: Use Smale's theorem of §6A to find regular values $b \in Y$. Then $\varphi^{-1}(b)$ is a finite set and the orientations permit an algebraic count (independent of the choice of b) giving d_φ. This was the context in which Elworthy **[25]** and Tromba **[64]** developed degree, following a special case due to Smale **[59]**. *The degree is an invariant of proper C^1-homotopies through maps respecting the oriented Fredholm structures* **[64]**.

An advantage of Mukherjea's cohomological description is that we can obtain analogues of Hopf's theorems **[30]**:

Take any $\eta \in H_{\mathscr{C}}^{\infty-i}(Y; Q)$ and follow the square

$$\begin{array}{ccc} H_{\mathscr{C}}^{\infty-i}(X;Q) & \xleftarrow{\varphi^*} & H_{\mathscr{C}}^{\infty-i}(Y;Q) \\ \downarrow{\scriptstyle \mathscr{D}_X} & & \downarrow{\scriptstyle \mathscr{D}_Y} \\ H_i(X;Q) & \xrightarrow{\varphi_*} & H_i(Y;Q); \end{array}$$

we obtain $\mathscr{D}_Y^{-1}\varphi_*\mathscr{D}_X\varphi^*(\eta) = d_\varphi\eta$. In particular, *if $d_\varphi \neq 0$, then $\varphi_*: H_i(X; Q) \to H_i(Y; Q)$ is an epimorphism. Furthermore, $\varphi_*\pi_1(X)$ has finite index j in $\pi_1(Y)$, and j divides d_φ. In particular, if $d_\varphi = 1$, then $\varphi_*: \pi_1(X) \to \pi_1(Y)$ is surjective.*

Application: *Let $\varphi: X \to Y$ be a proper $\Phi_0 C^\infty$-map of oriented Fredholm manifolds, and suppose $d_\varphi = 1$. If X is contractible, then so is Y.*

There is also **[25]** a degree theory for proper Φ_k-maps ($k \geq 0$), based on framed cobordism, following ideas of Pontrjagin-Thom; in that situation X should be parallelizable. A further generalization, for closed bounded subsets of Banach spaces, has been given by Gęba **[26]**.

(B) Example 1 **[14]**, **[25]**, **[57]**,**[64]**. Let E and F be complex Banach spaces, and D a closed domain in E containing the origin in its interior. Let $\varphi: (D, \text{bdy } D, 0) \to (F, F - 0, 0)$ be a holomorphic proper Φ_0-map. Then the Leray–Schauder degree of φ relative to 0 is strictly positive. As an application, *if $\varphi: X \to Y$ is a holomorphic Φ_0-map of complex analytic manifolds which is light (i.e., for which $\varphi^{-1}(y)$ is discrete for all $y \in Y$), then φ is an open map.*

Example 2 **[25]**, **[28]**, **[35]**. This is a form of Borsuk's antipodal points theorem: Let X be a closed symmetric domain containing the origin in a Banach space E. Assume that $\tilde{K}_R(X) = 0$. Let $\varphi: (X, \text{bdy } X) \to (F, F - 0)$ be a proper $\Phi_0 C^2$-map into a Banach space F. If φ is odd (i.e., $\varphi(-x) = -\varphi(x)$ for all $x \in X$), then $d_\varphi \equiv 1 \bmod 2$.

Example 3 **[25]**. Let $X = E = Y$ be the separable infinite dimensional real Hilbert space and $\varphi \in GC(E)$ an orientation reversing operator. φ and the identity map I are properly homotopic through diffeomorphisms, and yet $d_\varphi = -1$, $d_I = +1$.

(C) The following generalization of the Lefschetz Coincidence Theorem is due to Mukherjea **[46]**:

Let X and Y be Fredholm C^∞-manifolds modeled on C^∞-smooth separable Banach spaces. Let $\varphi: X \to Y$ be an oriented proper $\Phi_0 C^\infty$-map and $\psi: X \to Y$ a compact

C^∞-map. Let $\theta_i(\varphi, \psi): H_i(X; Q) \to H_i(X; Q)$ be defined by the composition $\theta_i(\varphi, \psi) = \mathscr{D}_X \circ \varphi^* \circ \mathscr{D}_Y^{-1} \circ \psi_*$:

$$\begin{array}{ccc} H_{\mathscr{C}}^{\infty-i}(X; Q) & \xleftarrow{\varphi^*} & H_{\mathscr{C}}^{\infty-i}(Y; Q) \\ \downarrow \mathscr{D}_X & & \downarrow \mathscr{D}_Y \\ H_i(X; Q) & \xrightarrow{\psi_*} & H_i(Y; Q) \end{array}$$

Then the Lefschetz coincidence number $\Lambda(\varphi, \psi)$ is an integer defined by $\Lambda(\varphi, \psi) = \sum_{i \in Z} (-1)^i \operatorname{Trace} \theta_i(\varphi, \psi)$.

If $\Lambda(\varphi, \psi) \neq 0$, *then there exists a point $x \in X$ such that $\varphi(x) = \psi(x)$.* Furthermore, if the Fredholm structures of both X and Y are oriented and if the coincidence points (x_k) are isolated and regular, then $\Lambda(\varphi, \psi)$ has the analytical expression

$$\Lambda(\varphi, \psi) = \sum_k \operatorname{sign} (\det (\varphi_i(x_k) - \psi_i(x_k))).$$

In particular, taking $\psi_b: X \to b \in Y$ gives $\Lambda(\varphi, \psi_b) = \operatorname{Trace} \theta_0 = d_\varphi$:

If $\varphi: X \to Y$ is a proper Φ_0-map of connected oriented Fredholm manifolds of degree $d_\varphi \neq 0$, then φ is surjective. (That conclusion can also be derived directly from degree theory.)

EXAMPLE. Let $P(K)$ be the projective space of the separable Hilbert space E over K with some oriented Fredholm structure. If X is an oriented Fredholm manifold and $\varphi: X \to P(K)$ a proper Φ_0-map with $d_\varphi \neq 0$, then φ has coincidence points with any compact map $\psi: X \to P(K)$.

BIBLIOGRAPHY

1. R. Abraham, *Lectures of Smale on differential topology*, Notes at Columbia University, New York, 1962–63.

2. R. Abraham and J. Robbin, *Transversal mappings and flows*, Benjamin, New York, 1967.

3. R. D. Anderson, D. W. Henderson and J. E. West, *Negligible subsets of infinite dimensional manifolds*, Compositio Math. (To appear).

4. M. F. Atiyah, *K-theory*, Benjamin, New York, 1967.

5. M. F. Atiyah and R. Bott, *Notes on the Lefschetz fixed point theorem for elliptic complexes*, Harvard University, Cambridge, Mass., 1964. *A Lefschetz fixed point formula for elliptic complexes*. I, Ann. of Math. (2) **86** (1967), 374–407.

6. R. A. Bonic, *Four brief examples concerning polynomials on certain Banach spaces*, J. Differential Geometry (To appear).

7. A. Borel, *Topics in the homology theory of fibre bundles*, Lecture Notes in Mathematics, no. 36, Springer-Verlag, Berlin, 1967.

8. R. Bott, *The stable homotopy of the classical groups*, Ann. of Math. **70** (1959), 313–337.

9. N. Bourbaki, *Éléments de mathématique*. Fasc. X. Premiere partie. Livre III: *Topologie générale*. Chaiptre 10: *Espaces fonctionnelles*, Actualities Sci. Indust., no. 1084, Hermann, Paris, 1949; 2nd ed., 1961; English transl., Hermann, Paris and Addison-Wesley, Reading, Mass., 1966.

10. F. E. Browder, *Families of linear operators depending upon a parameter*, Amer. J. Math. **87** (1965), 752–758.

11. J. W. Calkin, *Two sided ideals and congruences in the ring of bounded operators in Hilbert space*, Ann. of Math. (2) **42** (1941), 839–873.

12. J. Callahan, *Intersection theory for infinite dimensional manifolds*, Thesis, New York University, 1967.

13. H. Cartan, *Topologie algébraique*, Séminaire Ecole Norm Sup. 1950–51.

14. J. Cronin, *Analytic functional mappings*, Ann. of Math. (2) **58** (1953), 175–181.

15. J. Delsarte, *Les groupes de transformations linéaires dans l'espace de Hilbert*, Mémor. Sci. Math. **57** (1932).

16. A. Douady, *Un espace de Banach dont le groupe linéaire n'est pas connexe*, Indag. Math. **68** (1965), 787–789.

17. J. Dugundji, *An extension of Tietze's theorem*, Pacific J. Math. **1** (1951), 353–367.

18. N. Dunford and J. T. Schwartz, *Linear operators*, Interscience, New York, 1958.

19. J. Eells, *On the geometry of function spaces*, Sympos. Internat. Topologia Algebraica, Mexico City, 1958, pp. 303–308.

20. ——— *Alexander–Pontrjagin duality in function spaces*, Proc. Sympos. Pure Math. **3**, Amer. Math. Soc., Providence, R.I., 1961, 109–129.

21. ———, *A setting for global analysis*, Bull. Amer. Math. Soc. **72** (1966), 751–807.

22. J. Eells and K. D. Elworthy, *On the differential topology of Hilbertian manifolds*, Proc. Sympos. Pure Math, vol. 2, Amer. Math. Soc., Providence, R.I., 1969.

23. J. Eells and J. H. McAlpin, *An approximate Morse–Sard theorem*, J. Math. Mech. **17** (1968), 1055–1064.

24. J. Eells and N. H. Kuiper, *Homotopy negligible subsets*, Compositio Math. (To appear).

25. K. D. Elworthy, *Fredholm maps and $GL_c(E)$-structures*, Bull. Amer. Math. Soc. **74** (1968), 582–586.

26. K. Gęba, *Fredholm σ-proper maps of Banach spaces*, Habilitacyjna; Gdańsk 1968. Fund. Math. (To appear).

27. K. Gęba and A. Granas, *Algebraic topology in linear normed spaces*. III; IV, Bull. Acad. Polon. Sci. Sér. Sci. Math. Astronom. Phys. **15** (1967), 137–143; 145–152.

28. A. Granas, *The theory of compact vector fields and some of its applications to topology of functional spaces*. I, Rozprawy Mat. **30** (1962), 93 pp.

29. F. Hirzebruch, *Elliptische Differentialoperatoren auf Mannigfaltigkeiten*, Arbeitsgemeinschaft für Forschung des Landes Nordrhein-Westfalen, Natur-, Ingenieur- und Gesellschaftswissenschaften, Heft 157, Westdeutscher Verlag, Cologne, 1966, pp. 33–60.

30. H. Hopf, *Zur Algebra der Abbildungen von Mannigfaltigkeiten*, J. Reine Angew. Math. **163** (1930), 171–188.

31. K. Jänich, *Vektorraumbündel und der Raum der Fredholm-Operatoren*, Math. Ann. **161** (1965), 129–142.

32. J. L. Kelley, *General topology*, Van Nostrand, Princeton, N.J., 1955.

33. U. Koschorke, *Die characteristischen Klassen von Douady*, Diplomarbeit Bonn, 1966.

34. ———, *Infinite dimensional K-theory and characteristic classes of Fredholm bundle maps*, Thesis, Brandeis University, 1968.

35. M. A. Krasnosel'skiĭ, *Topological methods in the theory of nonlinear integral equations*, Macmillan, New York, 1964.

36. N. Krikorian, *A note concerning the fine topology on function spaces*, Compositio Math. (To appear).

37. ———, Thesis, Cornell University.

38. N. H. Kuiper, *The homotopy type of the unitary group of Hilbert space*, Topology **3** (1965), 19–30.

39. J. Leray, *La théorie des points fixe set ses applications en analyse*, Proc. Internat. Congr. Math., Cambridge, Mass., vol. 2, 1950, pp. 202–208. Amer. Math. Soc., Providence, R.I., 1952.

40. J. Leray and J. Schauder, *Topologie et équations fonctionnelles*, Ann. E.N.S. **51** (1934), 45–78.

41. L. A. Lusternik, *Intersections dans les espaces localement linéaires*, Dokl. Akad. Nauk. SSSR **27** (1940), 771–774.

42. ———, *Structure topologique d'un espace fonctionnelle*, Dokl. Akad. Nauk SSSR **27** (1940), 775–777.

43. A. D. Michal, *Affinely connected function space manifolds*, Amer. J. Math. **50** (1928), 473–517.

44. ———, *General differential geometries and related topics*, Bull. Amer. Math. Soc. **45** (1939), 525–563.

45. J. Morava, *Algebraic topology of Fredholm maps*, Thesis, Rice University, 1968.

46. K. K. Mukherjea, *Coincidence theorems for infinite dimensional manifolds*, Bull. Amer. Math. Soc. **74** (1968), 493–496.

47. ———, *Fredholm structures and cohomology*, Thesis, Cornell University, 1968.

48. G. Neubauer, *Der Homotopietyp der Automorphismengruppe in den Raüme l_p und c_0*, Math. Ann. **174** (1967), 33–40.

49. ———, *On a class of sequence spaces with contractible linear group*, Notes, Berkeley, 1967.

50. R. S. Palais, *On the homotopy type of certain groups of operators*, Topology **3** (1965), 271–279.

51. ———, *Homotopy type of infinite dimensional manifolds*, Topology **5** (1966), 1–16.

52. ———, *Foundations of global non-linear analysis*, Mathematics Lecture Note Series, Benjamin, New York, 1968.

53. F. Quinn, *Transversal approximation on Banach manifolds*, Proc. Sympos. Pure Math., vol. 2, Amer. Math. Soc., Providence, R.I., 1969.

54. R. Schatten, *Norm ideals of completely continuous operators*, Ergebnisse der Mathematik und ihrer Grenzgebiete, Heft 27, Springer-Verlag, Berlin, 1960.

55. J. R. Schue, *Hilbert space methods in the theory of Lie algebras*, Trans. Amer. Math. Soc. **95** (1960), 69–80.

56. ———, *Cartan decompositions for L^* algebras*, Trans. Amer. Math. Soc. **98** (1961), 334–349.

57. J. T. Schwartz, *Compact analytic mappings of B-spaces and a theorem of Jane Cronin*, Comm. Pure Appl. Math. **16** (1963), 253–260.

58. W. Shih, *Fiber cobordism and the index of a family of elliptic differential operators*, Bull. Amer. Math. Soc. **72** (1966), 984–991.

59. S. Smale, *An infinite dimensional version of Sard's theorem*, Amer. J. Math. **87** (1965), 861–866.

60. D. C. Spencer, *Deformations of structures on manifolds defined by transitive continuous pseudogroups*, Ann. of Math. (2) **76** (1962), 306–445.

61. ———, *de Rham theorems and Neumann decompositions associated with linear partial differential equations*, Ann. Inst. Fourier (Grenoble) **14** (1964), 1–20.

62. N. E. Steenrod, *The topology of fibre bundles*, Princeton Univ. Press, Princeton, N.J., 1951.

63. A. S. Švarc, *The homotopic topology of Banach spaces*, Dokl. Akad. Nauk SSSR **154** (1964), 61–63; English transl., Amer. Math. Soc. Transl. **5** (1964), 57–59.

64. A. J. Tromba, *Degree theory on Banach manifolds*, Thesis, Princeton University, 1968.

65. M. M. Vainberg, *Variational methods for the study of nonlinear operators*, Holden-Day, San Francisco, Calif., 1964.

66. K. Vala, *On compact sets of compact operators*, Ann. Acad. Sci. Fenn. Ser. A I No. 351 (1965), 9 pp.

67. J. H. C. Whitehead, *Combinatorial homotopy*. I, Bull. Amer. Math. Soc. **55** (1949), 213–245.

CORNELL UNIVERSITY

DEGREE THEORY ON BANACH MANIFOLDS

K. D. Elworthy and A. J. Tromba

We will be discussing an extension of the Leray-Schauder degree theory using some simple techniques of differential topology and following the ideas of Smale in [**14**]. Although the resulting degree can probably be obtained in a much more general situation using the theory of F. Browder and R. Nussbaum in [**4**] it nevertheless seems to us that the differential approach is worth considering because of its geometric simplicity. An elementary discussion of the finite-dimensional theory can be found in Milnor's book [**11**]. We will only be mentioning results here which seem closely related to the qualitative theory of nonlinear partial differential equations; for a survey of the topological aspects in infinite dimensions and of related results see Eells' article in these Proceedings [**9**].

Let $L(E)$ denote the Banach algebra of bounded linear opeators on the Banach space E, and $L_c(E)$ the subset consisting of elements of the form $I + K$ where K is compact. $GL(E)$ will denote the group of invertible operators in $L(E)$, and $GL_c(E)$ the subgroup of $GL(E)$ consisting of those elements which lie in $L_c(E)$. If E is real this subgroup has two components: the identity component, or group of "orientation preserving" elements, will be denoted by $SL_c(E)$, and the set of orientation reversing elements by $SL_c^-(E)$.

Our manifolds will be assumed to be Hausdorff, paracompact and of class C^r, $1 \leq r \leq \infty$, and usually modelled on an infinite-dimensional Banach space.

Definition. A "C-structure" or "Fredholm-structure", M_c, modelled on E, on a manifold M consists of a maximal atlas $\{(U_i, \varphi_i)\}$ for M, $\varphi_i : U_i \to E$, such that when defined $D(\varphi_i \circ \varphi_j^{-1})(x) \in GL_c(E)$. The structure M_c is said to be "orientable" if there is a subatlas of $\{(U_i, \varphi_i)\}$ with the $D(\varphi_i \circ \varphi_j^{-1})(x)$ lying in $SL_c(E)$. A maximal such subatlas will be called an "orientation" of M_c.

Note that if M is connected any orientable C-structure on M will admit precisely two orientations. M will be called "completely orientable" if it admits a C-structure and if every such structure that it admits is orientable. Each C-structure M_c on M gives rise to a class $\nu(M_c)$ in the first singular cohomology group $H^1(M, Z_2)$ of M such that M_c is orientable iff $\nu(M_c) = 0$. It can be shown that if M admits C^r partitions of unity, $r \geq 3$, and $GL(E)$ is contractible then M is completely orientable iff $H^1(M, Z_2) = 0$.

Example. Let L denote the open Möbius band and E separable Hilbert space. Set $M = L \times E$, given the product manifold structure. Then the natural C-structure on M is not orientable. However M can be embedded as an open tubular neighborhood of any smooth embedding of L in E and this will induce a C-structure on M which is orientable (in fact "flat").

The oriented degree. A map $f:M \to N$ is a "proper" map if the inverse image of any compact set in N is compact; it is "σ-proper" if M can be written as a countable union of closed subsets $M = \bigcup_j M_j$ with $f \mid M_j$ a proper map. Recall that a Fredholm map is locally proper; thus a Fredholm map with separable domain is necessarily σ-proper. We shall rely strongly on Smale's infinite-dimensional version of Sard's theorem [**14**]. The following generalization is due to F. Quinn:

THEOREM. *Suppose that $f:M \to N$ is a C^r, σ-proper, Fredholm map between the Banach manifolds M, N, where $r > \max(\operatorname{ind} f, 0)$. Then the set of regular values of f is a Baire subset of N. If f is proper they are open and dense.*

Suppose that M, N are manifolds with C-structures M_c, N_c, modelled on E. A C^r map $f:M \to N$ will be called a $C(I)$-map from M_c to N_c if

$$D(\psi_i \circ f \circ \varphi_j^{-1})(\varphi_j(x))$$

lies in $L_c(E)$ for all $x \in M$ and charts ψ_i, φ_j of M_c, N_c for which it is defined.

If $r \geq 2$ and M_c, N_c have orientations and if f is proper, since a $C(I)$ map is a Φ_0-map (Fredholm of index zero), we may apply Smale's theorem to obtain an oriented degree for f just as in the finite-dimensional case; namely take a regular value y of f in N and let $\deg f$ be the algebraic number of points in $f^{-1}(y)$:

$$\deg f = \sum_{x \in f^{-1}(y)} \operatorname{sgn} T_x f$$

where $\operatorname{sgn} T_x f = \pm 1$ depending on whether $D(\psi_i \circ f \circ \varphi_j^{-1})(\varphi_j(x))$ lies in $SL_c(E)$ or $SL_c^-(E)$ for oriented charts ψ_i, φ_j at $f(x)$, x. This degree gives an invariant of f under proper C^r homotopies through proper $C(I)$-maps of M_c into N_c.

The following theorem shows that Φ_0-maps are often $C(I)$-maps for some C-structure.

THEOREM 1 (PULL BACK THEOREM). (i) *A Φ_0-map $f:M \to E$ induces a unique C-structure $\{M, f\}_c$ on M, modelled on E, with respect to which f is a $C(I)$-map into E with its trivial structure.*

(ii) *If M_c is a C-structure on M modelled on E and if M admits C^r partitions of unity there is a C^r Φ_0-map $f:M \to E$ with $\{M, f\}_c = M_c$.*

PROOF. (i) Let (U, φ) be a chart about a point p of M. Then $D(f \circ \varphi^{-1})(\varphi(p))$ is Fredholm of index zero and so there is an operator R of finite rank with $D(f \circ \varphi^{-1})(\varphi(p)) + R$ a linear isomorphism onto E. By the inverse function theorem there is a neighborhood W of p in U such that $f \circ \varphi^{-1} + R$ sends $\varphi(W)$ diffeomorphically onto an open subset of E.

Define $\bar{\varphi}: W \to E$ by $\bar{\varphi}(x) = f + R \circ \varphi$ to obtain a chart $(W, \bar{\varphi})$ of M. Then

$$f = \bar{\varphi} - R \circ \varphi$$

giving

$$f \circ \bar{\varphi}^{-1} = I - R \circ \varphi \circ \bar{\varphi}^{-1},$$

showing that f is a $C(I)$-map with respect to $(W, \bar{\varphi})$.

Do this for each point p of M, to obtain an atlas for M consisting of such charts.

Suppose now that at some point p in M there is a neighborhood V and charts $\varphi_0, \varphi_1 : V \to E$ such that

$$f \circ \varphi_0^{-1} = 1 + k_0, \qquad f \circ \varphi_1^{-1} = 1 + k_1$$

are $C(I)$-maps. Then

$$(I + k_0)\varphi_0 = (I + k_1)\varphi_1$$

giving

$$(I + k_0)\varphi_0\varphi_1^{-1} = (I + k_1)$$

or

$$\varphi_0\varphi_1^{-1} = I + k_1 - k_0\varphi_0\varphi_1^{-1}$$

which is a $C(I)$-map.

This proves that the atlas we have constructed gives a C-structure on M, and also proves the uniqueness.

(ii) Let $\{(W_i, \varphi_i)\}_i$ be a locally finite atlas for M_c and $\{\mu_i\}_i$ a C^r partition of unity subordinate to the cover $\{W_i\}_i$.

Define

$$f : M \to E \quad \text{by } f(x) = \sum_i \mu_i(x)\varphi_i(x).$$

It is easy to check that f is a $C(I)$-map on M_c. It follows by the uniqueness part of (i) that $M_c = \{M, f\}_c$.

Theorem 1 has several extensions, in particular the range space E of f in (i) could be replaced by any manifold with C-structure. However, for simplicity we shall be mainly restricting ourselves to the case of maps whose range is a Banach space.

It follows from Theorem 1 that one may obtain an oriented degree for a proper $C^r\Phi_0$-map $f : M \to E$, $r \geq 2$, by considering it as a $C(I)$ map on $\{M, f\}_c$, provided the latter is orientable. This degree will not, however, be an invariant of proper homotopy through Φ_0-maps. For example, suppose E is an infinite-dimensional Hilbert space and $T \in SL_c^-(E)$. Then $\deg T = -1$, although $GL(E)$ is connected and so T is homotopic in $GL(E)$ to the identity map, which has degree $+1$. To get a proper homotopy invariant through Φ_0-maps one has to take the absolute value of the degree.

A more direct extension of the Leray-Schauder degree occurs when one has a closed domain B of E (or M), whose boundary will be denoted by ∂B, together with a point y of E and a proper $C^r\Phi_0$-map $f : B, \partial B \to E, E - \{y\}$, $r \geq 2$, which extends differentiably over a neighborhood A of B. In the same way as before, given orientability, we can define an integer $\deg(f, \partial B, y)$ by looking at the inverse image of a regular value of f lying in the component of y in $E - f(\partial B)$. This is possible since $f(\partial B)$ will be closed because a proper map is a closed map. This degree will depend only on the component of y in $E - f(\partial B)$, and its absolute value will be invariant through proper C^r homotopies of Φ_0-maps $f_t: B, \partial B \to E, E - \{y\}$. If f is a compact field and B is bounded in E it reduces to the Leray-Schauder degree.

Special cases. (i) *Complex analytic maps.* The inverse function theorems still hold in the complex analytic case and so we may talk about complex analytic C-structures induced on complex manifolds by complex analytic Fredholm maps. If E is a complex Banach space $GL_c(E)$ is connected. Hence, as in the finite-dimensional situation, any complex analytic C-manifold has a natural orientation; also, when defined, the degree of a complex analytic Φ_0-map will be nonnegative. (Here the degree is obtained by forgetting about the complex structures but using the induced orientations.) That deg $(f, \partial B, 0)$ is positive in this situation when there is an isolated zero is essentially proven by J. Cronin in [**8**]. Theorems 2 and 3 below give an extension of this:

THEOREM 2. *Let $f: M \to N$ be a proper Φ_0 analytic map between complex analytic manifolds, with N a connected C-manifold. Then* deg f *is defined and nonnegative. Moreover $f(M)$ is either all of N or nowhere dense in N.*

PROOF. By the Smale-Sard theorem the regular values of f are open and dense. If f assumes a regular value deg $f > 0$, thus f must be onto N.

THEOREM 3. *Let E and F be complex Banach spaces and B a closed bounded domain in E containing the origin in its interior. Let $f: B, \partial B \to F, F - \{0\}$ be a complex analytic proper Φ_0-map with $f(0) = 0$. Then* deg $(f, \partial B, 0) > 0$.

PROOF. Adding a sufficiently small complex linear map of finite rank onto f we may assume that $Df(0)$ is an isomorphism. Then, by the inverse function theorem, there is a connected neighborhood U of 0, inside B, such that $f \mid U$ is a diffeomorphism of U onto an open neighborhood V of 0 in $F - f(\partial B)$. By Smale's theorem there is a regular value P of f in V. Since P also lies in the image of f, deg $(f, \partial B, P) > 0$. But P is connected to 0 in V and so

$$\deg (f, \partial B, P) = \deg (f, \partial B, 0).$$

COROLLARY 1. *Let E, F be complex Banach spaces and B a closed domain in E. Let $g: B, \partial B \to F, F - \{y_0\}$ be a complex analytic proper Φ_0-map, where y_0 is some point of F. Then if y_0 lies in the image of g,* deg $(g, \partial B, y_0) > 0$. *Consequently, there is a neighborhood of y_0 contained in the image of the interior of B.*

PROOF. Suppose $g(x_0) = y_0$, $x_0 \in B$. For x in $B - \{x_0\}$ (the translate of B) set $f(x) = g(x_0 + x) - y_0$ and apply the theorem to obtain deg $(g, \partial B, y_0) > 0$.

Let V be a connected neighborhood of y_0 in $F - f(\partial B)$. Then, if $y \in V$,

$$\deg (f, \partial B, y) = \deg (f, \partial B, y_0) > 0.$$

Hence y lies in the image of the interior of B.

COROLLARY 2. *Let $f: M \to N$ be a complex analytic Φ_0-map of complex analytic manifolds. If the inverse image of each point of N is discrete then f is an open map.*

PROOF. Given $x_0 \in M$, let $f(x_0) = y_0$. Since a Φ_0-map is locally proper we may take a chart about y_0 in N and apply Corollary 1 to a sufficiently small chart about x_0 in M. This shows that f sends each element of a base for the topology of M to an open subset of N. Hence f is an open map.

COROLLARY 3. *Let $f:E \to F$ be a complex analytic Φ_0-map of complex Banach spaces. Suppose that for each y_0 in F there is a compact component of $f^{-1}(y_0)$. Then $f(E)$ is open in F.*

PROOF. Take $y_0 \in f(E)$. Let K be a compact component of $f^{-1}(y_0)$ in E. Since f is locally proper there is a neighborhood U of K in E with $f|\bar{U}$ proper. Moreover we may take U sufficiently small so that $f(\partial \bar{U}) \in F - \{y_0\}$. Corollary 1 with $B = \bar{U}$ completes the proof.

(ii) *Odd maps.* One of the most useful results in the Leray-Schauder theory has been Borsuk's theorem on the mod 2 degree of odd maps [**10**]. The following is an extension of this to our situation:

THEOREM 4. *Let B be a closed symmetric domain of E containing the origin, and let $f:B, \partial B \to F, F - \{0\}$ be a proper Φ_0-map of class C^2 into a Banach space F. Then if $KO^{\sim}(B) = \{0\}$ and if f is an odd map (i.e. $f(-x) = -f(x)$, $x \in B$) the degree of f with respect to 0, $\deg(f, \partial B, 0)$, is odd.*

The condition $KO^{\sim}(B) = \{0\}$ means here that all maps of B into the classifying space of the infinite orthogonal group are inessential. This will be true, for example, if B is the closed unit ball of E. A particular result of this condition is that B will be completely orientable and so the orientable degree of f will be defined.

The present proof depends strongly on some algebraic topology and will not be given here. The method is to reduce the problem to the finite-dimensional situation of a compact N-dimensional manifold X^N with boundary, and stably trivial tangent bundle, an involution J on X^N with a unique fixed point, and a map $g:X^N \to R^N$ with $g(Jx) = -g(x)$ for all x in X^N. The manifold X^N is obtained as the inverse image under f of a suitably chosen N-dimensional disc in F.

Nonlinear elliptic problems. In this section we give some examples to show how the degree theory can be applied to nonlinear elliptic partial differential equations.

Let Ω be an open bounded region in Euclidean n-space R^n, with $\partial\Omega$ a smooth submanifold of R^n. Let $C^{k+\alpha}(\bar{\Omega})$ denote the space of functions which are of class C^k in Ω and whose kth derivatives satisfy a uniform Hölder condition with exponent α, $0 < \alpha < 1$, with the usual norm. Define $\bigwedge^{k+\alpha}(\bar{\Omega})$ to be the closure of the space of C^∞ functions in $C^{k+\alpha}(\bar{\Omega})$. This space seems more convenient for topological results than the traditional $C^{k+\alpha}(\bar{\Omega})$ because it is linearly isomorphic to the sequence space c_0, and hence admits C^∞ partitions of unity and has contractible general linear group. This fact was conjectured by A. Tromba, and a very general proof can be found in the paper of Bonic, Frampton, and Tromba [**2**]. Also $\bigwedge^{k+\alpha}(\bar{\Omega})$ is separable whereas $C^{k+\alpha}(\bar{\Omega})$ is not.

Let J^k be the trivial k-jet bundle of $\bar{\Omega}$:

$$J^k = \bar{\Omega} \times L(R^n, R) \times L_s^2(R^n, K) \times \cdots \times L_s^k(R^n, R).$$

If $u \in C^k(\Omega)$ define $j_k u : \bar{\Omega} \to J^k$ by

$$(j_k u)(x) = (x, u(x), Du(x), \ldots, D^k u(x)).$$

By a nonlinear differential operator on $\bar{\Omega}$ we shall mean a map $F: J^k(\bar{\Omega}) \to R$. This will be called elliptic if $D_{p_k} F(x, p_0, \ldots, p_k)$ exists for all $(x, p_0, \ldots, p_k) \in J^k$ and satisfies

$$D_{p_k} F(x, p_0, \ldots, p_k)q \geq C \|q\|$$
$$(x_0, p_0, \ldots, p_k) \in J^k \quad \text{and all} \quad q \in L_s^k(R^n, R).$$

Let $\Lambda_0^{2m+\alpha}(\bar{\Omega})$ denote the subspace of $\Lambda^{2m+\alpha}(\bar{\Omega})$ consisting of those functions whose first m derivatives all vanish on $\partial\Omega$. The pertinent fact that we need about elliptic operators is that if $F: J^{2m} \to R$ is of class C^r then F induces a map $f: \Lambda_0^{2m+\alpha}(\bar{\Omega}) \to \Lambda^{0+\alpha}(\bar{\Omega})$, for all $\alpha \in (0, 1)$, defined by

$$f(u)(x) = F(j_{2m} u(x))$$

which is a Φ_0-map of class C^{r-1}. This fact is essentially "well known", for proofs see [**5**], [**3**].

To apply the degree theory we need to know some properness conditions on f. This turns out very satisfactorily:

LEMMA. *Suppose $i: E_1 \to E_0$, $j: F_1 \to F_0$ are dense linear embeddings of Banach spaces and $f_0: E_0 \to F_0$, $f_1: E_1 \to F_1$, Fredholm maps of the same index such that the following diagram commutes*

$$\begin{array}{ccc} E_0 & \xrightarrow{f_0} & F_0 \\ \uparrow i & & \uparrow j \\ E_1 & \xrightarrow{f_1} & F_1. \end{array}$$

Then if i is a compact map f_1 is proper on every closed bounded subset of E_1.

PROOF. Suppose $\operatorname{Ind} f_0 = \operatorname{Ind} f_1 = 0$. By a careful examination of the inverse function theorem and of the proof of Theorem 1 one can show that each x_0 in E_0 has a neighborhood U_0 with a diffeomorphism φ_0 of U_0 onto an open subset V_0 of F_0 such that $f_0 \circ \varphi_0^{-1}$ is of the form $x \mapsto x + k(x)$, where k has bounded range in a finite-dimensional subspace of $j(F_1)$. Moreover φ_0 "restricts" to give a diffeomorphism φ_1 of $U_1 = i^{-1}(U_0)$ onto the open subset $V_1 = j^{-1}(V_0)$ of F_1 with the same property. The map f_1 will be proper on any closed subset of U_1. Since i is compact any closed bounded subset of F_1 may be covered by a finite number of subsets U_1 obtained in this way.

THEOREM 5. *If $F: J^{2m} \to R$ is of class C^r and elliptic the induced map $f: \Lambda_0^{2m+\alpha}(\bar{\Omega}) \to \Lambda^{0+\alpha}(\bar{\Omega})$, $0 < \alpha < 1$, is a C^{r-1} Φ_0-map which is proper on every closed bounded subset of $\Lambda_0^{2m+\alpha}(\bar{\Omega})$.*

PROOF. Choose β with $0 < \beta < \alpha < 1$. Then the natural injections

$$\Lambda_0^{2m+\alpha}(\bar{\Omega}) \to \Lambda_0^{2m+\beta}(\bar{\Omega}), \quad \text{and} \quad \Lambda^{0+\alpha}(\bar{\Omega}) \to \Lambda^{0+\beta}(\bar{\Omega})$$

are compact maps with dense images. Thus we may apply the lemma to the maps induced by F on $\Lambda_0^{2m+\alpha}(\bar{\Omega})$ and $\Lambda_0^{2m+\beta}(\bar{\Omega})$.

REMARKS. (i) The same method may be applied to the spaces $C_0^{2m+\alpha}(\bar{\Omega})$, for although $C_0^{2m+\alpha}(\bar{\Omega})$ is not dense in $C_0^{2m+\beta}(\bar{\Omega})$, $0 < \beta < \alpha < 1$, we may apply the lemma to $C_0^{2m+\alpha}(\bar{\Omega}) \subset \Lambda_0^{2m+\beta}(\bar{\Omega})$ which is a dense compact embedding.

(ii) The properness on closed bounded subsets can be proved for quasilinear elliptic operators by a direct analytical method, and for this we need only assume that F lies in $\Lambda^{0+\alpha}(J^{2m})$.

THEOREM 6. *Let F, $G: J^{2m}(\bar{\Omega}) \to R$ be elliptic and of class C^3. For some $\alpha \in (0, 1)$ suppose that there is a constant K such that for $0 \leq t \leq 1$ any solution $u \in \Lambda_0^{2m+\alpha}(\bar{\Omega})$ of*

$$tf(u) + (1 - t)g(u) = 0$$

satisfies

$$\|u\|_{2m+\alpha} < K.$$

Then if g is an odd function of u the equation $f(u) = 0$ has a solution in $\Lambda_0^{2m+\alpha}(\bar{\Omega})$.

PROOF. Let B denote the closed ball of radius K about the origin in $\Lambda_0^{2m+\alpha}(\bar{\Omega})$. Then $f: B \to \Lambda^{0+\alpha}(\bar{\Omega})$ is a proper Fredholm map of index zero, and $\deg(f, \partial B, 0)$ is defined. By the homotopy property of degree

$$\deg(f, \partial B, 0) = \deg(g, \partial B, 0).$$

By Borsuk's theorem the right-hand side is nonzero and thus so is $\deg(f, \partial B, 0)$.

In the following paragraphs we describe how degree theory can be applied to give an elegant proof of the classic Leray-Schauder existence theorem for second order quasilinear strongly elliptic equations.

This theorem has been extended by Nirenberg [**12**] and to the n-dimensional case by Cordes [**6**]. The techniques of our proof of the Leray-Schauder theorem can also be applied to give new proofs of the Nirenberg-Cordes results.

THEOREM 7 (LERAY-SCHAUDER). *Let Ω be a bounded convex domain in R^2 such that its boundary Γ is a C^4 submanifold of R^2 represented by $x = x(s)$, $y = y(s)$ and such that Γ has positive curvature everywhere. Let $A: \Omega \times R^3 \to R$, $B: \Omega \times R^3 \to R$, and $C: \Omega \times R^3 \to R$ be defined and continuous for all points in $\Omega \times R^3$. Assume further that A, B, C are of class C^3. Consider the quasilinear partial differential equation given by*

$$\Phi(z)(x, y) = 0$$

where

$$\begin{aligned}\Phi(z)(x, y) &= A(x, y, z, z_x, z_y)z_{xx} + B(x, y, z, z_x, z_y)z_{xy} + C(x, y, z, z_x, z_y)z_{yy} \\ &= A(z)z_{xx} + B(z)z_{xy} + C(z)z_{yy}.\end{aligned}$$

In addition suppose that there is a positive number m such that for all real ξ *and* η *and all points in* $\Omega \times \boldsymbol{R}^3$

$$A\xi^2 + B\xi\eta + C\eta^2 \geq m(\xi^2 + \eta^2).$$

This is the ellipticity condition on Φ. *Suppose* $\phi \in \Lambda^{3+\alpha}(\Gamma)$. *Then* $\Phi(z) = 0$ *has at least one solution* $z \in \Lambda_\phi^{2+\alpha}(\bar{\Omega})$. ($\Lambda_\phi^{2+\alpha}(\bar{\Omega})$ *is the affine subspace of* $\Lambda^{2+\alpha}(\bar{\Omega})$ *consisting of those functions which agree with* ϕ *on* $\partial\Omega$.)

The proof of this theorem is based on the following estimate due to Schauder [**13**], which, in turn, was an extension of results by S. Berstein.

PROPOSITION 8. *Under the hypothesis of theorem F, there is a constant K dependent only on* $|\phi|_{3+\alpha}^{\Gamma}$ *so that if* $\Phi(z) = 0$, $|z|_{2+\alpha} < K$.

PROOF OF THEOREM 7. Let $V \subset \Lambda_\phi^{2+\alpha}(\bar{\Omega})$ be the intersection of the open ball B of radius K in $\Lambda^{2+\alpha}(\bar{\Omega})$ with $\Lambda_\phi^{2+\alpha}(\bar{\Omega})$. The map $\Phi: V \to \Lambda^{0+\alpha}(\bar{\Omega})$ has a Leray-Schauder degree deg $(\Phi, \partial V, 0)$. Consider the homotopy Φ_λ given by

$$\Phi_\lambda(z) = A(\lambda z)z_{xx} + B(\lambda z)z_{xy} + C(\lambda z)z_{yy}$$

where $A(\lambda z) = A(x, y, \lambda z, \lambda z_x, \lambda z_y)$ and similarly for $B(\lambda z)$ and $C(\lambda z)$.

The map $(\lambda, z) \rightsquigarrow \Phi_\lambda(z)$ for $(\lambda, z) \in I \times \bar{V}$ is a proper map. By Proposition 8, we can apply the homotopy property of degree to conclude that deg $(\Phi_1, \partial V, 0) =$ deg $(\Phi_0, \partial V, 0)$. By the Schauder existence theorem [7], $\Phi_0: \Lambda_0^{2+\alpha}(\bar{\Omega}) \to \Lambda^{0+\alpha}(\bar{\Omega})$ is an isomorphism and so deg $(\Phi_0, \partial V, 0) = 1$. Hence deg $(\Phi_1, \partial V, 0) = 1$ and thus there is a $z \in V$ with $\Phi(z) = 0$.

In fact, we have shown that there is a ball $\mathcal{O}$, about 0 in $\Lambda^{0+\alpha}(\bar{\Omega})$ so that for $f \in \mathcal{O}$ there is a $z \in \Lambda_\phi^{2+\alpha}(\bar{\Omega})$ with $\Phi(z) = f$.

REMARK. If the coefficients A, B, C are independent of z then the solution u of $\Phi(u) = 0$ is unique (e.g. see Bers [**1**]). To prove this in our context, we observe that if the coefficients are independent of z, then the map Φ is a local diffeomorphism. This follows from the Schauder existence theorem and the inverse function theorem. Thus Φ induces an orientable C-structure on V so that sgn $T_x\Phi = 1$ for x and 0 is a regular value for Φ. Since deg $(\Phi, \partial V, 0) = 1$, it follows that there can only be one point in $\Phi^{-1}(0)$.

In order to define deg $(\Phi, \partial V, y)$ we required that Φ be C^2-smooth. Since the degree of a map (up to sign) is an invariant of the proper homotopy class it extends naturally in this context to the topological category, and we give a brief outline on how to make this extension in the quasilinear case.

Again let $J^{2m-1}(\bar{\Omega})$ be the trivial $2m - 1$ jet bundle over $\bar{\Omega}$, with

$$j^{2m-1}: \Lambda_0^{2m+\alpha}(\bar{\Omega}) \to J^{2m-1}(\bar{\Omega})$$

a restriction of the usual $2m - 1$ jet extension map. Suppose $a, a_\beta \in \Lambda^{0+\alpha}(J^{2m-1}(\bar{\Omega}))$. Let $\Phi: \Lambda_0^{2m+\alpha}(\bar{\Omega}) \to \Lambda^{0+\alpha}(\bar{\Omega})$ be the nonlinear differential operator given by

$$\Phi(u) = \sum_{|\beta|=2m} a_\beta(j^{2m-1}(u))D^\beta u + a(j^{2m-1}(u)).$$

If

$$\sum_{|\beta|=2m} a_\beta(j^{2m-1}(u)(p))\xi^\beta \geq c\,|\xi|^{2m}$$

for some $c > 0$ and for all $p \in \bar{\Omega}$, then Φ is a quasilinear elliptic operator and proper on bounded sets.

Let $y \in \Lambda^{0+\alpha}(\bar{\Omega})$ and let V be a bounded open subset of $\Lambda_0^{2m+\alpha}(\bar{\Omega})$ with $\Phi(\partial V) \cap \{y\} = \phi$. By approximating a_β, a by a sequence of C^∞ functions a_β^n, a^n we obtain a sequence of quasilinear elliptic operators Φ^n which are of class C^∞. If N is sufficiently large, then for $n > N$, $\Phi^n(\partial V) \cap \{y\} = \phi$, and if $m > N$

$$(t\Phi^n + (1-t)\Phi^m)(\partial V) \cap \{y\} = \phi.$$

Thus for $n > N$, deg $(\Phi^n, \partial V, y)$ is defined and independent of n. Set

$$\deg(\Phi, \partial V, y) = \lim_{n\to\infty} \deg(\Phi^n, \partial V, y).$$

It is an easy matter to check that deg $(\Phi, \partial V, y)$ has all the properties of Leray-Schauder degree.

REMARK. Since Nirenberg [**12**] did not assume that A, B, C were smooth functions, this is exactly the extension of degree one needs to prove his result in this setting.

BIBLIOGRAPHY

1. L. Bers, *Linear partial differential equations*, New York, 1964.

2. R. A. Bonic, J. Frampton and A. J. Tromba, *Λ-manifolds*, J. Functional Analysis (to be published).

3. ———, *Λ-manifolds and Fredholm maps*, J. Functional Analysis (to be published).

4. F. Browder and R. D. Nussbaum, *The topological degree for noncompact nonlinear mappings in Banach spaces*, Bull. Amer. Math. Soc. **74** (1968), 671–676.

5. F. Browder, *Topological methods for non-linear elliptic equations of arbitrary order*, Pacific J. Math. **17** (1966), 17–31.

6. H. O. Cordes, *Über die erste Randwertaufgabe bei quasilinearen Differentialgleichungen zweiter Ordnung in mehr als zwei Variablen*, Math. Ann. **131** (1956), 278–373.

7. R. Courant, *Methods of mathematical physics.* Vol. II: *Partial differential equations*, Interscience, New York, 1962.

8. J. Cronin, *Analytic functional mappings*, Ann. of Math. **58** (1953), 175–181.

9. J. Eells, Jr., *Fredholm structures*, these Proceedings, 1968.

10. M. A. Krasnosel'skiĭ, *Topological methods in the theory of nonlinear integral equations*, Macmillan, New York, 1964.

11. J. Milnor, *Topology from the differentiable viewpoint*, Univ. Press of Virginia, Charlottesville, 1965.

12. L. Nirenberg, *On nonlinear elliptic partial differential equations and Hölder continuity*, Comm. Pure Appl. Math. **12** (1959), 623–727.

13. J. Schauder, *Über das Dirichletsche problem in Grossen für nichtlineare elliptische Differentialgleichungen*, Math. Z. **34** (1933), 623–634.

14. S. Smale, *An infinite dimensional version of Sard's theorem*, Amer. J. Math. **87** (1965), 861–866.

UNIVERSITY OF MANCHESTER, ENGLAND
STANFORD UNIVERSITY

ON THE EXTENSION OF CONTRACTIONS ON NORMED SPACES

D. G. deFigueiredo and L. A. Karlovitz[1]

Introduction. Let X and Y be two normed linear spaces. A mapping T defined on a subset S of X with values in Y is said to be a *contraction* if $\|Tx - Ty\| \leqslant \|x - y\|$ for all $x, y \in S$. If S is a proper subset of X, we are concerned with the existence of contractions $\tilde{T}$ defined on all of X such that $Tx = \tilde{T}x$ for all $x \in S$.

In the case $X = Y = R^n$ the extension problem was first considered by Kirszbraun [**12**], where the following result is proved.

KIRSZBRAUN LEMMA. *Let $\{B(x_i; r_i): i = 1, \ldots, m\}$ be a set of m closed balls with centers $x_i \in R^n$ and with radii r_i. If the intersection of these balls is nonempty and $x'_i \in R^n$ such that $\|x'_i - x'_j\| \leqslant \|x_i - x_j\|$ for all $i, j = 1, \ldots, m$, then the intersection of the balls $B(x'_i, r_i)$ is also nonempty.*

(A simple proof of Kirszbraun lemma is the object of a paper by Schoenberg [**18**].) With the help of this lemma one proves that a contraction defined on an arbitrary subset of R^n can be extended as a contraction to all of R^n. In order to make this precise, we introduce the following concepts: A pair (X, Y) of normed linear spaces is said to have *Property* (K) if the following holds:

(K) For all $x_i \in X$, $y_i \in Y$, $r_i > 0$, $i \in I$, (where I is an arbitrary index set) such that $\|y_i - y_j\| \leqslant \|x_i - x_j\|$, $i, j \in I$, and such that the intersection of the balls $B(x_i; r_i)$ in X is nonempty, it follows that the intersection of the balls $B(y_i; r_i)$ in Y is also nonempty.

Following Schönbeck [**19**] we say that a pair (X, Y) of normed spaces has the *contraction-extension* property, abbreviated *Property* (E), if:

(E) Whenever $S \subset X$ and T is a contraction defined on S with values in Y, there exists a contraction $\tilde{T}$ defined on all of X with values in Y such that $Tx = \tilde{T}x$ for all $x \in S$.

It is easy to prove, Valentine [**22**], that the two properties above are equivalent. It follows readily from the Kirszbraun lemma and Helly's theorem that (R^n, R^n) has Property (K), and consequently Property (E).

In the case that X and Y are arbitrary Hilbert spaces, it has been proved by

[1] The research of the second author was supported in part by Atomic Energy Commission Grant AEC AT(40-1)3443, and in part by National Science Foundation Grant NSF-GP 6631.

Valentine [**22**] (see also Mickle [**14**], Minty [**15**], Browder [**2**]) that the pair (X, Y) has Property (E).

In these and other early works of Browder and Minty on the theory of monotone operators in Hilbert spaces, the extension problem for contractions played a central role. Subsequently the theory of monotone operators was extended to Banach spaces using new techniques (Galerkin approximations) to prove the basic theorems. (Browder [**3**], Minty [**16**].)

Let us now turn to the extension problem for contractions in the case that at least one of X and Y is not a Hilbert space. It is easy to show that (l^p, l^p), $p \neq 2$, does not have Property (K) (see J. Schwartz [**21**, p. 27]). The following result has been proved by Grünbaum [**10**] (see also Danzer-Grünbaum-Klee [**6**, p. 15]).

THEOREM (GRÜNBAUM). *Let X be a two-dimensional normed linear space. Then (X, X) has Property* (E) *if and only if X is either a Hilbert space or the unit sphere of X is a parallelogram.*

For normed linear spaces of dimension higher than 2 the extension problem was first considered by Schönbeck [**19**]. For a strictly convex normed space Y, he proved the following theorem.

THEOREM (SCHÖNBECK). *Let X be a Banach space and Y a strictly convex Banach space Y with dimension $Y > 1$. Then (X, Y) has Property* (E) *if and only if both X and Y are Hilbert spaces.*

Edelstein and Thompson [7] prove, in a very elegant fashion, a stronger version of this theorem for the case $X = Y$.

Recently Schönbeck has extended the Grünbaum theorem to all finite-dimensional spaces X. We state an infinite dimensional version of this.

THEOREM 1. *Let X be a Banach space such that (X, X) has Property* (E). *Suppose that X contains a subspace M, of finite dimension not less than 2, which is the range of a linear projection of norm 1. Then either X is a Hilbert space or M is a P_1 space (i.e., M is isometric to $C(K)$, where K is an extremally disconnected compact Hausdorff space).*

REMARK. If X is finite dimensional we can take $X = M$ and this is precisely the result of Schönbeck [**20**, Theorem 2.4]. Assuming his result we give a proof of Theorem 1 in §2.

We will give elsewhere a complete discussion about the problem of extending contractions defined on arbitrary subsets of infinite-dimensional Banach spaces.

One might imagine that the rather negative conclusion of Theorem 1 stems from the arbitrariness of the domain S of the given contraction T. It is an important question to consider the extension problem only for contractions defined in *closed convex subsets* S of X. In many applications, Browder-Petryshyn [**4**], Moreau [**17**], Lions-Stampacchia [**13**], Zabreiko-Kachurovsky-Krasnoselsky [**24**], this is the case. With that in mind we give the following definition. A normed linear space X is said to have the *extension property for contractions defined in convex sets*, abbreviated *Property* (EC), if:

(EC) Whenever $C \subset X$ is a closed convex subset of X, every contraction $T: C \to X$ can be extended to a contraction $\tilde{T}: X \to X$ defined in all of X with the additional property that $\tilde{T}(X)$ is contained in the convex closure of $T(C)$.

It is easy to see that Property (EC) is equivalent to *Property* (P) defined by:

(P) For any closed convex subset C of X there exists a *contractive retraction* $P: X \to C$, i.e. there exists a contraction P such that $P(X) = C$ and such that $Px = x$ for $x \in C$.

The existence of contractive retractions has been the object of works [**8**] and [**9**] by the authors. In a Hilbert space, Property (P) holds. In fact, for each closed convex subset C, a contractive retraction is given by the *proximity mapping* $P: X \to C$ defined by $Px = x_0$ where x_0 is the uniquely defined point in C such that $\|x - x_0\| = \inf\{\|x - y\| : y \in C\}$, see Cheney-Goldstein [**5**]. However, proximity mappings are not contractions in general Banach spaces, as shown by the authors. In [**8**] we considered the radial retraction of a normed space over its unit ball defined as follows $Px = x$, if $\|x\| \leqslant 1$, and $Px = x/\|x\|$ if $\|x\| > 1$. Such a mapping P is obviously a proximity mapping of X onto the unit ball. But is shown [**8**] that: *For a normed space X of dimension not less than* 3, *the radial retraction P is a contraction if and only if X is an inner-product space.* In §2 we consider also contractive retractions which are not necessarily proximity mappings. In particular we prove the following result.

THEOREM 2. *Let X be a reflexive Banach space, and X^* its dual space. Let C be a closed convex subset of X which contains the origin in its interior. Suppose that C satisfies the following condition.*

(σ) *C has a unique support hyperplane $H + y_0$ at one of its boundary points y_0 such that the linear subspace H is not the range of a linear projection of norm* 1.

Then there exists no contractive retraction P from X onto C.

Consequently X does not have Property (EC).

REMARK. For the case that X is finite dimensional, Theorem 2 has been proved by the authors in [**9**].

The following corollaries are also obtained.

COROLLARY 1. *Let X be a finite-dimensional, strictly convex Banach space of dimension not less than* 3. *If X has dimension* 3, *the assumption of strict convexity may be omitted. Then there exists a contractive retraction over the unit ball if and only if X is a Hilbert space.*

Consequently X has Property (EC) *if and only if X is a Hilbert space.*

COROLLARY 2. *There is no contractive retraction over the unit ball of the space $L^p(G)$, $1 < p < \infty$, $p \neq 2$, where G is an open subset of R^n.*

Consequently these spaces do not have Property (EC).

1. **Proof of Theorem 2 and related results.** Suppose that C is a closed convex subset of a normed linear space X, and that the origin belongs to the interior of C. Let X^* be the dual space of X, and ∂C the boundary of C. The *support (or Minkowski) functional* of C is a real valued mapping ρ defined on all of X by $\rho(x) = \inf\{\lambda > 0 : x \in \lambda C\}$. It can be proved readily that ρ is subadditive and positively homogeneous. A linear functional $f \in X^*$ is said to be tangent to C at $x_0 \in \partial C$ if $f(x_0) = \sup\{f(x) : x \in C\}$. The set $H + x_0$, where $H = \{x : f(x) = 0\}$, is called a *tangent hyperplane* to C at x_0. It can be proved that C has a unique tangent hyperplane at x_0 if and only if the support functional ρ of C is Gâteaux differentiable at x_0; that is, $\lim \lambda^{-1}[\rho(x_0 + \lambda y) - \rho(x_0)]$, as λ goes to zero, exists for all $y \in X$. Moreover, one can prove in this case that the Gâteaux differential of ρ at x_0; i.e., the mapping $D\rho(x_0)$ defined on X by $D\rho(x_0)y = \lim \lambda^{-1}[\rho(x_0 + \lambda y) - \rho(x_0)]$, is a bounded linear functional on X.

In view of the preceding remarks, condition (σ) in the statement of Theorem 2 is equivalent to:

(σ') The support functional ρ of the set C is Gâteaux differentiable at one point $y_0 \in \partial C$ and the Gâteaux differential is such that the closed hyperplane $H = \{y \in X : D\rho(y_0)y = 0\}$ is not the range of a linear projection of norm 1.

We note that if X is a reflexive Banach space which is smooth (i.e., the norm is Gâteaux differentiable at all points of X) and if C is the unit ball then condition (σ) is equivalent to the existence of a closed linear subspace H of codimensional 1 which is not the range of a linear projection of norm 1. We do not know if every smooth, reflexive Banach space, which is not a Hilbert space, contains such a subspace H.

Before proving Theorem 2 we establish a lemma, which is essentially contained in Kakutani [**11**].

LEMMA 1. *Let X be a reflexive Banach space. Let H be a closed linear subspace of codimension 1 which is not the range of a linear projection of norm 1. Then given $y_0 \notin H$, there are points $x_1, \ldots, x_n$ in H such that*

$$\bigcap_{j=1}^{n} B(x_j; \|y_0 - x_j\|) \cap H = \varnothing .$$

PROOF. Consider the sets $B(x, \|y - x\|) \cap H$, for all $x \in H$. These sets are convex, closed, bounded, and consequently weakly compact. If the assertion of the lemma is not true they satisfy the finite intersection property. So there exists a point y^* in the intersection

$$\bigcap_{x \in H} B(x, \|y_0 - x\|) \cap H.$$

Since every point $z \in X$ can be uniquely written as $z = x + \alpha y_0$, $x \in H$, the mapping $Pz = x + \alpha y^*$, of X onto H is well defined. It is easy to see that P is linear and $P^2 = P$. We claim that $\|P\| = 1$. Once this is proved we have arrived at a contradiction and our lemma is thus proved. If $\alpha = 0$, $\|Pz\| = \|z\|$. If $\alpha \neq 0$ we have

$$\|Pz\| = \|x + \alpha y^*\| = |\alpha|\, \|-\alpha^{-1}x - y^*\| \leqslant |\alpha|\, \|-\alpha^{-1}x - y_0\| = \|x + \alpha y_0\|$$
$$= \|z\|.$$

REMARK. In the case of a finite-dimensional space X, the number n of points in the previous lemma can be taken to be equal to the dimension of the space. This follows readily from Helly's theorem.

PROOF OF THEOREM 2. By hypothesis (σ) and by Lemma 1 the linear subspace H contains points $x_1, \ldots, x_n$ such that

$$\bigcap_{j=1}^{n} B(x_j, \|y_0 - x_j\|) \cap H = \varnothing. \tag{1}$$

The set $B = \bigcap_{j=1}^{n} B(x_j, \|y_0 - x_j\|)$ is weakly compact by virtue of the reflexivity of X. Let f be the continuous linear functional such that $H = \{x: f(x) = 0\}$ and $f(y_0) = 1$. On the set B, f achieves its infimum at a point $w_0 = \delta_0 y_0 + x_0$, where $x_0 \in H$ and, in view of (1), $\delta_0 > 0$. Let $z_0 = \delta_0 y_0$ and $y_j = x_j - x_0$. Then

$$z \in \bigcap_{j=1}^{n} B(y_j, \|z_0 - y_j\|) \Rightarrow z + x_0 \in \bigcap_{j=1}^{n} B(x_j, \|y_0 - x_j\|). \tag{2}$$

For if $z \in \bigcap_{j=1}^{n} B(y_j, \|z_0 - y_j\|)$ then, $\|z + x_0 - x_j\| = \|z - y_j\| \leqslant \|z_0 - y_j\| = \|z_0 + x_0 - x_j\| = \|w_0 - x_j\| \leqslant \|y_0 - x_j\|$, $j = 1, \ldots, n$. It follows from (2) and the definition of δ_0 that

$$z = \gamma y_0 + x, \quad x \in H, \qquad z \in \bigcap_{j=1}^{n} B(y_j, \|z_0 - y_j\|) \Rightarrow \gamma \geqslant \delta_0. \tag{3}$$

Now for each λ, $0 < \lambda < 1$, let $w_\lambda = (1 + \lambda\delta_0)y_0$ and $z_{j,\lambda} = y_0 + \lambda y_j$, $j = 1, \ldots, n$. From (3) it follows that

$$z = \mu y_0 + x, \quad x \in H, \qquad z \in \bigcap_{j=1}^{n} B(z_{j,\lambda}, \|w_\lambda - z_{j,\lambda}\|) \Rightarrow \mu \geqslant \lambda\delta_0 + 1. \tag{4}$$

Since $H + y_0$ is a tangent hyperplane of C it follows that $\rho(y) \geqslant 1$ for all $y \in H + y_0$. Thus if z is as in relation (4) we find

$$\rho(z) = \rho(\mu y_0 + x) = \mu\rho(y_0 + \mu^{-1}x) \geqslant \mu \geqslant \lambda\delta_0 + 1 = \rho(w_\lambda). \tag{4'}$$

Let us now define the real-valued functions

$$h(\lambda) = \rho(w_\lambda) = 1 + \lambda\delta_0,$$
$$g_j(\lambda) = \rho(z_{j,\lambda}) = \rho(y_0 + \lambda y_j), \qquad j = 1, \ldots, n,$$

for $0 \leqslant \lambda \leqslant 1$. Clearly $h(0) = g_j(\lambda) = 1, j = 1, \ldots, n$. Moreover, $h'(0) = \delta_0 > 0$ and $g_j'(0) = D\rho(y_0)y_j = 0$. Consequently we can choose λ^*, $0 < \lambda^* < 1$, such that

$$\rho(z_{j,\lambda^*}) = g_j(\lambda^*) < h(\lambda^*) = \rho(w_{\lambda^*}),$$

$j = 1, \ldots, n$. Let $\alpha = \max \{g_j(\lambda^*) : j = 1, \ldots, n\}$ and put $u_j = \alpha^{-1} z_{j,\lambda^*}$ and $v = \alpha^{-1} w_{\lambda^*}$. Then $\rho(u_j) \leqslant 1$ and $\rho(v) > 1$; that is, $u_j \in C$, $j = 1, \ldots, n$, and $v \notin C$. Now suppose that P is a retraction of X onto C. Let $z = Pv$. If P is also a contraction then

$$\|Pv - Pu_j\| = \|z - u_j\| \leqslant \|v - u_j\|,$$

$j = 1, \ldots, n$. This is equivalent to

$$\|\alpha z - z_{j,\lambda^*}\| \leqslant \|w_{\lambda^*} - z_{j,\lambda^*}\|,$$

$j = 1, \ldots, n$, i.e. $\alpha z \in \bigcap_{j=1}^{n} B(z_{j,\lambda^*}, \|w_{\lambda^*} - z_{j,\lambda^*}\|)$. By virtue of inequality (4′) this implies that

$$\rho(\alpha z) \geqslant \rho(w_{\lambda^*}).$$

Consequently $\rho(z) \geqslant \alpha^{-1} \rho(w_{\lambda^*}) > 1$. This however contradicts the fact that $z \in C$; and hence P is not a contraction. Since P was arbitrarily chosen, the proof is finished.

The following two paragraphs show that condition (σ) and reflexivity are both necessary for the validity of Theorem 2.

On the necessity of condition (σ). Let X be an arbitrary normed linear space and let $f_1, \ldots, f_n \in X^*$, n finite. If the subspaces $H_j = \{x : f_j(x) = 0\}$ are ranges of linear projections R_j of norm 1, then $C = \{x : f_j(x) \leqslant 1, j = 1, \ldots, n\}$ is the range of a contractive retraction P. For using R_j we can readily find contractive retractions P_j of X onto $C_j = \{x : f_j(x) \leqslant 1\}$; and then the desired contractive retraction is given by $P = P_1 \cdots P_n$. Clearly C fails to satisfy condition (σ). This shows that Theorem 2 fails to hold without hypothesis (σ).

On the necessity of reflexivity. Let $X = C[0, 1]$. Let $C =$ closed unit ball in X. There exist points $x \in X$, $\|x\| = 1$ such that C has a unique tangent hyperplane at x. (Indeed the set of all such points was characterized by Banach [**1**, p. 168].) Moreover we show in Proposition 2 below that there exists no closed hyperplane in X which is the range of a linear projection of norm 1. Hence C satisfies condition (σ). On the other hand, the truncation mapping P defined on X by

$$\begin{aligned}(Px)(t) &= 1 &&\text{if } x(t) \geqslant 1,\\ &= x(t) &&\text{if } |x(t)| \leqslant 1,\\ &= -1 &&\text{if } x(t) \leqslant -1,\end{aligned}$$

is readily seen to be a contractive retraction of X onto C. This shows that Theorem 2 fails to hold without the hypothesis of reflexivity.

PROOF OF COROLLARY 1 OF THEOREM 2. We consider only the case where X is not a Hilbert space. Suppose that X contains a closed linear subspace H of codimension 1 which is not the range of a linear projection of norm 1. Then, by virtue of the compactness of C, there exists $y_0 \in \partial C$ such that $H + y_0$ is the tangent hyperplane to C at y_0. Hence condition (σ) is satisfied and the result follows from Theorem 2. Thus we need only show the existence of such an H. In the case that X has dimension 3 this is a direct consequence of the hypothesis that X is not a

Hilbert space. (Kakutani [**11**, Theorem 4].) For higher dimensions the existence of H is insured by the following proposition.

PROPOSITION 1. *Let X be a strictly convex Banach space of finite-dimension n. If X is not a Hilbert space, then there exist linear subspaces of all dimensions v, $2 \leqslant v \leqslant n-1$, which are not the ranges of linear projections of norm* 1.

PROOF OF PROPOSITION 1. We first show that if E_1 and E_2 are linear subspaces which are ranges of linear projections P_1 and P_2 of norm 1, then so is $E_1 \cap E_2$. To see this we consider for each $x_0 \in X$ the sequence

$$x_n = n^{-1} \sum_{i=1}^{n} (P_1P_2)^i x_0, \qquad n \geq 1.$$

Since P_1 and P_2 have norm 1, the sequence $\{x_n\}$ is also bounded by 1; and it contains a convergent subsequence. Hence by the Kakutani-Yosida mean ergodic theorem, the whole sequence converges to some y and $P_1P_2y = y$. Clearly $y \in E_1$. If $y \notin E_2$, then the strict convexity of X implies that $\|P_2y\| < \|y\|$, and hence $\|P_1P_2y\| < \|y\|$. This contradicts $P_1P_2y = y$, and so we conclude that $y \in E_1 \cap E_2$. In view of this, it is readily observed that the mapping Q defined on X by

$$Qx = \lim_{n\to\infty} n^{-1} \sum_{i=1}^{n} (P_1P_2)^i x$$

is a linear projection, of norm 1, of X onto $E_1 \cap E_2$.

Now we prove Proposition 1 by contradiction. To this end, suppose that all subspaces of a fixed dimension v, $2 \leqq v \leqq n-1$, are ranges of projections of norm 1. Then by the above all of their intersections have the same property. In particular, all linear subspaces of dimension $v - 1$ have this property. Repeating the process we find that all linear subspaces of dimension 2 are ranges of linear projections of norm 1. This contradicts the assumption that X is not a Hilbert space and the proof is finished.

PROOF OF COROLLARY 2 OF THEOREM 2. It is well known that the closed unit ball C of the space $L^p(G)$, $1 < p < \infty$, has a unique tangent hyperplane at each one of its boundary points. Furthermore we show in Proposition 3 below that every closed hyperplane in $L^p(G)$, for $1 < p < \infty$, $p \neq 2$, fails to be the range of a linear projection of norm 1. Hence condition (σ) is satisfied and the result follows from Theorem 2.

We conclude this section with two propositions, used above, which describe properties of projections in the spaces L^p and $C[0, 1]$. The proofs of both depend on the following lemma.

LEMMA 2. *Let X be a normed space. Let H be a closed linear subspace of codimension* 1, *and let P be a linear projection of X onto H. We denote the* 1-*dimensional null space of P by $N(P)$. Then P has norm* 1 *if and only if $I - P$ is a proximity mapping onto $N(P)$, i.e. for each $x \in X$ and $y \in N(P)$, $\|x - (I - P)x\| \leqslant \|x - y\|$.*

Thus if P has norm 1, there exists a linear proximity mapping onto $N(P)$.

PROOF. Suppose that P is a linear projection, of norm 1, onto H. Then for each $x \in X$ and $y \in N(P)$

$$\|x - (I - P)x\| = \|Px\| = \|P(x - y)\| \leqslant \|x - y\|.$$

This shows that $I - P$ is a proximity mapping. Conversely, if $I - P$ is a proximity mapping then for each $x \in X$

$$\|Px\| = \|x - (I - P)x\| \leqslant \|x - 0\| = \|x\|.$$

Hence P has norm 1. The proof is finished by noting that if P is linear so is $I - P$.

PROPOSITION 2. *No closed linear subspace of codimension* 1 *of* $C[0, 1]$ *is the range of a linear projection of norm* 1.

PROOF. By virtue of Lemma 2, it suffices to show that no 1-dimensional linear subspace of $C[0, 1]$ is the range of a linear proximity mapping. In other words, it is sufficient to show that given $y \in C[0, 1]$, $y \neq 0$, there exists no linear mapping T defined for all $x \in C[0, 1]$ by

$$Tx = \lambda(x)y, \qquad \lambda(x) \text{ real},$$

with the additional property

$$\|\lambda(x)y - x\| \leqslant \|\lambda y - x\|, \qquad \text{for all real } \lambda.$$

The proof is by contradiction. After possible replacement of y by $-y$, we can choose $t_0 \in [0, 1]$ such that $y(t_0) = \|y\|$. Let $[\alpha, \beta]$ be a subinterval of $[0, 1]$ containing t_0 and such that $y(t) > 0$ for all $t \in [\alpha, \beta]$. Let $t_1 < t_2 < t_3$ be three distinct points in the interior of $[\alpha, \beta]$. Now let us define w_1 and $w_2 \in C[0, 1]$ as follows. Both w_1 and w_2 are identically zero outside $[\alpha, \beta]$, and both are linear in the intervals $[\alpha, t_1]$, $[t_1, t_2]$, $[t_2, t_3]$, $[t_3, \beta]$. Finally, $w_1(t_1) = w_2(t_1) = 1$, $w_1(t_2) = -1$, $w_2(t_2) = 0$, $w_1(t_3) = 0$ and $w_2(t_3) = -1$. It is easy to see that $\lambda(w_1) = \lambda(w_2) = 0$. However if $\lambda > 0$ is chosen so that $\|\lambda y\| < 1$ then $\|w_1 + w_2 - \lambda y\| < 2$. Since $\|w_1 + w_2\| = 2$ this shows that $\lambda(w_1 + w_2) \neq 0$. Consequently $\lambda(w_1) + \lambda(w_2) \neq \lambda(w_1 + w_2)$. This finishes the proof.

REMARK. It has been recently shown by Wulbert [**23**] that no closed linear subspace of $C[0, 1]$, with finite codimension, is the range of a linear projection of norm 1.

PROPOSITION 3. *Let X be the space L^p, $1 < p < \infty$, $p \neq 2$, of p-summable real-valued functions defined on some σ-finite measure space (S, Σ, μ), with the property that each subset A of S with positive measure $\mu(A)$ contains a subset of any positive measure less than $\mu(A)$. Then no closed linear subspace of codimension* 1 *in X is the range of a linear projection of norm* 1.

In particular the space $L^p(G)$, G an open subset of R^n, has this property.

REMARK. The condition on (S, Σ, μ) is necessary for the validity of the theorem. To see this we note that l^p contains hyperplanes which are ranges of linear projections of norm 1.

PROOF. By virtue of Lemma 2, it suffices to prove that for each $y \in X$, $y \neq 0$, the proximity mapping T, from X onto the 1-dimensional subspace generated by y, defined (uniquely) for every x, by

$$Tx = \lambda(x)y \quad \text{and} \quad \|x - \lambda(x)y\| \leqslant \|x - \lambda y\|, \qquad \text{all real } \lambda,$$

is not linear. We propose to exhibit w_1 and w_2 so that $\lambda(w_1 + w_2) \neq \lambda(w_1) + \lambda(w_2)$. To this end, let $A^+ = \{t, t \in S, y(t) > 0\}$. After possible replacement of y by $-y$, we have $\mu(A^+) > 0$. Let $A^+ = A_1 \cup A_2 \cup A_3$ be a decomposition of A^+ into pairwise disjoint sets of positive measure. We denote $A_1 \cup A_2$ by A_0. For $i = 0, 1, 2$, we let $w_i = \chi_{A_i} y$, where χ_{A_i} is the characteristic function of the set A_i. Minimizing $f_i(\lambda) = \|w_i - \lambda y\|$ by differentiation we find that

$$\lambda(w_i) = (1 + [\beta_i + \gamma/\alpha_i]^{q-1})^{-1},$$

where $q^{-1} + p^{-1} = 1$, $\alpha_i = \int_{A_i} |y|^p \, d\mu$, $\beta_i = \int_{A^+ - A_i} |y|^p \, d\mu$, and $\gamma = \int_{S-A^+} |y|^p \, d\mu$. It is easy to see that if $\alpha_1 = \alpha_2$ and α_1 is sufficiently small, then $\lambda(w_1) + \lambda(w_2) \neq \lambda(w_0)$. Since $w_0 = w_1 + w_2$ this proves our assertion.

2. **Proof of Theorem 1.** According to Schönbeck [**20**, Theorem 2.4] this result is true if X is finite dimensional. Therefore M is either a Hilbert space or a P_1 space. We consider the cases: (1) X is strictly convex, and (2) X is not strictly convex. If case (1) occurs then the result follows from the theorem (Schönbeck) stated in the Introduction. If X is not strictly convex, we can choose linearly independent x_1, x_2 in X so that $\|x_1\| = \|x_2\| = 1$ and $\|x_1 + x_2\| = 2$. Let $x_3 = (-x_1 - x_2)/2$. From $\|x_1 + x_2\| = 2$ it follows that $\|x_i - x_3\| = 2$, $i = 1, 2$. In M choose linearly independent y_1, y_2 such that $\|y_i\| = 1$, $i = 1, 2$, and $\|y_1 - y_2\| < \|x_1 - x_2\|$. Let $y_3 = -\lambda y_1 - (1 - \lambda)y_2/\|-\lambda y_1 - (1 - \lambda)y_2\|$, where $0 \leqslant \lambda \leqslant 1$ and $\|y_1 - y_3\| = \|y_2 - y_3\|$. By the strict convexity of M, $\|y_1 - y_3\|$, $\|y_3 - y_2\| < 2$. Finally choose $\delta > 1$ so that $\delta\|y_1 - y_2\| < \|x_1 - x_2\|$ and $\delta\|y_1 - y_3\| < 2$. Let $z_i = \delta y_i$, $i = 1, 2, 3$. Since M is the range of a linear projection of norm 1 it follows that

$$\min_{x \in X} \max_{1 \leqslant i \leqslant 3} \|z_i - x\| = \min_{x \in M} \max_{1 \leqslant i \leqslant 3} \|z_i - x\|.$$

If M is not a P_1 space then it is a Hilbert space and it follows that

$$\min_{x \in M} \max_{1 \leqslant i \leqslant 3} \|z_i - x\| = \min_{x \in \operatorname{span}[z_1, z_2]} \max_{1 \leqslant i \leqslant 3} \|z_i - x\|.$$

It is readily seen that

$$\min_{x \in \operatorname{span}[z_1, z_2]} \max_{1 \leqslant i \leqslant 3} \|z_i - x\| = \delta.$$

Altogether,

$$\min_{x \in X} \max_{1 \leqslant i \leqslant 3} \|z_i - x\| = \delta > 1.$$

On the other hand, by the choice of x_1, x_2, x_3,

$$\min_{x \in X} \max_{1 \leqslant i \leqslant 3} \|x_i - x\| = 1.$$

Since $\delta > 1$, it follows that the contraction T defined on $\{x_1, x_2, x_3\}$ by $Tx_i = z_i$ cannot be extended to the origin. This contradicts the hypothesis (X, X) has Property (E). Hence the assumption that X is not strictly convex is untenable, and the proof is finished.

References

1. S. Banach, *Operations linéaires*, Chelsea, New York, 1955.
2. F. E. Browder, *The solvability of non-linear functional equations*, Duke Math. J. **30** (1963), 557–566.
3. ———, *Existence and uniqueness theorems for solutions of nonlinear boundary value problems*, Proc. Sympos. Appl. Math., vol. 17, Amer. Math. Soc., Providence, R.I., 1965, pp. 24–49.
4. F. E. Browder and W. V. Petryshyn, *Construction of fixed points of nonlinear mappings in Hilbert spaces*, J. Math. Anal. Appl. **20** (1967), 197–228.
5. E. W. Cheney and A. A. Goldstein, *Proximity maps for convex sets*, Proc. Amer. Math. Soc. **10** (1959), 448–450.
6. L. Danzer, B. Grünbaum and V. Klee, *Helly's theorem and its relatives*, Proc. Sympos. Pure Math., vol. 7, Amer. Math. Soc., Providence, R.I., 1963, pp. 101–180.
7. M. Edelstein and A. C. Thompson, *Contractions, isometries and some properties of inner-product spaces*, Nederl. Akad. Wetensch. Proc. Ser. A **70** (1967), 326–332.
8. D. G. deFigueiredo and L. A. Karlovitz, *On the radial projection in normed spaces*, Bull. Amer. Math. Soc. **73** (1967), 364–368.
9. ———, *On the projection onto convex sets and the extension of contractions*, Proc. Conference on Projections and Related Topics (preliminary edition) Clemson University, Clemson, S.C., 1967.
10. B. Grünbaum, *On a theorem of Kirszbraun*, Bull. Res. Counc. Israel, **17F** (1958), 129–132.
11. S. Kakutani, *Some characterizations of Euclidean space*, Japan J. Math. **16** (1939), 93–97.
12. M. D. Kirszbraun, *Über die zusammenziehende und Lipschitzsche Transformationen*, Fund. Math. **22** (1934), 77–108.
13. J. L. Lions and G. Stampacchia, *Variational inequalities*, Comm. Pure Appl. Math. **20** (1967), 493–519.
14. E. J. Mickle, *On the extension of a transformation*, Bull. Amer. Math. Soc. **55** (1949), 160–164.
15. G. J. Minty, *Monotone (non-linear) operators in Hilbert space*, Duke Math. J. **29** (1962), 341–346.
16. ———, *On a "monotonicity" method for the solution of nonlinear equations in Banach spaces*, Proc. Nat. Acad. Sci. U.S.A. **50** (1963), 1038–1041.
17. J. J. Moreau, *Proximité et dualité dans un espace Hilbertien*, Bull. Soc. Math France **93** (1965), 273–299.
18. I. J. Schoenberg, *On a theorem of Kirszbraun and Valentine*, Amer. Math. Monthly **60** (1953), 620–622.
19. S. O. Schönbeck, *Extension of nonlinear contractions*, Bull. Amer. Math. Soc. **72** (1966), 99–101.
20. ———, *On the extension of Lipschitzian maps*, Ark. Mat. **7** (1967), 201–209.
21. J. Schwartz, *Nonlinear functional analysis*, Lecture Notes, New York University, New York, 1963/64.
22. F. A. Valentine, *A Lipschitz condition preserving extension for a vector function*, Amer. J. Math. **67** (1945), 83–93.
23. D. E. Wulbert, *Projections of norm 1 on C(X)*, Notices Amer. Math. Soc. **15** (1968), 362.
24. P. P. Zabreiko, R. I. Kachurovsky and M. A. Krasnoselsky, *On a fixed point principle for operators in Hilbert spaces*, Functional Anal. i Priložen. **1** (1967), 168–169.

UNIVERSITY OF ILLINOIS AT CHICAGO CIRCLE,
UNIVERSITY OF MARYLAND

ON SOME NONEXISTENCE AND NONUNIQUENESS THEOREMS FOR NONLINEAR PARABOLIC EQUATIONS[1]

Hiroshi Fujita

1. **Introduction.** In this paper we study some pathological phenomena occurring to solutions of initial value problems for nonlinear partial differential equations. Preferring straightforwardness to generality, we restrict our consideration to a certain class of quite simple semilinear parabolic equations, which is well represented by

$$\partial u/\partial t = \Delta u + f(u) \qquad (t > 0,\, x \in \Omega). \tag{1.1}$$

Here Ω is a domain in R^m and f is a given continuous function from R^1 to R^1. By IVP we denote the initial value problem composed of (1.1), the initial condition

$$u|_{t=0} = a(x) \qquad (x \in \Omega), \tag{1.2}$$

and the homogeneous Dirichlet boundary condition

$$u|_{\partial\Omega} = 0. \tag{1.3}$$

Among others we shall derive a sufficient condition for nonexistence of global solutions of IVP, i.e. a sufficient condition that the solution blow up in a finite time. Also a sufficient condition for the solution to be nonunique will be derived. These conditions are to be specified as we proceed. Here we just note that in either condition f is assumed to be increasing. This assumption is not too strange when we recall that IVP is well posed if f is decreasing and some subsidiary assumptions on a and $\partial\Omega$ are made, as has been known since quite a few years ago. For instance, see Pini [**17**] and Murakami [**14**].

Taking account of the principal subject of the present symposium, it would be appropriate to refer to the relationship between our study and the recent theory of nonlinear evolution equations with monotone nonlinear generators developed by Browder [**4**], Kato [**11**], Kōmura [**12**] and others. The monotonicity of the generator in the abstract theory corresponds to the decreasing property of f in IVP. Therefore, IVP with an increasing f stays just outside the scope of application of the abstract theory. Nevertheless, we hope that our result can be a contribution to the abstract theory to the effect that it supports the basic assumption of

[1] During the period of preparation for the present paper, the author was partly supported as a Visiting Member at the Courant Institute of Mathematical Sciences, by the National Science Foundation, Grant NSF-GP-8114.

monotonicity of generators by warning that even a relatively slight violation of the required monotonicity may cause seriously ill-conditioned situations.

Another aspect of our result is a parallelism between IVP and the following initial value problem,

$$d\mu/dt = f(\mu), \qquad \mu(0) = \mu_0. \tag{1.4}$$

Actually, from our theorems one can see that pathological characters of the ordinary differential equation (1.4) are inherited by IVP to a notable extent. In addition, our method of proof often leans on elementary facts concerning (1.4).

The present paper has two sections apart from this section. In §2 we shall discuss the nonexistence and existence of global solutions, i.e., solutions for all time. There we first deal with the case $\Omega = R^m$ (the Cauchy problem) and prove theorems generalizing a previous result in [8]. Then we proceed to the case of bounded Ω. §3 is devoted to the nonuniqueness and uniqueness consideration, where we mainly describe what has been obtained recently by S. Watanabe and the author [**9**].

Before concluding the present section we make some standing assumptions and definitions. In order to avoid unessential complications in the statements and the proofs of theorems we make generous smoothness assumptions for Ω and a in IVP. Although Ω may be bounded or unbounded, we always assume that $\partial\Omega$ is sufficiently smooth. Also we suppose that a is bounded and uniformly continuous in $\bar{\Omega}$.

A function $u = u(t, x)$ is said to be a solution of IVP in $[0, T] \times \bar{\Omega}$, T being a positive number, if the following (i) and (ii) hold:

(i) u is bounded and uniformly continuous in $[0, T] \times \bar{\Omega}$. The initial and boundary conditions are satisfied in the usual sense.

(ii) The equation (1.1) is satisfied by u in the distribution sense in $(0, T) \times \Omega$.

Furthermore, we say that a solution u in $[0, T] \times \bar{\Omega}$ is classical if u_t, $\nabla_x u$, $\nabla_x \nabla_x u$ are continuous in $(0, T) \times \Omega$. As is well known, every solution is classical if f is locally Hölder continuous. A global solution is a function defined in $[0, \infty) \times \bar{\Omega}$ whose restriction to $[0, T] \times \bar{\Omega}$ is a solution there for every $T > 0$.

Finally we remark that a bounded function u is a solution of IVP if and only if it satisfies the integral equation

$$u = Ha + \Phi_f u, \tag{1.5}$$

where H and Φ_f are integral operators such that

$$(Ha)(t, x) = \int_\Omega U(t, x, y)a(y)\, dy \tag{1.6}$$

and

$$(\Phi_f u)(t, x) = \int_0^t ds \int_\Omega U(t - s, x, y) f(u(s, y))\, dy. \tag{1.7}$$

Here $U = U(t, x, y)$ is the Green function of the heat equation under the boundary condition (1.3).

2. **Blowing up of solutions.** Throughout this section we assume that f in IVP satisfies the following

CONDITION (f.1). f is locally Lipschitz continuous.

Thus IVP has a unique solution which may be local in t. We first deal with the Cauchy problem, putting $\Omega = R^m$. Then the following criterion can be easily obtained by a well-known comparison theorem. Let $\bar{\mu} = \bar{\mu}(t)$ and $\underline{\mu} = \underline{\mu}(t)$ be the solutions of (1.4) subject to the initial conditions $\bar{\mu}(0) = \sup_x a(x)$ and $\underline{\mu}(0) = \inf_x a(x)$, respectively. If $\bar{\mu}$ and $\underline{\mu}$ both exist for all t, then the solution u of IVP is global and satisfies $\underline{\mu} \leq u \leq \bar{\mu}$ for all t and x. On the other hand, if either $\bar{\mu} \to -\infty$ in a finite time or $\underline{\mu} \to +\infty$ in a finite time, then u cannot be global. Although this criterion is sometimes useful, it does not give any significant information in many problems which are typical in its structure or important in application. An example is $f(r) = r^{1+\alpha}$. For such a case we need more sophisticated treatments. Actually the following theorem was proved in [**8**].

THEOREM 2.1. *Let $\Omega = R^m$ and let $f(r) = r^{1+\alpha}$ $(r \geq 0)$, α being a positive number. Assume $0 \leq a(x)$, which implies the solution u of IVP is nonnegative where it exists. Then* (i) *and* (ii) *hold.*

(i) *Let $0 < m\alpha < 2$ and suppose that $a(x_0) > 0$ for some x_0. Then the solution u of IVP is not global. That is, every positive solution of IVP blows up in a finite time if $0 < m\alpha < 2$.*

(ii) *Let $2 < m\alpha$. Then there are many positive solutions of IVP for all time. To be precise, for any $\kappa > 0$ we can choose $\delta > 0$ such that IVP has a global solution whenever $0 \leq a(x) \leq \delta \exp(-\kappa |x|^2)$.*

A generalization of this theorem is stated in two theorems below, for which we need the following conditions concerning f.

CONDITION (f.2). (i) $f(0) \geq 0$ and $f(r) > 0$ for $r > 0$.

(ii) $1/f$ is integrable at $r = +\infty$.

(iii) f is convex in $[0, \infty)$.

CONDITION (f.3). For some $M > 0$ there exists a continuous function $g = g(r)$ on $[0, M]$ with the properties (i) and (ii).

(i) $|f(r)| \leq g(|r|)$ $(|r| \leq M)$.

(ii) g is Lipschitz continuous on $[0, M]$ and $h(r) \equiv g(r)/r$ is increasing in $(0, M]$.

THEOREM 2.2. *Let $\Omega = R^m$ and suppose that f satisfies Conditions* (f.1) *and* (f.2). *Furthermore, assume that $0 \leq a(x)$ for all x and $a(x_0) > 0$ for some x_0. If*

$$F(r) \equiv \int_r^\infty \frac{d\lambda}{f(\lambda)} = o(r^{-2/m}) \quad \text{as } r \to +0, \tag{2.1}$$

then the solution u of IVP blows up in a finite time.

THEOREM 2.3. *Let $\Omega = R^m$ and suppose that f satisfies Conditions* (f.1) *and* (f.3). *Furthermore, assume that*

$$c_M \equiv \int_0^M r^{-2-2/m} g(r)\, dr < +\infty. \tag{2.2}$$

Then, for any $\kappa > 0$ there exists a $\delta_0 > 0$ such that the solution u of IVP is global whenever

$$|a(x)| \le \delta_0 \exp(-\kappa |x|^2) \qquad (x \in R^m).$$

OUTLINE OF PROOF OF THEOREM 2.2. Suppose that u is a global solution. Taking an arbitrary $T > 0$, we put

$$J_T = J_T(s) = \int_{R^m} U(T - s, 0, y) u(s, y)\, dy \qquad (0 \le s \le T), \tag{2.3}$$

where U is the Green function of the heat equation. By means of (1.1) and by integration by parts we have

$$\frac{dJ_T}{ds} = \int_{R^m} U(T - s, 0, y) f(u(s, y))\, dy,$$

whence follows

$$\frac{dJ_T}{ds} \ge f(J_T) \tag{2.4}$$

by Jensen's inequality which is here applicable by virtue of the convexity of f and the identity

$$\int_{R^m} U(T - s, 0, y)\, dy = 1. \tag{2.5}$$

Using $F(r)$ of (2.1), we have

$$F(J_T(0)) \ge T \tag{2.6}$$

from (2.4). On the other hand, we can easily show

$$J_T(0) \equiv \int_{R^m} U(T, 0, y) a(y)\, dy \ge c_1 T^{-m/2} \tag{2.7}$$

for $T \ge 1$, where c_1 is a positive constant. Substitution of (2.7) into (2.6) yields

$$F(c_1 T^{-m/2}) \ge T \tag{2.8}$$

since F is a decreasing function. However, this is impossible because of (2.1). Thus we have the theorem.

REMARK 2.4. A generalization of Theorem 2.2 is to replace Δ in (1.1) by an elliptic operator

$$L_0 = \sum_{j,k} \frac{\partial}{\partial x_j} a_{jk} \frac{\partial}{\partial x_k}$$

with a_{jk} subject to a certain smoothness and boundedness condition. Actually, an estimate from below of the Green function necessary for (2.7) has been given by Aronson [**1**].

OUTLINE OF PROOF OF THEOREM 2.3. We consider an auxiliary problem for each fixed $\kappa > 0$

$$\frac{\partial v}{\partial t} = \Delta v + g(v) \qquad (t > 0, x \in R^m), \tag{2.9}$$

$$v\big|_{t=0} = \delta \exp(-\kappa |x|^2).$$

It is enough to show that (2.9) has a global solution if $\delta > 0$ is sufficiently small, for then the *a priori* estimate $|u(t, x)| \leq v(t, x)$ guarantees the global solvability of IVP. We write the integral equation

$$v = \delta H a_0 + \Phi_g v \tag{2.10}$$

equivalent to (2.9), using H in (1.6) and Φ_g which is equal to Φ_f in (1.7) with f replaced by g. In (2.10) a_0 stands for $\exp(-\kappa |x|^2)$. We want to show that (2.10) is solvable by the iteration

$$v_0 = \delta H a_0, \qquad v_{n+1} = v_0 + \Phi_g v_n \qquad (n = 0, 1, \ldots), \tag{2.11}$$

provided that δ is sufficiently small. We put

$$\|v\| = \sup_{t>0, x\in R^m} \frac{|v(t, x)|}{\rho(t, x)}$$

where $\rho = H a_0$. When $\|v\| \leq M$ and $v \geq 0$, we can estimate $\|\Phi_g v\|$ in terms of $\|v\|$ as

$$\|\Phi_g v\| \leq \|v\|^{1+2/m} c_M / 4m\kappa$$

after a straightforward calculation making use of

$$g(v(s, y)) \leq \|v\| \, \rho(s, y) h(\|v\| \, \rho(s, 0)).$$

Thus we have

$$\|v_{n+1}\| \leq \delta + c'_M \|v_n\|^{1+2/m} \qquad (n = 0, 1, \ldots) \tag{2.12}$$

with $c'_M = c_M / 4m\kappa$ if we assume $v_n \geq 0$ and $\|v_n\| \leq M$ for all n. Conversely, we can show the following: If $\delta > 0$ is chosen so small that the numerical sequence $\{\nu_n\}$ defined by $\nu_0 = \delta$ and $\nu_{n+1} = \delta + c'_M \nu_n^{1+2/m}$ $(n = 0, 1, \ldots)$ has an upper bound $K(\delta) \leq M$, then $\|v_n\| \leq K(\delta) \leq M$ $(n = 0, 1, \ldots)$ automatically holds. Convergence of $\{v_n\}$ follows from this boundedness since $\{v_n\}$ forms an increasing sequence. The limit of v_n is the required global solution and we have the theorem.

REMARK 2.5. The assertion of Theorem 2.3 still holds if we replace Δ in (1.1) by L_0 mentioned in Remark 2.4.

We now proceed to the case of bounded Ω. We introduce the smallest eigenvalue λ_0 of $-\Delta$ and the corresponding eigenfunction ϕ_0, for which we may assume that $\phi_0 > 0$ in Ω:

$$(2.13)\qquad \begin{aligned} -\Delta\phi_0 &= \lambda_0\phi_0 \quad \text{in } \Omega, \\ \phi_0|_{\partial\Omega} &= 0. \end{aligned}$$

The following theorem may be regarded as a counterpart of Theorem 2.2.

THEOREM 2.6. *Let Ω be bounded and suppose that f satisfies Conditions* (f.1) *and* (f.2). *Assume that $a(x) \geq 0$ in Ω and that $f(r) - \lambda_0 r > 0$ for $r > \int_\Omega a_0\phi_0\,dx$, where ϕ_0 is now normalized as*

$$\int_\Omega \phi_0(x)\,dx = 1.$$

Then the solution of IVP blows up in a finite time.

OUTLINE OF PROOF. Suppose that u is a global solution of IVP and put

$$J = J(t) = \int_\Omega u(t, x)\phi_0(x)\,dx.$$

Then we have

$$(2.14)\qquad dJ/dt \geq -\lambda_0 J + f(J) \qquad (t \geq 0)$$

by means of (1.1) and Jensen's inequality. However, (2.14) implies that $J \to +\infty$ in a finite time, because

$$\int_{J(0)}^{\infty} \frac{dJ}{f(J) - \lambda_0 J} < +\infty$$

under the assumption of the theorem. Thus we have the theorem.

REMARK 2.7. In the special case of $f(r) = r^{1+\alpha}$ $(r \geq 0)$ the preceding theorem asserts that the solution blows up if

$$\int_\Omega a(x)\phi_0(x)\,dx \geq \lambda_0^{1/\alpha}.$$

REMARK 2.8. Generalization of Theorem 2.6 to IVP with Δ replaced by a general elliptic operator of the second order and with the Dirichlet boundary condition replaced by the boundary condition of the second or the third kind is easily achieved, making use of the positive eigenfunction of "the adjoint eigenvalue problem".

If $f(r) = r^{1+\alpha}$ $(\alpha > 0)$ and Ω is bounded, then the solution of IVP is global whenever $a = a(x)$ is sufficiently small, for we have Theorem 2.8 below which is a refined form of a special case of a stability theorem due to Bellman [**3**], Narasimhan [**15**] and Friedman [**5**].

THEOREM 2.8. *Suppose that* $|f(r)| \leq \lambda_0 |r|$ $(|r| \leq M)$ *for some* $M > 0$ *and assume that* $|a(x)| \leq M\phi_0(x)$ *in* Ω, *where* λ_0 *and* ϕ_0 *are those in* (2.13), ϕ_0 *being normalized as* $\max_x \phi_0(x) = 1$. *Then the solution* u *exists and satisfies* $|u| \leq M\phi_0$ *for all time.*

OUTLINE OF PROOF. $v(t, x) = M\phi_0(x)$ solves the majorizing initial value problem

$$\frac{\partial v}{\partial t} = \Delta v + \lambda_0 v,$$

$$v\big|_{t=0} = M\phi_0.$$

Hence follows the theorem.

3. **Nonuniqueness of solutions.** Suppose that $a \equiv 0$ in Ω and $f(0) = 0$. Then IVP obviously has a trivial solution $u \equiv 0$. Is this a unique solution? The answer is no in general according to Theorem 3.1 below which is a special case of a nonuniqueness theorem in [**9**]. We prepare the following

CONDITION (f.4). For some $M > 0$ the conditions (i)–(iv) hold.

(i) $f(0) = 0$ and $f(r) > 0$ in $0 < r \leq M$.

(ii) f is increasing on $[0, M]$.

(iii) $1/f$ is integrable near $+0$, i.e.,

$$\int_0^M \frac{dr}{f(r)} < +\infty. \tag{3.1}$$

(iv) f is concave in $[0, M]$.

THEOREM 3.1. *Suppose that* $a \equiv 0$ *and* f *satisfies Condition* (f.4). *Then the solution of IVP is not unique. Actually there exists, in addition to the trivial solution, a solution* $\bar{u}$ *in* $[0, T] \times \bar{\Omega}$ *for some* $T > 0$ *such that* $\bar{u}(t, x) > 0$ *in* $(0, T] \times \Omega$.

OUTLINE OF PROOF. (For details, see [**9**].) The general case can be reduced to the case of bounded Ω. Choose $T > 0$ such that $f(M)T \leq M$. We define $\bar{u}$ as the maximum solution of IVP in the class $S = S_{T,M}$ of continuous functions u subject to $0 \leq u \leq M$ in $[0, T] \times \bar{\Omega}$. As a matter of fact, $\bar{u}$ can be constructed simply by the iteration

$$u_0 \equiv M \quad \text{and} \quad u_{n+1} = \Phi_f u_n \qquad (n = 0, 1, \ldots). \tag{3.2}$$

For Φ_f recall (1.7). $\bar{u}$ satisfies $u \leq \bar{u}$ in $[0, T] \times \Omega$ for any solution $u \in S$ of IVP. Furthermore, $\bar{u}$ is shown to satisfy $w \leq \bar{u}$ for any solution $w \in S$ of the initial value problem

$$\begin{aligned} &\frac{\partial w}{\partial t} = \Delta w + q(t, x) f(w), \\ &w\big|_{t=0} = 0, \\ &w\big|_{\partial\Omega} = 0, \end{aligned} \tag{3.3}$$

provided that $0 \leq q(t, x) \leq 1$. Now we define w_ε by

$$w_\varepsilon(t, x) = \mu(\varepsilon t)\phi_0(x) \qquad (0 \leq t \leq M, x \in \bar{\Omega})$$

for small $\varepsilon > 0$, where $\mu = \mu(t)$ is a function given by the relation

$$\int_0^\mu \frac{dr}{f(r)} = t,$$

and where ϕ_0 is the nonnegative eigenfunction in (2.13) normalized as $\max_x \phi_0(x) = 1$. Putting

$$q = (\partial w_\varepsilon/\partial t - \Delta w_\varepsilon)/f(w_\varepsilon),$$

we see that $0 \leq q \leq 1$ for sufficiently small ε, noting that under Condition (f.4) $h(r) = r/f(r)$ is an increasing function in $(0, M]$ and decreases to 0 as $r \to +0$. With this q in (3.3) $w = w_\varepsilon$ is a solution of (3.3). $w_\varepsilon \in S$ holds for sufficiently small ε. Thus we have $\tilde{u} \geq w_\varepsilon > 0$ in $(0, T] \times \Omega$, whence follows the theorem immediately.

REMARK 3.2. As for a more general form of Theorem 3.1 we refer to [**9**].

In comparison with the preceding nonuniqueness theorem the following easy uniqueness theorem might be interesting. To state it, we need

CONDITION (f.5). There exists an increasing continuous function g on $[0, \infty)$ such that

$$|f(r_1) - f(r_2)| \leq g(|r_1 - r_2|) \qquad (r_1, r_2 \in R^1)$$

and $1/g(r)$ is not integrable near $r = +0$, i.e.

$$\int_{+0} \frac{dr}{g(r)} = +\infty.$$

THEOREM 3.3. *Suppose that f satisfies Condition* (f.5). *Then the solution of IVP is unique.*

OUTLINE OF PROOF. (For details of some parts, see [**9**].) Let u and v be two possible solutions of IVP and set $w = u - v$. Then w solves

$$\text{(3.4)} \qquad \begin{aligned} &\frac{\partial w}{\partial t} = \Delta w + F(t, x, w), \\ &w\big|_{t=0} = 0, \\ &w\big|_{\partial\Omega} = 0, \end{aligned}$$

where $F(t, x, \lambda) = f(\lambda + v(t, x)) - f(v(t, x))$. We can choose positive numbers M and T such that $|u| \leq M/2$ and $|v| \leq M/2$ in $[0, T] \times \bar{\Omega}$ and $Tg(M) \leq M$. Then $|w| \in S_{M,T}$. On the other hand, we can construct a solution $\bar{\mu} = \bar{\mu}(t)$ with $0 \leq \bar{\mu} \leq M$ of $d\bar{\mu}/dt = g(\bar{\mu})$, $\bar{\mu}(0) = 0$ by the iteration

$$\mu_0(t) = M \quad \text{and} \quad \mu_{n+1}(t) = \int_0^t g(\mu_n(s))\, ds.$$

It is easy to verify by induction that $|w(t, x)| \leq \mu_n(t)$ $(n = 0, 1, \ldots)$ in $0 \leq t \leq T$ and $x \in \bar{\Omega}$, since we have $|F(t, x, \lambda)| \leq g(|\lambda|)$ and g is an increasing function. Thus we have $|w| \leq \bar{\mu}$. However, $\bar{\mu}$ must vanish identically by Osgood's criterion for the ordinary differential equation. Hence we have the theorem.

BIBLIOGRAPHY

1. D. G. Aronson, *Bounds for the fundamental solution of a parabolic equation*, Bull. Amer. Math. Soc. **73** (1967), 890–896.

2. ———, *Non-negative solutions of linear parabolic equations*, Ann. Scuola Norm. Sup. Pisa **22** (1968), 607–694.

3. R. Bellman, *On the existence and boundedness of solutions of nonlinear partial differential equations of parabolic type*, Trans. Amer. Math. Soc. **64** (1948), 21–44.

4. F. E. Browder, *The solvability of non-linear functional equations*, Duke Math. J. **30** (1963), 557–566.

5. A. Friedman, *Convergence of solutions of parabolic equations to a steady state*, J. Math. Mech. **8** (1959), 57–76.

6. ———, *Partial differential equations of parabolic type*, Prentice-Hall, Englewood Cliffs, N.J., 1964.

7. ———*Remarks on nonlinear parabolic equations*, Proc. Sympos. Appl. Math., vol. 17, Amer. Math. Soc., Providence, R.I., 1965, pp. 3–23.

8. H. Fujita, *On the blowing up of solutions of the Cauchy problem for* $u_t = \Delta u + u^{1+\alpha}$, J. Fac. Sci. Univ. Tokyo. Sect. I **13** (1966), 109–124.

9. H. Fujita and S. Watanabe, *On the uniqueness and non-uniqueness of solutions of initial value problems for some quasi-linear parabolic equations*, Comm. Pure Appl. Math. **21** (1968), 631–652.

10. S. Itô, *Fundamental solutions of parabolic differential equations and boundary value problems*, Japan. J. Math. **27** (1957), 55–102.

11. T. Kato, *Nonlinear semigroups and evolution equations*, J. Math. Soc. Japan **19** (1967), 508–520.

12. Y. Kōmura, *Nonlinear semigroups in Hilbert space*, J. Math. Soc. Japan **19** (1967), 493–507.

13. J. L. Lions, *Équations différentielles opérationelles et problèmes aux limites*, Springer-Verlag, Berlin, 1961.

14. H. Murakami, *Semi-linear partial differential equations of parabolic type*, Funkcial. Ekvac. **3** (1960), 1–50.

15. R. Narasimhan, *On the asymptotic stability of solutions of parabolic differential equations*, J. Rational Mech. Anal. **3** (1954), 303–313.

16. S. Kaplan, *On the growth of solutions of quasi-linear parabolic equations*, Comm. Pure Appl. Math. **16** (1963), 305–330.

17. B. Pini, *Sul primo problema di valori al contorno per l'equatione parabolica non lineare del secondo ordine*, Rend. Sem. Mat. Univ. Padova **153** (1957), 149–161.

UNIVERSITY OF TOKYO AND
NEW YORK UNIVERSITY

A GENERAL FIXED-POINT THEOREM

Benjamin R. Halpern

The Shauder-Tychonoff Theorem states that a continuous function $f: V \to V$ from a compact convex subset $V \subset X$ of a locally convex topological linear space X into itself must have a fixed point. The requirements that V be convex, f single-valued, and $f(V) \subset V$ can be simultaneously weakened. A fixed point theorem is obtained here for compact subsets W of a locally convex topological linear space which satisfies a property roughly described as having a convex intersection with all translates of some subspace of finite codimension, for continuous set-valued functions $F: W \to X$ such that for each $x \in W$, $F(x)$ is a compact star-shaped set or the union of two compact convex sets and the directions from x to the points of $F(x) \subset X$ satisfy a certain "inwardness" condition.

This result extends to infinite-dimensional spaces the fixed point theorems for set-valued function of Eilenberg-Montgomery [**5**] and O'Neill [**9**]. It also generalizes the fixed point theorem for "inward" map of the author and Bergman [**7**]. In the case of continuous set-valued maps this result extends the fixed-point theorem of Glicksberg [**6**] and is more general in some but not all respects than the fixed point theorem for inward maps of Browder [**2**]. As a corollary of the main theorem a generalization of the Mackov-Kakutani fixed point theorem [**4**] for a commuting family of continuous linear maps is proved. It is also noted that the main theorems lead to extensions of some theorems of H. H. Schaefer [**10**] on nonlinear positive operators.

1. **Notations.** We shall use the following notations. Let $\boldsymbol{R}^n$ denote Euclidean n-dimensional space with $\boldsymbol{R}^1 = \boldsymbol{R} =$ the real numbers and $\boldsymbol{R}^0 =$ the singleton $\{0\}$. By $\boldsymbol{R}^\infty$ we will mean the space of all infinite sequences of real numbers with the product topology. When convenient we will identify $\boldsymbol{R}^n$ with $\{(x_1, x_2, \ldots) \in \boldsymbol{R}^\infty \mid x_i = 0 \text{ for } i > n\}$. H will denote the subspace of R^n consisting of all $x \in \boldsymbol{R}^n$ for which the norm $\|x\| = [\sum_1^\infty x_i^2]^{1/2}$ is finite. That is H is the Hilbert space of square summable sequences of real numbers. The projections p_n, P_n and R_n are defined by

$$p_n(x_1, x_2, \ldots) = x_n, \qquad P_n(x_1, x_2, \ldots) = (x_1, x_2, \ldots, x_n),$$
$$R_n(x_1, x_2, \ldots) = (0, 0, \ldots, 0, x_{n+1}, x_{n+2}, \ldots) \quad \text{for } (x_1, x_2, \ldots) \in \boldsymbol{R}^\infty.$$

If Y is a linear space and $x, y \in Y$ then we define $[x, y]$, $[x, y)$, and $(x, y\}$ to be $\{\alpha x + (1 - \alpha)y \mid 0 \leq \alpha \leq 1\}$, $\{\alpha x + (1 - \alpha)y \mid 0 < \alpha \leq 1\}$, and $\{x + \beta(y - x) \mid \beta > 0\}$ respectively. Note that $(x, x\} = \{x\}$. A subset $A \subset Y$ is starshaped provided there is a point $y \in A$ such that $[y, x] \subset A$ for all $x \in A$.

If X and Y are topological spaces, a *set-valued function* $F: X \to Y$ assigns to each point x of X a closed nonempty subset $F(x)$ of Y. If $F: X \to Y$ is a set-valued function, let $F^{-1}: Y \to X$ be the function such that $x \in F^{-1}(y)$ if and only if $y \in F(x)$. Then F is *upper* (*lower*) *semicontinuous* provided F^{-1} is closed (open). If both conditions hold, F is continuous. It should be noted that the composition of continuous (upper semicontinuous) set-valued functions is continuous (upper semicontinuous) and that if $f: X \to Y$ is a continuous single-valued function then the set-valued function $x \to \{f(x)\}$ is continuous.

If A is a subset of a topological space X then the closure of A will be denoted by cl A. We will mean by a "topological linear space" a Hausdorff topological linear space over the reals.

2. **Statement of main theorem.** Let X be a topological linear space. The main fixed point theorem of this study concerns the following sort of sets and maps.

2.1. DEFINITION. A nonempty subset $W \subset X$ is a *gw-set* with *slicing function* L and *retraction* r provided:

(i) L is a continuous linear map from X into some finite-dimensional Euclidean space $\boldsymbol{R}^N$, $L: X \to \boldsymbol{R}^N$.

(ii) r is a retraction of $\boldsymbol{R}^N$ onto $L(W)$ (i.e. $r: \boldsymbol{R}^N \to L(W)$, r is continuous and $r(x) = x$ for $x \in L(W)$).

(iii) For each $x \in \boldsymbol{R}^N$, $L^{-1}(x) \cap W$ is convex.

(iv) The following "continuity property" is satisfied:

For each $p \in W$ either

(a) $L^{-1}(L(p)) \cap W = \{p\}$ or

(b) there exists an $\epsilon > 0$ such that $\{y \in \boldsymbol{R}^N \mid \|y - L(p)\| = \epsilon$ and $[y, L(p)) \cap L(W) \neq \varnothing\} \subset L\{x \mid [x, p] \subset W\}$.

Let K be any subset of X and p a point of X. The *inward set* of p with respect to K is defined as the set of all points $x \in X$ such that $(p, x\} \cap K \neq \varnothing$. The *weakly inward set* of p is defined to be the closure of the inward set of p.

Now let W be a *gw*-set with slicing function L and retraction r.

2.2. DEFINITION. A set-valued function $F: W \to X$ is (weakly) inward relative to L and r if

(i) $L(F(p)) \cap (r^{-1}(L(p)) \sim \{L(p)\}) = \varnothing$ for all $p \in W$ and

(ii) for each $p \in W$ and $x \in F(p)$ either

(a) $L(x) \neq L(p)$ or

(b) $L(x) = L(p)$ and x is in the (weakly) inward set of p with respect to $L^{-1}L(p) \cap W$.

2.3. THEOREM. *Let X be a locally convex topological linear space and $W \subset X$ a compact gw-set with slicing function L and retraction r. Suppose $F: W \to X$ is a continuous set-valued function, weakly inward relative to L and r, and such that for each $x \in W$, $F(x)$ is a compact starshaped set or the union of two compact convex sets. Then there exists an $x \in W$ such that $x \in F(x)$.*

The proof of Theorem 2.3 is contained in §§3–8. Certain generalizations and extensions of Theorem 2.3 are proven in §§9 and 10. In particular a generalization of Theorem 2.3 concerning functions of the form $f \circ F$ where F is a set-valued map and f is a single-valued map is established in §9. In §10, Theorem 2.3 is extended to the case where W is not necessarily compact but $F(W)$ is relatively compact.

3. A Cube in H.

We start the proof of Theorem 2.3 by extending a fixed point theorem of O'Neill [**9**] to a cube in H.

3.1. THEOREM (O'NEILL). *Let F be a set-valued self-map of a compact polyhedron X and n an integer such that if $x \in X$, $F(x)$ is homologically trivial or consists of n homologically trivial components. If X is homologically trivial, then F has a fixed point.*

The homology referred to above is Čech homology theory with coefficients in a field. We will be concerned with convex and starshaped sets and since these sets are contractible they are always homologically trivial.

Let $F: W \to X$ be a set-valued map from a subset W of a topological linear space X into X.

3.2. DEFINITION. We will say that F is admissible if it is continuous and for each $x \in W$, $F(x)$ is either a compact starshaped set or the union of two compact convex sets.

3.3. NOTATION. Let a_i, $i = 1, 2, \ldots$, be a sequence of positive real numbers such that $\sum_{i=1}^{\infty} a_i^2 < \infty$. Set $H^0 = \{(x_1, x_2, \ldots) \in \boldsymbol{R}^\infty \mid |x_i| \leq a_i\}$. Note that $H^0 \subset H \subset \boldsymbol{R}^\infty$, H and $\boldsymbol{R}^\infty$ topologies agree in H^0, and H^0 is compact.

3.4. LEMMA. *If $F: H^0 \to H^0$ is an admissible set-valued function from H^0 into itself then F has a fixed point.*

PROOF. Let $H_n^0 = H^0 \cap R^n$. For each n consider the composite map $P_n \circ (F \mid H_n^0): H_n^0 \to H_n^0$. Since P_n is linear it carries convex and starshaped sets into convex and starshaped sets respectively. Noting that the union of two convex sets with nonnull intersection is a starshaped set we see that Theorem 3.1 applies to $P_n \circ (F \mid H_n^0): H_n^0 \to H_n^0$. Thus there is an $x_n \in P_n(F(x_n))$ for each n. From $F(x_n) \subset H^0$ it follows that $\rho(x_n, F(x_n)) = \inf\{\|x_n - y\| \mid y \in F(x_n)\} \leq [\sum_{m>n} a_m^2]^{1/2}$. Since H^0 is compact a subsequence of $\{x_n\}$ converges to a point $x \in H^0$, and from the continuity of F and the fact that $F(x)$ is closed we conclude that $x \in F(x)$.

Q.E.D.

4. Retraction lemma.

The following lemma will enable us to retract H^0 in a special way onto certain compact gw-sets contained in H^0.

4.1. LEMMA. *Suppose $W \subset H^0$ is a compact gw-set such that $L = P_N$ is a slicing function for W. If we set*

$$p(x) = \textit{the unique } y \in W \textit{ such that } P_N(x) = P_N(y)$$

$$\textit{and } \|y - x\| = \inf_{\substack{z \in W \\ P_N(z) = P_N(x)}} \|z - x\|$$

then p is a well-defined continuous map from $H^0 \cap P_N^{-1}(P_N(W))$ *into* W *such that* $p(x) = x$ *if* $x \in W$.

PROOF. p is well defined for $x \in H^0 \cap P_N^{-1}(P_N(W))$ because in this case $(P_N^{-1}(P_N(x))) \cap W$ is a nonempty compact convex set. It is obvious that $p(H^0 \cap (P_N^{-1}(P_N(W)))) \subset W$ and that $p(x) = x$ for all $x \in W$.

We shall now see that the "continuity condition" on W implies that p is continuous. Assume $x_n, x \in H^0 \cap P_N^{-1}P_N(W)$ and that x_n converges to x. Suppose that $p(x_n)$ did not converge to $p(x)$. Then since W is compact a subsequence of $p(x_n)$ converges to a point $z \in W$, $z \neq p(x)$. We may consequently assume that $\lim_n p(x_n) = z \neq p(x)$. Note that

$$(\star) \qquad P_N(z) = \lim_n P_N(p(x_n)) = \lim_n P_N(x_n) = P_N(x) = P_N(p(x)).$$

We consider the two cases (a) and (b) distinguished in the continuity condition on W at $p(x)$ (see Definition 2.1, iv). In case (a) we have $z = p(x)$ a contradiction. Next suppose case (b) holds at $p(x)$. We shall show that $\lim_n \|x_n - p(x_n)\| \leq \|x - p(x)\|$. Let $k^2 = \sum_{n>N} a_n^2 < \infty$. According to Definition 2.1, part iv(b) there exists an $\epsilon > 0$ such that $C = \{y \in \boldsymbol{R}^n \mid \|y - P_N(p(x))\| = \epsilon$ and there exists a β, $0 < \beta \leq 1$, such that

$$p(x) + \beta(y - p(x)) \in P_N\{(W)\} \subset P_N\{x' \mid [x', p(x)] \subset W\}.$$

If n is sufficiently large $\|P_N(x_n) - P_N(x)\| \leq \epsilon$. So we may assume that this holds for all n. Set $t_n = P_N(x_n) - P_N(x) = P_N(x_n) - P_N(p(x))$. Either $t_n = 0$ or $t_n \neq 0$. In the latter case

$$P_N(p(x)) + \epsilon t_n/\|t_n\| \in C \subset P_N\{x' \mid [x', p(x)] \subset W\}.$$

Consequently when $t_n \neq 0$ there exists an x_n' such that $[x_n', p(x)] \subset W$ and $P_N(x_n') = P_N(p(x)) + \epsilon t_n/\|t_n\|$. If $t_n = 0$ set $x' = p(x)$. Then in both cases, $t_n \neq 0$ and $t_n = 0$, we have

$$(\star\star) \qquad y_n' = p(x) + (\|t_n\|/\epsilon)(x_n' - p(x)) \in [x_n', p(x)] \subset W.$$

Note that $P_N(y_n') = P_N(x_n)$ and therefore

$$\|x_n - p(x_n)\| \leq \|x_n - y_n'\| = \|R_N(x_n) - R_N(y_n')\|.$$

Now from $(\star\star)$,

$$\|R_N(y_n') - R_N(p(x))\| = (\|t_n\|/\epsilon)\,\|R_N(x_n') - R_N(p(x))\| \leq \|t_n\|\, 2k/\epsilon.$$

Thus $R_N(y_n') \to R_N(p(x))$ as $n \to \infty$ and consequently $\|R_N(x_n) - R_N(y_n')\| \to \|R_N(x) - R_N(p(x))\| = \|x - p(x)\|$ as $n \to \infty$. We therefore have $\|x - z\| = \lim_n \|x_n - p(x_n)\| \leq \|x - p(x)\|$. Since $P_N(z) = P_N(p(x))$ (($\bigstar$) above) the uniqueness of $p(x)$ implies $z = p(x)$, a contradiction. Therefore p is continuous. Q.E.D.

5. **Subsets of $\boldsymbol{R}^\infty$.** In this section we prove Theorem 2.3 for *gw*-sets contained in $\boldsymbol{R}^\infty$ which have P_N as the slicing function. We need two preliminary lemmas.

5.1. LEMMA. *Let X and Y be two Hausdorff spaces and C a compact subset of $X \times Y$, $C \subset X \times Y$. Set $C_x = p_Y(p_X^{-1}(x) \cap C)$ for each $x \in X$ where p_X and p_Y are the projections of $X \times Y$ onto X and Y. If $x \in X$ and $C_x \subset \mathcal{O}$ where $\mathcal{O}$ is an open subset of Y then there is a neighborhood U of x such that $x' \in U$ implies $C_{x'} \subset \mathcal{O}$.*

PROOF. For each open set U containing x consider the sets $Y_U = (\mathrm{cl}\, U \times Y \cap C) \sim (X \times \mathcal{O})$. The Y_U's are closed subsets of C and because X is Hausdorff $\bigcap_{x \in U,\, U \text{ open}} Y_U = \varnothing$. Since C is compact we can find a finite number of neighborhoods $U_1, \ldots, U_n$ of x such that $\bigcap_{i=1}^n Y_{U_i} = \varnothing$. Since $V \subset V'$ implies $Y_V \subset Y_{V'}$ we have $Y_U \subset \bigcap_{i=1}^n Y_{U_i} = \varnothing$ where $U = \bigcap_{i=1}^n U_i$. It is easy to see that this U satisfies the conclusion of the lemma. Q.E.D.

5.2. LEMMA. *If $F: X \to Y$ is an upper semicontinuous set-valued function from a compact space X to a topological space Y such that $F(x)$ is compact for each $x \in X$ then $F(X)$ is compact.*

PROOF. Let $\mathcal{O}$ be an open cover of $F(X)$. Since $F(x)$ is compact for each $x \in X$ there is a finite subcollection $\mathcal{O}_x$ of $\mathcal{O}$, $\mathcal{O}_x \subset \mathcal{O}$, such that $F(x) \subset \bigcup \mathcal{O}_x$. For $A \subset Y$ set $F^*(A) = \{x \in X \mid F(x) \subset A\} = X \sim F^{-1}(Y \sim A)$. Then $\{F^*(\bigcup \mathcal{O}_x) \mid x \in X\}$ is an open cover of X and consequently there is a finite subcover $F^*(\bigcup \mathcal{O}_{x_1}), \ldots, F^*(\bigcup \mathcal{O}_{x_n})$. Then $\mathcal{O}_{x_1} \cup \cdots \cup \mathcal{O}_{x_n} \subset \mathcal{O}$ is a finite subcover of $F(X)$. Q.E.D.

5.3. LEMMA. *Let $W \subset \boldsymbol{R}^\infty$ be a compact gw-set with slicing function $L = P_N$ and retraction r. If $F: W \to \boldsymbol{R}^\infty$ is an admissible, weakly inward relative to L and r set-valued function then F has a fixed point.*

PROOF. Suppose $x \notin F(x)$ for all $x \in W$. $F(x)$ is compact and consequently there is an n such that $x \in P_n^{-1}(\mathcal{O}) \subset (\boldsymbol{R}^\infty \sim F(x))$ where $\mathcal{O}$ is an open subset of $\boldsymbol{R}^n$. Thus the sets $U_i = \{x \in W \mid P_i(x) \notin P_i F(x)\}$ form an open cover of W. Since W is compact and $U_i \subset U_j$ for $i \leq j$ we must have $W \subset U_n$ for some n. Therefore the continuous function $\rho(P_n(x), P_n F(x))$ is never zero. We can clearly assume it is ≥ 1 for $x \in W$.

By the assumption that F is weakly inward relative to $L = P_N$ and r, we can, for each $x \in W$ and $z \in F(x) \cap P_N^{-1} P_N(x)$, find a $y \in W$ and an $\alpha \geq 0$ such that $P_N(y) = P_N(x)$ and $\|P_n(z - ((1 - \alpha)x + \alpha y))\| < 2^{-n-3}$. Now it is clear that the y and α chosen for a given $z \in F(x) \cap P_N^{-1} P_N(x)$ will work for all z' in a neighborhood of z. Hence by the compactness of $F(x) \cap P_N^{-1} P_N(x)$ we can handle all points of $F(x) \cap P_N^{-1} P_N(x)$ by choosing y and α from some finite set S_x. If $F(x) \cap P_N^{-1} P_N(x) = \varnothing$ we will take $S_x = \varnothing$.

Let $x \in W$. We claim that there is a neighborhood U_x of x such that for each $x' \in U_x$ and $z' \in F(x') \cap P_N^{-1} P_N(x')$ there is a $y' \in W$ and $\alpha \in S_x$ such that $P_N(y') = P_N(x')$ and $\|P_n[z' - ((1 - \alpha)x' + \alpha y')]\| \leq 2^{-n-1}$. We will establish the claim by finding neighborhoods U_1, U_2 and U_3 of x in W and an $\epsilon > 0$ such that

(1) $x' \in U_1$ implies each $z' \in F(x') \cap P_N^{-1} P_N(x')$ is near some $z \in F(x) \cap P_N^{-1} P_N(x)$.

(2) Each $z \in F(x) \cap P_N^{-1}P_N(x)$ is near $(1-\alpha)x + \alpha y$ for some $y, \alpha \in S_x$.

(3) $x' \in U_2$, $\|P_n y' - P_n y\| < \epsilon$ implies $(1-\alpha)x + \alpha y$ is near $(1-\alpha)x' + \alpha y'$ for all $\alpha \in S_x$.

(4) $x' \in U_3$ and $y \in S_x$ implies there exists a $y' \in W$ such that $P_N(y') = P_N(x')$ and $\|P_n y' - P_n y\| < \epsilon$, where two points $a, b \in \boldsymbol{R}^\infty$ are said to be *near* each other provided $\|P_n a - P_n b\| < 2^{-n-3}$.

Once (1)–(4) are established the proof of the claim is easily completed by setting $U_x = U_1 \cap U_2 \cap U_3$.

Consider Proposition (1). Write $\boldsymbol{R}^\infty = \boldsymbol{R}^N \times R_N \boldsymbol{R}^\infty$ and apply Lemma 5.1 with $X = \boldsymbol{R}^N$, $Y = R_N \boldsymbol{R}^\infty$, $C = F(x)$ and

$$\mathcal{O} = \{q \in R_N \boldsymbol{R}^\infty \subset \boldsymbol{R}^\infty \mid \|P_n q - P_n R_N z\| < 2^{-n-4} \quad \text{for some} \quad z \in F(x) \cap P_N^{-1}P_N(x)\}.$$

We obtain a neighborhood U of $P_N(x)$ such that $(U \times R_N \boldsymbol{R}^\infty) \cap F(x) \subset U \times \mathcal{O}$. Let $\bar{U}$ be another neighborhood of $P_N(x)$ such that $\operatorname{cl} \bar{U} \subset U$. Then $F(x)$ is contained in the open set $Q = (U \times \mathcal{O}) \cup ((\boldsymbol{R}^N \sim \operatorname{cl} \bar{U}) \times R_N \boldsymbol{R}^\infty)$. Since F is upper semicontinuous there is a neighborhood U' of x such that $x' \in U'$ implies $F(x') \subset Q$.

Now for $x' \in U_1 \equiv U' \cap P_N^{-1}\bar{U} \cap \{x'' \in W \mid \|P_N x'' - P_N x\| < 2^{-n-4}\}$ we have $F(x') \cap P_N^{-1}P_N(x') \subset U \times \mathcal{O}$ and consequently $z' \in F(x') \cap P_N^{-1}P_N(x')$ implies there is a $z \in F(x) \cap P_N^{-1}P_N(x)$ such that $\|P_N z' - P_N z\| = \|P_N x' - P_N x\| < 2^{-n-4}$ and $\|P_n R_N z' - P_n P_N z\| < 2^{-n-4}$ from which it follows that $\|P_n z' - P_n z\| < 2^{-n-3}$. This proves (1). Note that the above argument holds even when $F(x) \cap P_N^{-1}P_N(x) = \varnothing$.

Proposition (2) was proven in the second paragraph. Proposition (3) follows from the continuity of the algebraic operations and the finiteness of S_x.

Now consider Proposition (4). We are given an $\epsilon > 0$ and $y \in S_x$. We may assume $\epsilon < 1$. Since we set $S_x = \varnothing$ when $F(x) \cap P_N^{-1}P_N(x) = \varnothing$ we must have $F(x) \cap P_N^{-1}P_N(x) \neq \varnothing$. If $W \cap P_N^{-1}P_N(x) = \{x\}$ then the inward set and consequently the weakly inward set of x with respect to $W \cap P_N^{-1}P_N(x)$ would be $\{x\}$ and thus any $z \in F(x) \cap P_N^{-1}P_N(x)$ would be a fixed point (F is weakly inward). But this would contradict our assumption that F is fixed point free. We can therefore conclude that $W \cap P_N^{-1}P_N(x) = W \cap P_N^{-1}P_N(y)$ is not a singleton and therefore alternative (b) of Definition 2.1 part (iv) must hold at y. Thus there is an $\epsilon' > 0$ such that

$$\begin{aligned} A = \{t \in R^N \mid \|t - P_N(x)\| = \epsilon' \quad &\text{and} \quad [t, P_N(x)) \cap P_N(W) \neq \varnothing\} \\ \subset P_N\{q \in W \mid [q, y] \subset W\}. \end{aligned}$$

Since W is compact there is a constant $K \geq 1$ such that $\|P_n a - P_n b\| \leq K$ for all $a, b \in W$. Set $U_3 = \{p \in W \mid \|P_N p - P_N x\| < \epsilon\epsilon'/K\}$ (note $\epsilon\epsilon'/K < \epsilon'$). Let $x' \in U_3$. Set $\alpha = \|P_N x' - P_N x\|/\epsilon' < \epsilon/K < 1$. If $\alpha = 0$ we may take $y' = y$. Assume now that $\alpha \neq 0$ and consider $t = P_N(x) + \alpha^{-1}(P_N(x') - P_N(x))$. Then $t \in A$ and consequently there is a $q \in W$ such that $P_N(q) = t$ and $[q, y] \subset W$. Set

$y' = y + \alpha(q - y)$. Then $y' \in W$ and $\|P_n y' - P_n y\| = \alpha \|P_n q - P_n y\| < \epsilon K/K = \epsilon$. Also $P_N(y') = P_N(y) + \alpha(P_N(q) - P_N(y)) = P_N(x) + \alpha(t - P_N(x)) = P_N(x')$. Thus y' has all the desired properties and Proposition (4) is established. As mentioned above the proof of the claim is clearly completed by setting $U_x = U_1 \cap U_2 \cap U_3$.

By compactness a finite number of the U_x, $x \in W$, cover W. From this we conclude that there is a finite set S such that for each $x \in W$ and $z \in F(x) \cap P_N^{-1}P_N(x)$ there is a point $u(x, z) = (1 - \alpha)x + \alpha y$ with $y \in W \cap P_N^{-1}P_N(x)$ and $\alpha \in S$ such that $\|P_n z - P_n u(x, z)\| < 2^{-n-1}$. Note that all the values of $u(x, z)$ are contained in the compact set $\bigcup_{\alpha \in S} ((1 - \alpha)W + \alpha W)$. Hence for each $i > 0$, we can find a real number a_i such that for each $x \in W$, $z' \in F(x)$ and $z \in F(x) \cap P_N^{-1}P_N(x)$ we have $|p_i(x)| < a_i$, $|p_i(z')| < a_i$ (see Lemma 5.2), and $|p_i z - p_i u(x, z)| < a_i$. Multiplication of each coordinate by an independent nonzero constant is a linear homeomorphism and preserves all the structure we are considering. Consequently, we may assume $a_i \leq 2^{-i}$ for $i > n$. (We have already put conditions on the first n coordinates.)

We now have $\sum_{i=1}^{\infty} a_i^2 < \infty$, $W \subset H^0 = \{(x_1, \ldots) \mid |x_i| \leq a_i \text{ for all } i\} \subset H$, and $F(W) \subset H^0$. Note that for all $x \in W$ and $z \in F(x) \cap P_N^{-1}P_N(x)$, $u(x, z)$ is at a distance of less than 1 from z, since $|p_i z - p_i u(x, z)| < 2^{-i}$ for i both $\leq n$ and $> n$. On the other hand, x is at a distance at least 1 from z (see first paragraph). Hence we have a point $u(x, z) = (1 - \alpha)x + \alpha y$, $y \in W \cap P_N^{-1}P_N(x)$ and $\alpha \in S$, on the ray drawn from x through some other point, y, of the convex set $W \cap P_N^{-1}P_N(x)$ (if $y = x$ then $u(x, z) = x$ and $1 < \|x - z\| = \|u(x, z) - z\| < 1$ which is impossible) and $u(x, y)$ is closer to z than x is. Thus there is a point $x' \in [y, x) \subset W \cap P_N^{-1}P_N$ closer to z than x is. Consequently $p(z) \neq x$ where p is as defined in Lemma 4.1. Again referring to the definition of p we see that $x \notin pF(x)$ for all $x \in W$ (where $pF(x) \equiv p(F(x) \cap P_N^{-1}P_N(W))$).

Consider $H^0 = P_N(H^0) \times R_N(H^0)$. Define $q: H^0 \to P_N^{-1}P_N(W)$ by $q(a, b) = (r(a), b)$ for all $(a, b) \in P_N(H^0) \times R_N(H^0) = H^0$ (r is the retraction associated with W). Since r is continuous q is also continuous. Invoking Lemma 4.1 we have $p \circ q: H^0 \to W$, $p \circ q$ is continuous and $p \circ q(x) = x$ for all $x \in W$. The composite function $F \circ p \circ q: H^0 \to H^0$ thus satisfies the hypothesis of Lemma 3.4 and so there is an $x \in H^0$ such that $x \in F(p(q(x)))$.

From $x \in F(p(q(x)))$ we have $P_N(x) \in P_N(F(p(q(x))))$. But then part (i) of Definition 2.2 implies $P_N(x) \notin r^{-1}P_N(p(q(x))) \sim \{P_N(p(q(x)))\}$. Since $rP_N(x) = P_N(q(x)) = P_N(p(q(x)))$ we must have $P_N(x) = P_N(p(q(x))) = P_N(q(x))$. It now follows that $x = q(x)$ because we always have $R_N(x) = R_N(q(x))$.

Thus $x \in F(p(x))$ and consequently $p(x) \in p(F(p(x)))$. Since $p(x) \in W$ this contradicts $y \notin pF(y)$ for all y which was established at the end of the third paragraph above. Therefore there must be an $x \in W$ such that $x \in F(x)$. Q.E.D.

6. **Lemma on set-valued functions.** We present here an alternative way of looking at set-valued functions which will prove useful in the following section. Let X be a metric space with metric d. If $\epsilon > 0$ is a real number and A is a subset of X then we shall use the notation $\epsilon A = \{y \in X \mid d(y, x) < \epsilon \text{ for some } x \in A\}$.

We make the following

6.1. Definition. If $X = (X, d)$ is a metric space we shall denote by X' the metric space (X', d') where $X' = \{C \mid C \text{ is a compact subset of } X\}$ and d' is the Hausdorff metric given by $d'(C_1, C_2) = \inf\{\epsilon \mid C_1 \subset \epsilon C_2 \text{ and } C_2 \subset \epsilon C_1\}$.

It is not difficult to show that d' is indeed a metric for X' (see [**8**]).

6.2. Definition. If $F: Y \to X$ is a set-valued function with $F(y)$ compact for each $y \in Y$ we shall let F' denote the single-valued function $F': Y \to X'$ given by $F'(y) = F(y) \in X'$ for all $y \in Y$.

6.3. Lemma. *With F and F' as in Definition 6.2, F is continuous if and only if F' is continuous.*

Proof. Assume F is continuous. We will show that F' is continuous at an arbitrary point $y \in Y$. Let $\epsilon > 0$. Since $\epsilon F(y)$ is open so is

$$\mathcal{O} = Y \sim F^{-1}(X \sim \epsilon F(y)).$$

Furthermore $y \in \mathcal{O}$, and $y' \in \mathcal{O}$ implies $F(y') \subset \epsilon F(y)$. In order to obtain the reverse relation, cover $F(y)$ with a finite number of open balls of radius $\epsilon/2$, $B_{\epsilon/2}(x_1), \ldots, B_{\epsilon/2}(x_n)$, each of which has nonempty intersection with $F(y)$. Then $y \in \mathcal{O}' = \bigcap_{i=1}^n F^{-1}B_{\epsilon/2}(x_i)$ and $\mathcal{O}'$ is an open set with the property that $y' \in \mathcal{O}'$ implies $F(y') \cap B_{\epsilon/2}(x_i) \neq \varnothing$ for $i = 1, \ldots, n$. Since the $B_{\epsilon/2}(x_i)$ cover $F(y)$ we have $F(y) \subset \epsilon F(y')$ for all $y' \in \mathcal{O}'$. Thus $y' \in \mathcal{O} \cap \mathcal{O}'$ implies $d'(F(y'), F(y)) < \epsilon$ and so F' is continuous at y. Since y was arbitrary F' is continuous.

Now assume F' is continuous. First we will show that F^{-1} is open. Let $\mathcal{O} \subset X$ be an open set. To show that $F^{-1}(\mathcal{O})$ is open it is sufficient to show that each $y \in F^{-1}(\mathcal{O})$ is an interior point. From $y \in F^{-1}(\mathcal{O})$ it follows that there is a $z \in F(y) \cap \mathcal{O}$. Since $\mathcal{O}$ is open there exists an $\epsilon > 0$ such that the open ball of radius ϵ about z is contained in $\mathcal{O}$, $B_\epsilon(z) \subset \mathcal{O}$. Set $U = F'^{-1}\{C \in X' \mid d'(C, F(y)) < \epsilon\}$. Because F' is continuous U is open. Moreover $y \in U$. If $y' \in U$ then $z \in F(y) \subset \epsilon F(y')$ and consequently $F(y') \cap B_\epsilon(z) \neq \varnothing$. Hence $F(y') \cap \mathcal{O} \neq \varnothing$ which shows that $y' \in F^{-1}(\mathcal{O})$. Thus $y \in U \subset F^{-1}(\mathcal{O})$ and so y is an interior point of $F^{-1}(\mathcal{O})$ as we wished to show.

Finally we will show that F^{-1} is closed. Let $C \subset X$ be closed. We must show that $F^{-1}(C)$ is closed, i.e. that $Y \sim F^{-1}(C) = \{y \in Y \mid F(y) \subset X \sim C\}$ is open. Let $y \in Y$ be such that $F(y) \subset X \sim C$. Since $F(y)$ is compact and $X \sim C$ is open there is an $\epsilon > 0$ such that $\epsilon F(y) \subset X \sim C$. Let $U = F'^{-1}\{A \mid d(A, F(y)) < \epsilon\}$ and note that U is open and $y' \in U$ implies $F(y') \subset \epsilon F(y) \subset X \sim C$. Thus $y \in U \subset Y \sim F^{-1}(C)$ which shows that $Y \sim F^{-1}(C)$ is a neighborhood of each of its points and is thus open. Q.E.D.

7. **Fibering lemma.** Theorem 2.3 will be obtained from Lemma 5.3 through using the "Fibering Lemma" below. This lemma is a strengthened form of the argument used in the Dunford-Schwartz lemma [**4**, V. 10.4], the analogous step in the proof of the Shauder-Tychonoff fixed point theorem. Lemma 7.1 also generalizes the fibering lemma, Lemma 3.2, of Halpern and Bergman [**7**].

7.1. FIBERING LEMMA. *Let X be a locally convex topological linear space and $W \subset X$ a compact subset of X. Suppose $F: W \to X$ is a continuous set-valued function such that $F(x)$ is compact for each $x \in W$ and $l_n: X \to \boldsymbol{R}$, $n = 1, \ldots, N'$, are continuous linear maps. Set $K = W \cup F(W)$ and suppose $|l_n(K)| \leq a_n$ for $n = 1, \ldots, N'$. Let $H^0 = \{(x_1, \ldots) \in R^\infty \mid |x_n| \leq a_n$ for $1 \leq n \leq N'$ and $|x_n| \leq 1/n$ for $n \geq N' + 1\}$. Then there exists a continuous linear transformation $\mathscr{F}: X \to \boldsymbol{R}^\infty$ such that $\mathscr{F}(K) \subset H^0$, $l_n = p_n \circ \mathscr{F}$ for $1 \leq n \leq N'$, and there exists a continuous set-valued function $\bar{F}: \mathscr{F}(W) \to H^0$ such that $\mathscr{F} \circ F = F \circ \mathscr{F}$.*

PROOF. According to Lemma 5.2, $F(W)$ is compact and consequently $K = F(W) \cap W$ is also compact. It follows that the X^* topology for K and its original topology coincide since local convexity implies that the X^* topology is Hausdorff.

We will say that a finite sequence $g_1, \ldots, g_n$ of continuous linear functionals is determined by a set of continuous linear functionals $\mathscr{H} = \{h\}$ if for each $\epsilon > 0$ there exists a neighborhood of the origin $N(\gamma, \delta) = \{x \in X \mid |h(x)| < \delta,\ h \in \gamma\}$, where γ is a finite subset of $\mathscr{H}$, with the property that if $p, q \in W$ and $p - q \in N(\gamma, \delta)$, then $d'(g(F(p)), g(F(q))) < \epsilon$ where $g = g_1 \times \cdots \times g_n: X \to \boldsymbol{R}^n$. A set of continuous linear functionals $\mathscr{G} = \{g\}$ is determined by another set $\mathscr{H} = \{h\}$ if each finite sequence $g_1, \ldots, g_n \in \mathscr{G}$ is determined by $\mathscr{H}$. It is clear that provided $\mathscr{G}$ is determined by $\mathscr{H}$, $p, q \in W$ and $h(p) = h(q)$, $h \in \mathscr{H}$, implies that for each finite sequence $g_1, \ldots, g_n \in \mathscr{G}$, $g(F(p)) = g(F(q))$ where $g = g_1 \times \cdots \times g_n: X \to \boldsymbol{R}^n$.

We will show that each finite sequence $g_1, \ldots, g_n$ of continuous linear functionals is determined by some denumerable set of functionals $\mathscr{H}$. Let $g = g_1 \times \cdots \times g_n: X \to \boldsymbol{R}^n$. Since W is compact we can conclude that $(g \circ F)': X \to \boldsymbol{R}^{n'}$ is uniformly continuous in the sense that for each $\epsilon > 0$ there exists a neighborhood $N(\gamma_\epsilon, \delta_\epsilon) = \{x \in X \mid |h(x)| < \delta_\epsilon, h \in \gamma_\epsilon\}$, where γ_ϵ is a finite subset of X^* such that $p, q \in W$ and $p - q \in N(\gamma_\epsilon, \delta_\epsilon)$ implies $d'(g(F(p)), g(F(q))) < \epsilon$. Set $\epsilon_m = 1/m$ and $\mathscr{H} = \bigcup_{m=1}^\infty \gamma_{\epsilon_m}$; then $g_1, \ldots, g_n$ is determined by $\mathscr{H}$.

Since the set of all finite sequences with elements taken from a denumerable set is itself denumerable it follows that each denumerable set $\mathscr{G} \subset X^*$ is determined by some denumerable set $\mathscr{H} \subset X^*$. We can even assert that each denumerable set $\mathscr{G}$ is contained in some denumerable self-determined set $\mathscr{H}$. For $\mathscr{G}$ is determined by a denumerable set $\mathscr{G}_1$, $\mathscr{G} \cup \mathscr{G}_1$ is determined by a denumerable set $\mathscr{G}_2$, and $\mathscr{G} \cup \mathscr{G}_1 \cup \mathscr{G}_2$ is determined by a denumerable set $\mathscr{G}_3$, etc. Set $\mathscr{H} = \mathscr{G} \cup \bigcup_{m=1}^\infty \mathscr{G}_n$; then $\mathscr{H}$ is self-determined and denumerable.

Now let $\mathscr{H} = \{h_n\}$ be a denumerable self-determined set of continuous linear functionals containing $\{l_1, \ldots, l_{N'}\}$. We may assume that $h_n = l_n$ for $n = 1, 2, \ldots, N'$. Repetitions in the sequence $h_1, h_2, \ldots$ are allowed. Since K is compact, $h_n(K)$ is bounded. Because we can multiply each h_n, $n > N'$, by an independent constant without changing the situation we may also assume $|h_n(K)| \leq 1/n$ for $n > N'$. Now define the mapping $\mathscr{F}: X \to R^\infty$ by $\mathscr{F}(x) = [h_n(x)] \in R^\infty$ for each $x \in X$. It is obvious that $\mathscr{F}$ is continuous and linear, $\mathscr{F}(K) \subset H^0$ and $l_n = p_n \circ \mathscr{G}$ for $n = 1, \ldots, N'$. According to a comment above

$p, q \in W$, $p, q \in \mathscr{F}^{-1}(x)$ implies that $h_1 \times \cdots \times h_M(F(p)) = P_M(\mathscr{F}(F(p))) = P_M(\mathscr{F}(F(q))) = h_1 \times \cdots \times h_M(F(q))$ for all M. The latter implies that $d'(\mathscr{F}(F(p)), \mathscr{F}(F(q))) \leq 2\epsilon_M$ where $\epsilon_M^2 = \sum_{M+1}^{\infty} (1/n^2)$ provided $M \geq N'$. Consequently $d'(\mathscr{F}(F(p)), \mathscr{F}(F(q))) = 0$ and $\mathscr{F}(F(p)) = \mathscr{F}(F(q))$. For each $x \in \mathscr{F}(W)$ set $\bar{F}(x) =$ the common value of $\mathscr{F}(F(p))$ for $p \in W$, $\mathscr{F}(p) = x$. Then $\bar{F} \circ \mathscr{F} = \mathscr{F} \circ F$ and $\bar{F}(\mathscr{F}(W)) = \mathscr{F}(F(W)) \subset \mathscr{F}(K) \subset H^0$.

Finally we shall show that $\bar{F}: \mathscr{F}(W) \to H^0$ is continuous. Given $\epsilon > 0$ choose $M \geq N'$ such that $\epsilon_M^2 = \sum_{M+1}^{\infty} (1/n^2) < \epsilon^2$. Since $\mathscr{H}$ is self-determined, there is a $\delta > 0$ and an m such that if $|h_j(p) - h_j(q)| < \delta$, $j = 1, \ldots, m$ and $p, q \in W$, then

$$(\star) \qquad d'(h_1 \times \cdots \times h_M(F(p)), h_1 \times \cdots \times h_M(F(q))) < \epsilon.$$

Thus if $b, b_0 \in \mathscr{F}(W)$ and $|b - b_0| < \delta$, and if p and q are any two points of W such that $b = \mathscr{F}(p) = [h_n(p)]$ and $b_0 = \mathscr{F}(q) = [h_n(q)]$, then $(\star)$ holds. Consequently using $(\star)$ and the triangular inequality we have

$$d'(\bar{F}(b), \bar{F}(b_0)) = d'(\mathscr{F}(F(p)), \mathscr{F}(F(q))) \leq \epsilon + 2\epsilon_M \leq 3\epsilon.$$

This shows that $\bar{F}$ is continuous and the proof is complete. Q.E.D.

8. **Proof of Theorem 2.3.** Consider the sets $C_l = \{x \in W \mid l(x) \in l(F(x))\}$ where l is any continuous linear transformation from X into some finite-dimensional Euclidean space, $l: X \to \boldsymbol{R}^M$. By using the continuity of l and F it is easily seen that C_l is closed. We shall show that the collection $\{C_l\}$ has the finite intersection property. Since

$$C_{l_1} \cap \cdots \cap C_{l_m} \supset C_{l_1 \times \cdots \times l_m}$$

it is sufficient to show that each C_l is nonempty. Let l be a continuous linear transformation $l: X \to \boldsymbol{R}^M$. Set $l_n = p_n \circ L$ for $n = 1, \ldots, N$, and $l_n = p_{n-N} \circ l$ for $n = N + 1, \ldots, N + M$. Since K (see Lemma 5.2) is compact there are real numbers a_n such that $|l_n(K)| \leq a_n$ for $1 \leq n \leq N + M$. Now apply Lemma 7.1 with $N' = N + M$.

Next we will show that Lemma 5.3 applies to $\bar{F}: \mathscr{F}(W) \to H^0 \subset \boldsymbol{R}^\infty$. Since $\mathscr{F}$ is continuous, $\mathscr{F}(W)$ is compact. The fact that $\mathscr{F}(W)$ is a gw-set with P_N as a slicing function and r as a retraction is easily checked with the aid of the following observations. For each $x \in \boldsymbol{R}^N$

$$P_N(\mathscr{F}(W)) = L(W),$$

$$P_N^{-1}(x) \cap \mathscr{F}(W) = \mathscr{F}(L^{-1}(x) \cap W)$$

and if $p = \mathscr{F}(q)$, $q \in W$, then

$$L\{x' \mid [x', q] \subset W\} = P_N \mathscr{F}\{x' \mid [x, q] \subset W\} \subset P_N\{y'[y'; p] \subset \mathscr{F}(W)\}$$

and

$$P_N(p) = L(q).$$

Next we shall see that $\bar{F}: \mathscr{F}(W) \to \boldsymbol{R}^\infty$ is inward relative to P_N and r. Let $q = \mathscr{F}(p) \in \mathscr{F}(W)$. Then $P_N \bar{F}(q) = P_N \circ \bar{F} \circ \mathscr{F}(p) = P_N \circ \mathscr{F} \circ F(p) = L(F(p))$ and

$P_N(q) = L(p)$. Thus Definition 2.2 part (i) is satisfied by $\bar{F}$, $\mathscr{F}(W)$ and q by virtue of its being satisfied by F, W and p. Consider part (ii) of Definition 2.2. Let $y \in \bar{F}(q) = \bar{F}\mathscr{F}(p) = \mathscr{F}F(p)$. Thus $y = \mathscr{F}(x)$ for some $x \in F(p)$. Since $P_N(y) = L(x)$, $P_N(q) = L(p)$ and the image under $\mathscr{F}$ of the inward set (and consequently the weakly inward set) of p with respect to $L^{-1}(L(p)) \cap W$ is contained in the inward set (weakly inward set) of q with respect to $P_N^{-1}P_N(q) \cap \mathscr{F}(W) \subset \boldsymbol{R}^\infty$, it follows that $\bar{F}$ satisfies part (ii) of Definition 2.2.

Because $\mathscr{F}$ is linear it takes convex sets into convex sets and starshaped sets into starshaped sets. Since $\mathscr{F}: X \to \boldsymbol{R}^\infty$ is continuous, for each $p \in W$, $\bar{F}(\mathscr{F}(p)) = \mathscr{F}(F(p))$ is either a compact starshaped subset of $\boldsymbol{R}^\infty$ or the union of two compact convex subsets of $\boldsymbol{R}^\infty$. Thus $\bar{F}: \mathscr{F}(W) \to \boldsymbol{R}^\infty$ satisfies all the hypotheses of Lemma 5.3 and so $x \in \bar{F}(x)$ for some $x \in \mathscr{F}(W)$. Let $P = p_{N+1} \times \cdots \times p_M$; then $l = P \circ \mathscr{F}$. Since $x \in \mathscr{F}(W)$, $x = \mathscr{F}(y)$ for some $y \in W$. Then $l(y) = P \circ \mathscr{F}(y) = P(x) \in P\bar{F}(x) = P \circ \bar{F} \circ \mathscr{F}(y) = P \circ \mathscr{F} \circ F(y) = lF(y)$. Thus $y \in C_l$ and $C_l \neq \varnothing$ as desired.

Because W is compact, each C_l is a closed subset of W, and $\{C_l\}$ has the finite intersection property, we can conclude that there exists a $p \in \bigcap_l C_l \subset W$.

The proof will be completed by showing $p \in F(p)$. Assume that $p \notin F(p)$. Then for each $x \in F(p)$ there exists, due to the local convexity of X, a continuous linear functional $l_x \in X^*$ which separates p and x in the sense that $l_x(p) \leq k_x$ and $l_x(x) > k_x$ for some real constant k_x. Obviously the same l_x and k_x work for all y in a neighborhood of x. Since $F(p)$ is compact there exists a finite set of linear functionals $l_{x_1}, \ldots, l_{x_n} \in X^*$ such that each $x \in F(p)$ is separated from p by at least one of them. Thus $l(p) \notin (F(p))$ where $l = l_{x_1} \times \cdots \times l_{x_n}$, and so $p \notin C_l$, a contradiction. Therefore $p \in F(p)$ and the proof is complete. Q.E.D.

9. **Generalizations of the main theorem.** In this section we will prove the following generalization of Theorem 2.3.

9.1. THEOREM. *Let X and Y be locally convex topological linear spaces and $W \subset X$ a compact gw-set with slicing function L and retraction r. Suppose $G: W \to Y$ is a continuous set-valued function such that for each $x \in W$, $G(x)$ is a compact starshaped set or the union of two compact convex sets and $f: G(W) \to X$ is a continuous single-valued function. If $F = fG: W \to X$ is weakly inward relative to L and r then there is an $x \in W$ such that $x \in F(x)$.*

Theorem 2.3 is the special case of the above where $Y = X$ and f is the identity on $G(W)$.

The proof of Theorem 9.1 will proceed by starting from a generalization of Theorem 3.1 and carrying this generalization through the proof of Theorem 2.3. The modifications are mostly straightforward and so we will be very brief.

We begin the development leading to a proof of Theorem 9.1 with a result from O'Neill [**9**]. Though not stated explicitly in O'Neill [**9**] this result follows immediately from the definitions, Lemma 8 and Theorem 9 of O'Neill [**9**].

9.2. THEOREM. *Let G be a set-valued function from a compact polyhedron X into itself and n an integer such that if $x \in X$ then $G(x)$ is homologically trivial or consists of n homologically trivial component. If $f: X \to X$ is a continuous map and X is homologically trivial then $F = f \circ G$ has a fixed point.*

Next we have the following generalization of Lemma 3.4.

9.3. LEMMA. *If $G: H^0 \to H^0$ is an admissible set-valued function and $f: H^0 \to H^0$ is a continuous single-valued map then $F = f \circ G$ has a fixed point.*

PROOF. Consider the functions $P_n \circ (G \mid H_n^0): H_n^0 \to H_n^0$ and $P_n \circ (f \mid H_n^0): H_n^0 \to H_n^0$ and proceed as in proof of Lemma 3.4. Q.E.D.

Our next task is to extend Lemma 5.3.

9.4. LEMMA. *Let $W \subset \mathbf{R}^\infty$ be a compact gw-set with slicing function $L = P_N$ and retraction r. Suppose $G: W \to \mathbf{R}^\infty$ is an admissible set-valued function and $f: G(W) \to \mathbf{R}^\infty$ is a continuous map. If $F = fG: W \to \mathbf{R}^\infty$ is weakly inward relative to L and r then F has a fixed point.*

PROOF. Proceed as in the proof of Lemma 5.3. When things have reached the stage where $W \subset H^0$, $G(W) \subset H^0$ and $f(G(W)) \subset H^0$ extend f to a continuous function defined on all of H^0 going into H^0. This is possible due to Dugundji's extension of Zeitze's theorem [**5**] which states "If A is a closed subspace of a metrizable space B and $h: A \to S$ is a continuous map of A into a locally convex topological linear space S, then there exists a continuous extension $H: B \to S$ of h such that $H(B)$ is contained in the convex hull of $h(A)$ in S". The rest of the proof of Lemma 5.3 can then be mimicked without any trouble. Q.E.D.

The proof of Theorem 9.1 can now be completed along the lines of §8 using the Fibering Lemma as it stands.

PROOF OF THEOREM 9.1. First we may assume $Y = X$, for we may always consider X and Y as subspaces of $X \times Y$. Clearly we may also assume $G(W) \cap W = \varnothing$. Now when the time comes to apply the Fibering Lemma as in §8 we will apply the lemma to the function $H: G(W) \cup W \to X$ given by

$$\begin{aligned} H(x) &= \{f(x)\} \quad \text{if } x \in G(W), \\ &= G(x) \quad \text{if } x \in (W). \end{aligned}$$

In this way we obtain an $\mathscr{F}$, $\bar{G}$ and $\bar{f}$ such that $\mathscr{F} \circ G = \bar{G} \circ \mathscr{F}$ and $\mathscr{F} \circ f = \bar{f} \circ \mathscr{F}$. Then with $\bar{F}$ defined as $\bar{f} \circ \bar{G}$ we have $\mathscr{F} \circ F = \mathscr{F} \circ f \circ G = \bar{f} \circ \mathscr{F} \circ G = \bar{f} \circ \bar{G} \circ \mathscr{F} = \bar{F} \circ \mathscr{F}$ and we can proceed as in §8. Q.E.D.

REMARK. O'Neill's Theorem 3.1 holds for upper semicontinuous set-valued functions when $n = 1$ [**5**], [**9**]. It is natural to try to extend this result to a theorem like Theorem 2.3. Glicksberg [**6**] and Browder [**2**] have obtained by methods different from those presented here some results in this direction in the case of a convex W and an F such that $F(x)$ is convex for each $x \in W$. The techniques presented here all go through except for the Fibering Lemma and the trouble there is that it is not apparent how to ensure that the set-valued function is "uniformly upper semicontinuous". In fact there is trouble in formulating this concept without implying continuity.

Let X be a locally convex topological linear space and $W \subset X$ a compact gw-set with slicing function $L: X \to \boldsymbol{R}^{\infty}$ and retraction r. Since L is linear $L(X)$ is a linear subspace of $\boldsymbol{R}^N$ and consequently homeomorphically isomorphic to $\boldsymbol{R}^{N'}$ for some $N' \leq N$. Thus we might as well assume L is onto. Let $e_1 = (1, 0, 0, \ldots, 0) \in \boldsymbol{R}^N$, $e_2 = (0, 1, 0, \ldots, 0) \in \boldsymbol{R}^N$, etc. Algebraically we may write $\varphi: X \cong L^{-1}(0) \oplus \boldsymbol{R}^N$ where $\varphi(x) = (\varphi_1(x), \varphi_2(x)) = (x - gLx, L(x))$ and the linear function g is (not uniquely) defined by picking an $x_i \in L^{-1}(e_i)$ for each $i = 1, \ldots, N$ and setting $g(e_i) = x_i : \varphi^{-1}$ is given by $\varphi^{-1}(a, b) = a + g(b)$. Since g is linear and has the finite-dimensional space $\boldsymbol{R}^N$ as domain it is continuous. Thus φ and φ^{-1} are also continuous and X is homeomorphically isomorphic to $L^{-1}(0) \oplus \boldsymbol{R}^N$.

Let $x \in X$ and $A \subset X$. A point $y \in X$ is said to be in the *(weakly) outward* set of y with respect to A provided $x - (y - x)$ is in the (weakly) inward set of x with respect to A. The definition of "a set-valued function $F: W \to X$ being (weakly) outward relative to L and r" is obtained by replacing "inward" by "outward" in Definition 2.2. We now can state the following extension of Theorem 2.3.

9.5. THEOREM. *Theorem* 2.3 *holds with "inward" replaced by "outward".*

PROOF. If F satisfies the hypothesis of Theorem 9.5 then

$$
\begin{aligned}
H(x) &= \varphi^{-1}(\varphi_1(x) - (\varphi_1(F(x)) - \varphi_1(x)), \varphi_2(x)) \\
&= \varphi^{-1}\{(\varphi_1(x) - (\varphi_1(y) - \varphi_1(x)), \varphi_2(x)) \mid y \in F(x)\}
\end{aligned}
$$

satisfies the hypothesis of Theorem 2.3 and x_0 is a fixed point for F if and only if it is a fixed point for H. Q.E.D.

This same technique can also be used to prove Theorem 9.1 for the "outward" case. But first we must replace $G: W \to Y$ and $f: G(W) \to X$ by $\bar{G}: W \to X \times Y$ and $\bar{f}: \bar{G}(W) \to X$ where $\bar{G}(x) = (x, G(x)) = \{(x, y) \mid y \in G(x)\}$ and $\bar{f}(x, y) = f(y)$. Then $F = \bar{f} \circ \bar{G}$ and we can compare F to H where $H = h \circ \bar{G}$ and $h(x, y) = \varphi^{-1}(\varphi_1(x) - (\varphi_1(\bar{f}(x, y)) - \varphi_1(x)), \varphi_2(x))$. Thus we have

9.6. THEOREM. *Theorem* 9.1 *holds with "inward" replaced by "outward".*

A set-valued function $F: W \to X$ from a subset $W \subset X$ of a topological linear space is said to be (weakly) inward if $F(x)$ is contained in the (weakly) inward set of x with respect to W for each $x \in W$. If W is a compact gw-set with slicing function L and retraction r such that $L(W)$ is convex and $r(p) =$ the unique point $q \in L(W)$ which minimizes $\|p - q'\|$ among $q' \in L(W)$ then an inward set-valued function $F: W \to X$ is inward relative to L and r. Thus an admissible inward set-valued function $F: W \to X$ from such a W into X (assumed locally convex) must have a fixed point.

If W is a convex subset of a locally convex topological linear space X then W can be made into a gw-set by taking $L: W \to \boldsymbol{R}^0$ to be the only such function—$L(W) = \{0\}$ and $r: \boldsymbol{R}^0 \to L(W)$ again the only such function—$r(0) = 0$. Then any weakly inward set-valued function $F: W \to X$ is weakly inward relative to L and

r. Thus any admissible weakly inward set-valued function $F: W \to X$ from a compact convex subset of a locally convex topological linear space into that space must have a fixed point. This specialization of Theorem 9.1 is a generalization of the main result of Halpern and Bergman [7] which in turn generalizes the Shauder-Tychonoff fixed point theorem.

We conclude this section with a generalization of the concept of *a gw*-set. This new kind of set we call a *ggw*-set and by adjusting the proofs a little the theorems concerning *gw*-sets may be established for *ggw*-sets.

9.7. DEFINITION. Let X be a topological linear space. A nonempty subset W of X is a *ggw-set* with *slicing function* L and *retraction* r provided

(i) $L: X \to \boldsymbol{R}^\infty$ is a continuous linear map.
(ii) $L(W) \subset H^\circ \subset \boldsymbol{R}^\infty$ (see 3.3).
(iii) $L^{-1}(t) \cap W$ is convex for each $t \in \boldsymbol{R}^\infty$.
(iv) $r: H^\circ \to L(W)$ is a continuous retraction.
(v) The following "continuity" condition is satisfied:

For each $x \in W$ either

1. $L^{-1}L(x)W = \{x\}$ or
2. There exists an $\epsilon > 0$ such that $\{y \in H \mid \|y - L(x)\| = \epsilon$ and $[y, L(x)) \cap L(W) \neq \varnothing\} \subset L\{Z \in W[Z, x] \subset W\}$.

The concept of *pw*-set, defined in §10, can be similarly generalized along with the related theorems.

10. Noncompact case. We now take up the problem of extending Theorem 9.1 to the situation where W is closed (but not necessarily compact) and cl $F(W)$ is compact. The basic idea is to consider F restricted to a compact subset $W' \subset W$ of W which is formed by intersecting W with an appropriate compact convex set containing $F(W)$. In order to insure that Theorem 9.1 applies to $F \mid W' = f(G \mid W'): W' \to X$ we will assume for W a stronger form of the "continuity condition" of Definition 2.1.

Let X be a topological linear space.

10.1. DEFINITION. A nonempty subset $W \subset X$ is a *pw-set* with *slicing function* L and *retraction* r **provided there is a compact set** $S \subset W$ **such that** $L(S) = L(W)$ and

(i) L is a continuous linear map from X into some finite-dimensional Euclidean space $\boldsymbol{R}^N$.
(ii) r is a retraction of $\boldsymbol{R}^N$ onto $L(W)$.
(iii) For each $x \in \boldsymbol{R}^N$, $L^{-1}(x) \cap W$ is convex.
(iv) The following "strong continuity property" is satisfied: For each $p \in W$ either

(a) $L^{-1}(L(p)) \cap W = \{p\}$ or
(b) There exists an $\epsilon > 0$ such that

$$\{y \in \boldsymbol{R}^N \mid \|y - L(p)\| = \epsilon \quad \text{and} \quad [y, L(p)) \cap L(W) \neq \varnothing\}$$
$$\subset L\{x \mid [x, p] \subset W \text{ and } x \in S\}.$$

Note that Definition 10.1 is Definition 2.1 with the parts (in bold type) added. Thus a *pw*-set is a *gw*-set with the same slicing function and retraction.

10.2. THEOREM. *Let X and Y be locally convex topological linear spaces with X complete and $W \subset X$ a closed pw-set with slicing function L and retraction r. Suppose $G: W \to Y$ is a continuous set-valued function such that for each $x \in W$, $G(x)$ is a compact starshaped set or the union of two compact convex sets and $f: G(W) \to X$ is a continuous single-valued function. If $F = f \circ G: W \to X$ is inward relative to L and r, and* cl *$F(W)$ is compact then there is an $x \in W$ such that $x \in F(x)$.*

PROOF. Let K be the closed convex hull of (cl $F(W)) \cup S$ where S is as in Definition 10.1. Since by hypothesis cl $F(W)$ and S are compact it follows that (cl $F(W)) \cap S$ is compact and then by a theorem of Bourbaki [**1**], K is also compact. The remaining part of the proof consists simply of showing that the hypotheses of Theorem 9.1 are fulfilled by $W' = W \cap K$, L, r, f, and $G \mid W'$.

First W' is a *pw*-set (and thus a *gw*-set) with slicing function L and retraction r. Since $L(S) = L(W)$ and $S \subset W' \subset W$ we have $L(W') = L(W)$ and thus r is a retraction of $\boldsymbol{R}^N$ onto $L(W')$. For each $x \in \boldsymbol{R}^N$, $L^{-1}(x) \cap W' = (L^{-1}(x) \cap W) \cap K$ is convex because $(L^{-1}(x) \cap W)$ and K are convex. It is easy to see that part (iv) of Definition 10.1 holds for W' and S.

Since $W' = W \cap K$ is the intersection of a compact set K with a closed set W, it is compact. The only nontrivial thing left to check is that $(F \mid W') = f \circ (G \mid W'): W' \to X$ is inward (and thus weakly inward) relative to L and r. Part (i) of Definition 2.2 is immediate. Consider part (ii) of Definition 2.2. Let $p \in W'$ and $x \in (F \mid W')(p) = F(p)$. Since F is inward relative to L and r condition (a) or (b) of Definition 2.2, part (ii) must hold for p, x, F and W. Condition (a) for p, x, F and W will imply condition (a) for p, x, $F \mid W'$ and W'. Assume now that condition (b) holds for p, x, F and W, that is, $L(x) = L(p)$ and x is in the inward set of p with respect to $L^{-1}L(p) \cap W$. Since $L^{-1}L(p) \cap W$ is convex this means that there is a $y \in [x, p) \cap W$. Since $p \in W' \subset K$ and $x \in F(p) \subset F(W) \subset K$, we have $y \in [x, p) \subset K$ by the convexity of K. Thus $y \in [x, p) \cap K \cap W = [x, p) \cap W'$ which shows that x is in the inward set of p with respect to $L^{-1}L(p) \cap W'$ (remember $L(x) = L(p)$). Thus condition (b) holds for p, x, $F \mid W'$ and W'.

Consequently, Theorem 9.1 applies and there is an $x \in W' \subset W$ such that $x \in (F \mid W')(x) = F(x)$. Q.E.D.

H. H. Schaefer [**10**] has applied the Schauder-Tychonoff fixed point theorem to the situation of nonlinear (or rather, not necessarily linear) maps defined on (or on a subset of) the "positive" cone in a partially ordered locally convex topological linear space. See [**10**] for definitions and statements of the theorems to be mentioned below. Morgenstein's theorem and Theorem 1 of [**10**] can be immediately extended, using Theorem 10.2 in place of the Schauder-Tychonoff theorem, to admissible set-valued T. (Note that the map $y \to cy/\|y\|$ used there actually preserves convex and starshaped subsets of C.) Theorem 2 of [**10**] can similarly be generalized to the case where T is a continuous set-valued function such that $T(x)$ consists of one or two points for each x.

11. A covering property for outward maps. Let X be a locally convex topological linear space. Consider a compact gw-set $W \subset X$ with slicing function $L: X \to \boldsymbol{R}^N$ and retraction $r: \boldsymbol{R}^N \to L(W)$ such that $L(W)$ is convex and $r(p)$ is the nearest point of $L(W)$ to p for each $p \in \boldsymbol{R}^N$. We will say that a set-valued function $F: W \to X$ is (*weakly*) *outward* if $F(x)$ is in the (weakly) outward set of x with respect to W for each $x \in W$. Since the (weakly) outward set of x consists of all y such that $x - (y - x)$ is in the (weakly) inward set of x (with respect to W) we see that F is (weakly) outward if and only if $G(x) = x - (F(x) - x)$ is (weakly) inward.

If A is a subset of a linear space Y then the center or star center of A is the set $\text{sc } A = \{x \in A \mid [x, y] \subset A \text{ for all } y \in A\}$. We will need the following lemma.

11.1. LEMMA. *Suppose a subset A of a topological linear space Y contains 0 in its center, $0 \in \text{sc } A \subset A \subset Y$. Then the inward set of any point $x \in X$ is closed under multiplication by $\frac{1}{2}$. Hence so is the weakly inward set.*

PROOF. See Figure 11.2 where z is an arbitrary point in the inward set of x and $y = \alpha z + (1 - \alpha)x$, $\alpha > 0$, is a point in $(x, z\} \cap A$. The reader can easily supply the numerical argument, getting $y' \in (x, \frac{1}{2}z\} \cap [0, y] \subset (x, \frac{1}{2}z\} \cap A$.

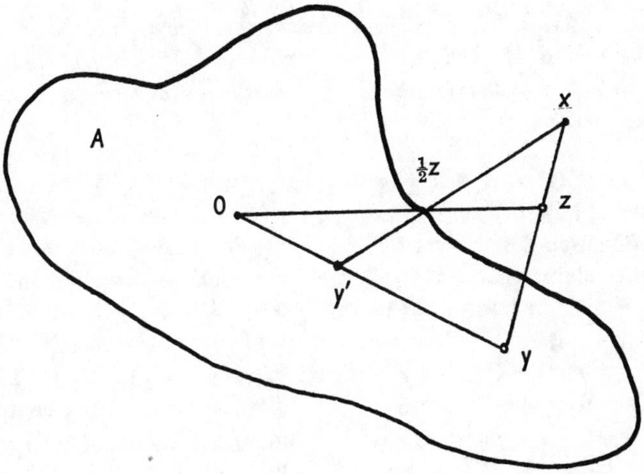

FIGURE 11.2

With W and X as above we have the following theorem.

11.2. THEOREM. *If $F: W \to X$ is a weakly outward admissible set-valued function then $x \in F(x)$ for some $x \in W$ and* sc $W \subset F(W)$.

PROOF. Consider $G(x) = x - (F(x) - x)$, then $G: W \to X$ is weakly inward and satisfies the hypothesis of Theorem 2.3. (See the discussion following Theorem 9.6.) Therefore G has a fixed point x_0 which is obviously also a fixed point for F. Now let $p \in \text{sc } W$. We may assume $p = 0$. Since $0 \in F(x)$ if and only if $x \in \frac{1}{2}G(x)$

we need only show that $\frac{1}{2}G$ has a fixed point. But it follows immediately from Lemma 11.1 that $\frac{1}{2}G$ is weakly inward. Consequently Theorem 2.3 guarantees that $\frac{1}{2}G$ has a fixed point and the proof is complete. Q.E.D.

The same technique used above can be used to obtain the same result for closed *pw*-sets W, and outward set-valued functions $F: W \to X$ such that X is complete and cl $G(W)$ is compact where $G(x) = x - (F(x) - x)$. We could also obtain these results in the same way for F's which factor $F = fG$ as in Theorem 9.1. Finally the same procedure shows that $W \subset F(W)$ when it is assumed (along with the usual assumptions) that $F(x)$ is in the weakly outward set of x with respect to sc W for all $x \in W$.

12. **Commuting linear maps.** The Markov-Kakutani fixed point theorem states

12.1. THEOREM. *If $\mathscr{F} = \{f\}$ is a commuting family of continuous linear self-maps of a compact convex subset $V \subset X$ of a topological linear space X then they have a common fixed point.* (*Dunford-Schwartz* [**4**].)

Using Theorem 2.3 we can prove the following extension of the Markov-Kakutani theorem in the case where X is locally convex.

12.2. THEOREM. *Let X be a locally convex topological linear space and $V \subset X$ a compact convex subset of X. If $\mathscr{F} = \{f\}$ is a commuting family of continuous linear maps from X into X such that $(f \mid V): \dot{V} \to X$ is inward for each $f \in \mathscr{F}$, then they have a common fixed point in V.*

PROOF. Let $K(f) = \{x \in X \mid f(x) = x\}$ and $R(f) = VK(f)$ for each $f \in \mathscr{F}$. We must show that $\bigcap_{f \in \mathscr{F}} R(f) \neq \varnothing$. Since each $f \in \mathscr{F}$ is continuous the sets $R(f)$ are closed. By Theorem 2.3 they are nonempty. (See discussion following Theorem 9.6.) We now claim that $\{R(f)\}_{f \in \mathscr{F}}$ has the finite intersection property. Let $f_1, \ldots, f_n \in \mathscr{F}$. We wish to show $\bigcap_{i=1}^n R(f_i) \neq \varnothing$, $^{(\star)}$. We will use induction on n. Obviously $^{(\star)}$ holds for $n = 1$. Now assume we have $^{(\star)}$ for n and $f_1, \ldots, f_{n+1} \in \mathscr{F}$. Since each $f \in \mathscr{F}$ is linear, $K(f)$ is convex. Thus $K = \bigcap_{i=1}^n K(f_i)$ is convex and closed and therefore $V' = V \cap K = \bigcap_{i=1}^n R(f_i)$ is a nonempty compact convex subset of V. If $x \in V'$ and $1 \leq i \leq n$ then $f_{n+1}(x) = f_{n+1}(f_i(x)) = f_i(f_{n+1}(x))$. Consequently $f_{n+1}(x) \in K(f_i)$ for $i = 1, \ldots, n$, and so $f_{n+1}(x) \in \bigcap_{i=1}^n K(f_i) = K$, which is convex. We will now show that $f_{n+1} \mid V'$ is inward. Since $f_{n+1} \mid V$ is inward and V is convex, for each $x \in V' \subset V$ there is a $y \in [f_{n+1}(x), x) \cap V$. Since $f_{n+1}(x) \in K$ and $x \in V' = V \cap K \subset K$ we have $[f_{n+1}(x), x) \subset K$ by the convexity of K. Thus $y \in [f_{n+1}(x), x) \cap K \cap V = [f_{n+1}(x), x) \cap V'$ which shows that $f_{n+1} \mid V'$ is inward. By Theorem 2.3 there is an $x_0 \in V'$ such that $f_{n+1}(x_0) = x_0$. Then $x_0 \in \bigcap_{i=1}^{n+1} R(f_i)$ and the induction step is complete.

Since V is compact $\bigcap_{f \in \mathscr{F}} R(f) \neq \varnothing$ which is what we wished to show. Q.E.D.

REMARKS. If we have a commuting family $\mathscr{G}$ of continuous linear maps such that for each $g \in \mathscr{G}$, $g \mid V$ is either inward or outward then we may obtain a commuting family $\mathscr{F}$ of continuous inward (i.e. the restrictions $f \mid V$ are inward for $f \in \mathscr{F}$) linear maps from $\mathscr{G}$ by replacing each outward $g \in \mathscr{G}$ by $\hat{g}(x) = x - (g(x) - x) = 2x - g(x)$. Furthermore, a common fixed point for $\mathscr{F}$ is a common fixed point for $\mathscr{G}$.

Next, note that the hypothesis of Theorem 12.2 can be weakened by allowing one $f \in \mathscr{F}$ to be possibly nonlinear but still inward and a $g \in \mathscr{F}$, $g \neq f$, to be any continuous linear map with a fixed point in V. This case is proven exactly as above except for the added precaution that when trying to show $R(f_1) \cap \cdots \cap R(f_{n+1}) \neq \varnothing$ we make sure that if f is among the f_i it is f_{n+1} and if g is among the f_i it is f_1.

13. **Examples of *gw*-sets and *pw*-sets.** If $X = X' \times R$ where X' is a topological linear space we may consider sets of the form $W = \{(x', t) \mid x' \in g(t)V, a \leq t \leq b\}$ where $V \subset X'$ and $g:[a, b] \to R$. We may insure that W is a compact *gw*-set by requiring V to be a compact convex subset of X with $0 \in V$ and $g:[a, b] \to \boldsymbol{R}$ a function satisfying a Lipschitz condition: $|g(t) - g(t')| \leq K\,|t - t'|$ for $t, t' \in [a, b]$ and some constant K. We take $L:X' \times \boldsymbol{R} \to \boldsymbol{R}$ to be given by $L(x', t) = t$ and $r:\boldsymbol{R} \to L(W) = [a, b]$ to be given by

$$\begin{aligned} r(t) &= a \quad \text{if } t < a, \\ &= t \quad \text{if } a \leq t \leq b, \\ &= b \quad \text{if } b < t. \end{aligned}$$

It is easy to check that W is a *gw*-set with slicing function L and retraction r. To prove that W is compact it is sufficient to show that it is a closed subset of $(\max_{a \leq t \leq b} |g(t)|)(V \cup (-V)) \times [a, b]$ which is compact by the Tychonoff theorem. We leave the verification of this to the reader. Actually W is a *pw*-set with $S = 0 \times [a, b]$.

Bibliography

1. N. Bourbaki, *Espaces vectoriels topologiques Éléments de Mathématique*, Chapters I, II, Actualités Sci. Indust. no. 1189, Hermann, Paris, 1953.
2. F. E. Browder, *The fixed point theory of multi-valued mappings in topological vector spaces*, Math. Ann. **117** (1968), 283–301.
3. J. Dugundji, *An extension of Tietze's theorem*, Pacific J. Math. **1** (1951), 353–367.
4. N. Dunford and J. T. Schwartz, *Linear operations*. I, Interscience, New York, 1964.
5. S. Eilenberg and D. Montgomery, *Fixed point theorems for multivalued transformations*, Amer. J. Math. **68** (1946), 214–222.
6. I. L. Glicksberg, *A further generalization of the Kakutani fixed point theorem with applications to Nash equilibrium points*, Proc. Amer. Math. Soc. **3** (1952), 170–174.
7. B. Halpern and G. Bergman, *A fixed-point theorem for inward and outward maps*, Trans. Amer. Math. Soc. **130** (1968), 353–358.
8. K. Kuratowski, *Topologie*. II, Hefner, New York, 1950.
9. B. O'Neill, *Induced homology homomorphisms for set-valued maps*, Pacific J. Math. **7** (1957), 1179–1184.
10. H. H. Schaefer, *On nonlinear positive operators*, Pacific J. Math. **9** (1959), 847–860.

University of California, Berkeley, California

ON CRITICAL SETS FOR NONLINEAR EQUATIONS OF EVOLUTION WITH NONUNIQUENESS[1]

G. Stephen Jones

This study concerns the existence of critical sets in set valued flows generated by nonlinear equations of evolution over locally convex linear Hausdorff spaces. Critical sets in turn relate to the problem of existence and location of critical points, periodic solutions, and pseudo-periodic solutions.

The evolution of many dynamical processes can be predicted, or at least partially so, by the study of initial value formulations of a standard type. That is, with X an appropriate space, $R^+ = [0, \infty)$, and $f: R^+ \times X \to X$ continuous, we consider a differential system of the form

$$\dot{x}(t) = f(t, x(t)) \tag{1}$$

together with the initial condition that $x(0) = x_0$. More generally a hereditary formulation may be required, so with E some appropriately defined function space over X and $g: R^+ \times E \to X$ continuous, we consider equations of evolution having the form

$$\dot{x}(t) = g(t, x(\alpha_t(x(t)))), \tag{2}$$

where $t \in \alpha_t(\xi) \subset (-\infty, t]$ and $x(\alpha_t(\xi))$ is the restriction of x to $\alpha_t(\xi)$. For an initial condition, a function $\phi: R \to X$ is specified and it is required that $x \mid (-\infty, 0] = \phi \mid (-\infty, 0]$.

Assuming conditions sufficient to assure the existence of solutions on all of R^+ starting from each initial value or initial function in a specified set of values or functions, we pass on to consider the problem of existence of solutions with special properties. In particular, we consider some general techniques for determining critical points, periodic solutions, and pseudo-periodic solutions. For these considerations, we switch our attention to the flows generated by equations of evolution. That is, functions $F: R^+ \times Y \to Y^*$ where $F(0, y) = y$ for y in Y, Y is an appropriately defined subset of either X or E, and Y^* is the space of nonempty compact convex subsets of Y. F in general is assumed to be upper semicontinuous. If initial data is assumed to determine a solution uniquely, then our flow equation simplifies to $F: R^+ \times Y \to Y$.

---

[1] This work was supported in part by Contract NSF-GP-7846, National Science Foundation.

It should be pointed out that the relationship between flows and equations of evolution we have expressed is strictly motivational and serves only to make natural the conditions imposed. It is otherwise nonessential to the results on flows presented. For example, if we simply start with a mapping $f: Y \to Y^*$, it is perfectly admissible to consider a flow

$$F(t; x) = ([t + 1] - t)F([t]; x) + (t - [t])f(F([t]; x)) \tag{3}$$

where $[t]$ denotes the largest integer $<t$ and $F(0; x) = x$.

For our considerations Y is assumed to be a closed convex subset of a locally convex complete linear Hausdorff space. A trajectory initiating at x_0 in Y and belonging to the flow $F: R^+ \times Y \to Y^*$ is a continuous function $\xi: R^+ \to Y$ such that $\xi(0) = x_0$ and $\xi(t) \in F(t; x_0)$ for all $t \in R^+$. A flow is referred to as single-valued if emanating from each initial point there exists precisely one trajectory.

Most of the classical theory developed for flows has been concerned with single-valued autonomous flows. In this paper we provide a theory applicable to the study of equations of evolutions where trajectories are not uniquely determined by initial data and which generate nonautonomous flows. Autonomy is replaced by the notions of weak autonomy and quasi-autonomy which are defined in the sequel.[2]

For each set $S \subset Y$ let $\Lambda(t; S)$ denote the set of all points $x \in S$ such that $x \in F(t; x)$. A compact minimal set $\Omega(S)$ such that $\Omega(S) \cap \Lambda(t; S)$ is nonvoid for all $t \in R^+$, we shall call a *critical set* of F.

In the classical sense, a single-valued flow is called an autonomous flow or dynamical system if $F(t_1; F(t_2; x)) = F(t_1 + t_2; x)$ for every pair t_1, t_2 in R^+ and every $x \in Y$. The same notion is, of course, meaningful for set valued flows. A simple proposition relating to autonomous flows and critical sets may be stated as follows.

PROPOSITION 1. *If F is an autonomous flow and has a critical set $\Omega(S)$ on $S \subset Y$, then F contains a constant trajectory (often referred to as a critical point).*

Proposition 1 is a special case of a more general proposition which may be stated as follows.

PROPOSITION 2. *Let F on $S \subset Y$ have a critical set $\Omega(S)$ and for each $t \in R^+$ let x_t denote a point in $F(t; S) \cap \Omega$. Suppose there exists a continuous function $\mu: R^+ \to R^+$ with $\mu(t) > 0$ for $t > 0$ and such that for all t_1 and t_2 in R^+, $x_{t_1} \in F(t_1; x_{t_1})$ implies $F(t_1; x_{t_1}) \subset F(t_1 + \mu(t_2); x_{t_1})$. Then F contains a constant trajectory.*

PROOF. Consider the sequence $\{x_{1/n}\}$, $n = 1, 2, \ldots$, in $\Omega(S)$. $\Omega(S)$ being compact implies $\{x_{1/n}\}$ has a limit point which we denote by x^*. We shall show that $x^* \in F(t; x^*)$ for all $t \in R^+$.

---

[2] Results for single-valued autonomous flows similar to those presented here are contained in references [**1**] and [**2**].

Assume $x^* \notin F(t^*; x^*)$ for some $t^* \in R^+$. Then there exist neighborhoods $N_1(x^*)$ and $N(F(t^*; x^*))$ of x^* and $F(t^*, x^*)$ respectively such that $N_1(x^*) \cap N(F(t^*; x^*)) = \varnothing$. By the upper semicontinuity of F, there exist $\delta > 0$ and a neighborhood $N_0(x^*) \subset N_1(x^*)$ such that $t \in [t^* - \delta, t^*]$ and $x \in N_0(x^*)$ imply $F(t; x) \subset N(F(t^*; x^*))$. Clearly we may choose n such that $1/n < \delta$, $\mu(1/n) < \delta$, and $x_{1/n} \in N_0(x^*)$. Now $x_{1/n} \in F(1/n; x_{1/n})$ and by hypothesis

$$F\left(\frac{1}{n}; F\left(\frac{1}{n}; x_{1/n}\right)\right) \subset F\left(\frac{1}{n} + \mu\left(\frac{1}{n}\right); x_{1/n}\right) \subset \cdots \subset F\left(\frac{1}{n} + k\mu\left(\frac{1}{n}\right); x_{1/n}\right)$$

where k is an arbitrary positive integer. Clearly for some k, $1/n + k\mu(1/n) \in [t^* - \delta, t^*]$ so it follows that $x_{1/n} \in N(F(t^*; x^*)) \cap N_0(x^*)$. But since $N_0(x^*) \subset N_1(x^*)$ this is contradictory, and we must conclude that t^* can not exist. Hence $\xi(t) = x^*$ is a constant trajectory for our flows and Proposition 2 is verified.

For arbitrary $S \subset Y$ and $t \in R^+$, we define $F^1(t; S) = \bigcup \{F(t; x) : x \in S\}$ and $F^n(t; S) = F^1(t; F^{n-1}(t; S))$ for $n = 2, 3, \ldots$. For purposes of defining a notion of weak autonomy, consider two not necessarily distinct open subsets S_0 and S_1 of Y such that $S_0 \subset S_1$. F is said to be *weakly autonomous* relative to S_0, S_1 if the following conditions are satisfied:

(1) *There exists a strictly increasing function* $\phi: R^+ \to R^+$ *such that*

$$F^1(t_1 + \phi(t_2); S_0) \subset F^1(t_1; F^1(t_2; S_1))$$

for all t_1 *and* t_2 *in* R^+.

(2) *For every pair* t, τ *in* R^+ *with* $\tau > t$ *and every* $x \in S_0$ *there exists a positive integer* n *such that* $F^n(t; x) \subset F^1(\tau^*, S_0)$ *for some* $\tau^* > \tau$.

A flow which is weakly autonomous relative to pair of sets S_0 and S_1 is said to be *quasi-autonomous* relative to S_0, S_1 if it has the following additional property:

(3) *For some continuous function* $\mu: R^+ \to R^+$ *with* $\mu(t) > 0$ *for* $t > 0$, $x_1 \in S_0$ *and* $x_1 \in F(t_1; x_1)$ *imply* $F(t_1; F(t_2; x_1)) \subset F(t_1 + \mu(t_2); x_1)$ *for all* t_1, t_2 *in* R^+.

It is clear that every autonomous flow is quasi-autonomous and that many perturbed or asymptotically autonomous flows are quasi-autonomous. It is also true that if we consider the period map of a periodic equation of evolution and construct a flow such as suggested by (3), then in many cases this resulting flow will be quasi-autonomous. The main theorem of this paper sets forth general conditions under which the existence of critical sets and critical points can be established for weakly autonomous and quasi-autonomous flows respectively. It may be stated as follows.

THEOREM 1. *Suppose* Y *is locally compact and* S_0 *and* S_1 *with* $S_0 \subset S_1$ *are convex sets open in* Y. *Let* F *be weakly autonomous relative to* $\bar{S}_0$, $\bar{S}_1$ *and suppose* $F^1(t_1; \bar{S}_1)$ *at some point* $t_1 \in R^+$ *is contained in a compact subset of* S_0. *Then* F *has a critical set* $\Omega(S_0)$. *If* F *is quasi-autonomous, then* F *has a constant trajectory.*

In the proof of Theorem 1 we shall make use of the following lemma which is of interest in itself.

LEMMA 1. *Let F be weakly autonomous relative to a pair of open sets S_0, S_1 such that $S_0 \subset S_1 \subset Y$ and suppose $F^1(t_1; \bar{S}_1)$ at some point $t_1 \in R^+$ is contained in a compact subset of S_0. Then $F^1(t; \bar{S}_1)$ is contained in a compact subset of S_0 for all t sufficiently large.*

PROOF. Since $F^1(t_1; \bar{S}_1)$ is contained in a compact subset of S_0, there exists $t_2 > t_1$ such that $F^1(t; \bar{S}_1) \subset U$ for all $t \in [t_0, t_1]$. Using this fact and the weak autonomy of F we shall show that $F^1(t; \bar{S}_1) \subset U$ for all

$$t \geqslant t_2 + \frac{\phi(t_2)^2}{\phi(t_2) - \phi(t_1)}.$$

For all $t \in [t_1, t_2]$ and all positive integers n, let us observe the following set ordering resulting from the weak autonomy of F.

$$\begin{aligned}\bar{S}_1 \supset S_0 \supset U \supset F^1(t; \bar{S}_1) &\supset F^1(t; F^1(t; S_1)) \\ &\supset F^1(t + \phi(t); \bar{S}_1) \supset F^1(t + \phi(t); F(t; \bar{S}_1)) \\ &\supset F^1(t + 2\phi(t); \bar{S}_1) \supset \cdots \supset F^1(t + n\phi(t); \bar{S}_1).\end{aligned}$$

Since ϕ is strictly increasing, we have that $F^1(t; \bar{S}_1) \subset U$ for all $t \in \bigcup_{n=1}^{\infty} [t_1 + n\phi(t_1), t_2 + n\phi(t_2)]$. Considering

$$n \geqslant \frac{\phi(t_1)}{\phi(t_2) - \phi(t_1)}$$

we observe the implication that

$$n(\phi(t_2) - \phi(t_1)) + (t_2 - t_1) \geqslant \phi(t_1)$$

and

$$t_2 + n\phi(t_2) \geqslant t_1 + (n + 1)\phi(t_1).$$

But this clearly implies that

$$[t_2 + n\phi(t_2), \infty) \subset \bigcap_{n=1}^{\infty} [t_1 + n\phi(t_1), t_2 + n\phi(t_2)].$$

We may choose $n \geqslant \phi(t_1)/(\phi(t_2) - \phi(t_1))$ and $\leqslant \phi(t_2)/(\phi(t_2) - \phi(t_1))$, so

$$\left[t_2 + \frac{\phi(t_2)^2}{\phi(t_2) - \phi(t_1)}, \infty\right) \subset \bigcup_{n=1}^{\infty} [t_1 + n\phi(t_1); t_2 + n\phi(t_2)].$$

Thus $F^1(t; \bar{S}_1) \subset U$ for all $t \geqslant t_2 + \phi(t_2)^2/(\phi(t_2) - \phi(t_1))$ and our lemma is proved.

PROOF OF THEOREM 1. By Lemma 1 it follows that there exists $\tau \in R^+$ such that $F(t; \bar{S}_1)$ is contained in a compact subset U of S_0. Since U is contained in a complete locally convex space and S_0 is convex, we can always choose U convex. Using an extension of the Kakutani fixed point theorem proved by Glicksberg in [**3**], it follows that for all $t \geqslant \tau$, there exists $x_t \in U$ such that $x_t \in F(t; x_t)$. Since Y is locally compact there exists a neighborhood N of U in S_0 whose closure is compact.

If F has no critical set $\Omega(S_0)$ then there exists some $t_0 \in (0, \tau]$ such that $x \notin F(t_0; x)$ for all $x \in S_0$. Then by a homotopy type argument for set valued mapping presented in [**4**], there must exist $t^* \in (t_0, \tau)$ such that for some $x^* \in \bar{N} \setminus N$, $x^* \in F(t^*; x^*)$. But $x^* \in F(t^*; x^*)$ implies $x^* \in F^n(t^*, x^*)$ for all positive integers n. Since F is weakly autonomous there exists n and $\tau^* > \tau$ such that $F^n(t^*; x^*) \subset F^1(\tau^*; \bar{S}_1) \subset U$ and we are lead to the conclusion that $x^* \in U$. Since this is contradictory, it follows that no such point x^* can exist. Therefore, no point t_0 can exist and we have established the existence of a critical set $\Omega(S_0)$ for F. If F is quasi-autonomous then it follows from Proposition 2 that F must have a constant trajectory and our proof is complete.

Our next theorem is an extension of Theorem 1 which is more involved to state but offers the advantage of not requiring Y to be locally compact.

THEOREM 2. *Let $Z \subset Y$ be closed convex and locally compact. Let $r: Y \to Z$ be a retraction, $F: R^+ \times Y \to Y^*$ a flow, and let $G: R^+ \times Y \to Y^*$ be defined by the formula $G(t; x) = F(t; r(x))$. Relative to open convex subsets $S_0 \subset S_1 \subset Y$ let G be weakly autonomous where $r(S_0)$ is contained in S_0 and open in Z. If $G^1(t_1; \bar{S}_1)$ at some point $t_1 \in R^+$ is contained in a compact subset of $r(S_0)$, then F has a critical set. If G is quasi-autonomous, then F has a constant trajectory.*

PROOF. By Theorem 1 G has a critical set $\Omega(r(S_0))$ and a constant trajectory if quasi-autonomous. But since $\Omega(r(S_0)) \subset r(S_0) \subset Z$ it follows that $G \mid \Omega(r(S_0)) = F \mid \Omega(r(S_0))$, so these conclusions also hold for F.

A flow F is said to be *pseudo-periodic* of period $w > 0$ relative to a neighborhood N of the origin if for all $t \in R^+$ and all integers n,

$$F^1(t; F^n(w; x)) \subset F(t + nw; x) + N.$$

If

$$F^1(t; F(w; x)) \subset F(t + w; x),$$

then F is said to be *strongly pseudo-periodic*. If F has this property and is single valued then it is, of course, periodic in the usual sense.

A trajectory $\xi: R^+ \to Y$ is said to be pseudo-periodic of period $w > 0$ relative to N if

$$\xi(t) \subset \xi(t + nw) + N$$

for all t and n. A trajectory ξ is strongly *pseudo-periodic* if $\xi(t) \subset \xi(t + w)$ for all t in R^+.

THEOREM 3. *Suppose Y is locally compact and S_0 and S_1 with $S_0 \subset S_1$ are open convex sets in Y. Let F be weakly autonomous relative to $\bar{S}_0$, $\bar{S}_1$ and pseudo-periodic of period w relative to a neighborhood N of the origin. If $F^1(t_1; \bar{S}_1)$ at some point $t_1 \in R^+$ is contained in a compact subset of S_0, then F has a pseudo-periodic trajectory of period w. If, in addition, F is strongly pseudo-periodic of period w, then F has a strongly pseudo-periodic trajectory of period w.*

PROOF. By Theorem 1 F has a critical set $\Omega(S_0)$. Hence there exists $x^* \in S_0$ such that $x^* \in F(w; x^*)$. Now F is pseudo-periodic so

$$F^1(t; F^n(w; x^*)) \subset F(t + nw; x^*) + N.$$

Since $x^* \in F^n(w; x^*)$ for all n, it follows that

$$F^1(t; x^*) \subset F(t + nw; x^*) + N$$

for all n. Thus any trajectory with initial data x^* is a pseudo-periodic trajectory of period w. If F is strongly pseudo-periodic, then $F^1(t; x^*) \subset F(t + nw; x^*)$ for all n and any trajectory with initial data x^* is strongly pseudo-periodic of period w. Hence Theorem 3 is established.

REFERENCES

1. G. S. Jones, *The existence of critical points in generalized dynamical systems*, Seminar on Differential Equations and Dynamical Systems, Springer-Verlag, New York, 1968, pp. 7–19.

2. G. S. Jones and J. A. Yorke, *The existence and nonexistence of critical points in bounded flows*, J. Differential Equations **5** (1969).

3. I. L. Glicksberg, *A further generalization of the Kakutani fixed point theorem, with application to Nash equilibrium points*, Proc. Amer. Math. Soc. **3** (1952), 170–175.

4. A. Cellina and G. S. Jones, *Fixed point theory and set valued mappings*, University of Maryland Report, 1969.

UNIVERSITY OF MARYLAND

ACCRETIVE OPERATORS AND NONLINEAR EVOLUTION EQUATIONS IN BANACH SPACES

Tosio Kato

1. **Introduction.** In this paper we are concerned with nonlinear differential equations of the form

$$du(t)/dt + A(t)u(t) = 0 \tag{1.1}$$

and related problems. Here the unknown u takes values in a Banach space X and $\{A(t)\}$ is a family of (possibly) nonlinear *accretive* operators in X. By definition A is accretive if $|(1 + \lambda A)u - (1 + \lambda A)v| \geq |u - v|$ for each $u, v \in D(A)$ and $\lambda > 0$.

In the previous paper [**5**] (in which the term "monotonic" was used instead of "accretive"), it was shown that (1.1) has a unique solution for each initial value $u(0) = a \in D(A(0))$ if $A(t)$ is *m-accretive* and depends on t smoothly in a certain sense and if X^* is uniformly convex. By definition A is m-accretive (m-monotonic in [**5**]) if it is accretive and the range of $1 + \lambda A$ is the whole of X for each (or, equivalently, for some) $\lambda > 0$.

But this condition is not necessary even when $A(t) = A$ is independent of t, as is seen from the following simple but typical example.

Example 1.1. Let $X = R^1$ and

$$Ax = 1 \quad \text{for} \quad x > 0, \quad Ax = 0 \quad \text{for} \quad x \leq 0. \tag{1.2}$$

Clearly A is accretive, but not m-accretive. A is discontinuous at $x = 0$. The solution of (1.1) with $A(t) = A$ and $u(0) = a$ is given by

$$\begin{aligned} u(t) &= \max\,(a - t, 0) \quad \text{if} \quad a > 0, \\ u(t) &= a \quad \text{if} \quad a \leq 0. \end{aligned} \tag{1.3}$$

If $a > 0$, $u(t)$ is not differentiable at $t = a$, but such an exception may be allowed in (1.1). But $A0 = 0$ is crucial. If we changed A by setting $A0 = 1$ so that A is right-continuous at $x = 0$ instead of left-continuous, no solution of (1.1) would exist for $u(0) = a = 0$ even locally (near $t = 0$), and the solution for $a > 0$ would exist only for $t < a$.

We shall give a sufficient condition, partly more general than the one given in [**5**], in order that (1.1) be solvable globally for each $u(0) = a \in D(A(0))$. The condition is stated in terms of *multiple-valued* m-accretive operators.

Thus we are led to consider multiple-valued operators A in X, for which Au is a subset of X rather than a point of X. We shall define accretiveness and m-accretiveness for such operators, which are straightforward generalizations of the same notions for single-valued operators. (Multiple-valued operators have been considered by many authors in connection with monotone operators from X to X^*, which coincide with accretive operators when X is a Hilbert space so that $X^* = X$. In particular, multiple-valued m-monotone operators in a Hilbert space have been studied in [7] in connection with nonlinear semigroups.)

The sufficient condition for the solvability of (1.1) mentioned above is that both X and X^* are uniformly convex and $A(t)$ is the *canonical restriction* of an m-accretive (multiple-valued) operator depending on t "smoothly". Here we say A is the canonical restriction of a multiple-valued operator B if Au is the set of all $x \in Bu$ such that $|x| = \inf_{y \in Bu} |y|$. (The A in Example 1.1 is exactly such an operator, where B is given by $Bx = \{0\}$ for $x < 0$, $Bx = \{1\}$ for $x > 0$, and $B0 = [0, 1]$.) Under the stated assumptions on X and X^*, it can be shown that $A(t)$ is single-valued and the differential equation (1.1) makes sense. Even if X is not uniformly convex, the result is still true if (1.1) is replaced by

$$du(t)/dt \in -A(t)u(t), \tag{1.4}$$

in which $A(t)$ is in general multiple-valued. ("Differential equations" of the form (1.4) were considered in [7], but there only weak solutions of a certain type were obtained.)

We shall study not only the existence of solutions for (1.4) but various a priori estimates related to it. One of the interesting results is that whenever u is a solution of (1.4) in which $A(t)$ is accretive (but not necessarily m-accretive), u satisfies a stronger equation of the same form with $A(t)$ replaced by its canonical restriction (under some mild additional conditions).

If $A(t) = A$ is m-accretive (or the canonical restriction of an m-accretive operator or any operator between the two) and independent of t, setting $u(t) = U(t)u(0)$ defines a semigroup $\{U(t)\}$ of nonlinear, nonexpansive operators on $D(A)$ into itself. It can be extended by continuity to a semigroup of such operators on the closure $[D(A)]$ of $D(A)$ into itself. $\{U(t)\}$ may be said to be generated by $-A$. In this paper, however, we do not go into the definition of the generator of a given semigroup. In any case it seems to the writer that the question is open whether $\{U(t)\}$ determines A uniquely.

There is another notion which we find it convenient to introduce. We say a multiple-valued operator A is *locally m-accretive* on $S \subset X$ if each $x \in S$ has a neighborhood contained in the ranges of $1 + \varepsilon_n A$ for a sequence $\varepsilon_n \downarrow 0$. It turns out, eventually, that an operator A locally m-accretive on $[D(A)]$, for example, is m-accretive if X^* is uniformly convex. But we have been able to prove this only after solving (1.3) with $A(t) = A$.

Since the notion of local m-accretiveness is formally weaker than that of m-accretiveness, it is convenient as a criterion for m-accretiveness (when X^* is uniformly convex). In particular it is useful in perturbation problems, in which one wants to show that $A + B$ is m-accretive if A is m-accretive, B is accretive and "small" relative to A in some sense or other. The last two sections of this paper are devoted to theorems of this type.

The writer wants to thank Professors F. E. Browder, M. Crandall, and A. Pazy for stimulating conversations.

(After the manuscript of this paper was almost completed, the writer was informed of similar or related results obtained by several authors, [**2**], [**3**], [**8**]. These papers contain, among others, answers to some questions raised in §9 below and results on the differentiability of nonlinear semigroups.)

2. **Multiple-valued operators.** Throughout this paper X is a real Banach space, with the dual space X^*, and with the pairing between $x \in X$ and $f \in X^*$ denoted by (x, f). The norms in X and X^* are denoted by $|\ \ |$.

For a subset $S \subset X$, we denote by $[S]$ its closure. For $S, T \subset X$, $S + T$ is the set of all $x + y$ with $x \in S$ and $y \in T$; $S + T$ is empty if at least one of S, T is empty. λS is the set of all λx with $x \in S$, where λ is a scalar. We write $S + y$ or $y + S$ for $S + \{y\}$.

We find it convenient to use the notation

$$\|S\| = \inf_{x \in S} |x| \tag{2.1}$$

for any nonempty $S \subset X$. It is easy to see that

$$\|S + T\| \le \|S\| + \|T\|, \qquad \|\lambda S\| = |\lambda|\ \|S\|. \tag{2.2}$$

$\|S\| \le \|S + T\| + \|T\|$ is in general not true, but it is true if T consists of a single element:

$$\|S\| \le \|S + y\| + |y|. \tag{2.3}$$

Let A be multiple-valued (nonlinear) operator in X. By this we mean that A assigns a $u \in X$ a subset Au of X. Au may be empty for some u. The *domain* $D(A)$ of A is by definition the set of all $u \in X$ for which Au is not empty. We define the *range* $R(A)$ of A as the union of the Au for all $u \in X$. Thus $\|Au\|$ is defined by (2.1) for all $u \in D(A)$.

A single-valued operator A with domain and range in X is regarded as a special case of a multiple-valued operator in X. In this case Au for $u \in D(A)$ may denote the value of A in the usual sense or the set consisting of this single element, and Au is the empty set if $u \notin D(A)$.

In what follows an operator means a multiple-valued operator in X unless otherwise stated.

Let A, A' be two operators. A' is an *extension* of A, and A is a *restriction* of A', in symbols $A' \supset A$, $A \subset A'$, if $Au \subset A'u$ for each $u \in X$. Thus $D(A) \subset D(A')$.

The sum $A + B$ of two operators A, B is the operator that sends u into $Au + Bu$. Thus $D(A + B) = D(A) \cap D(B)$. For any scalar λ, λA is the operator defined by $(\lambda A)u = \lambda(Au)$. Thus $D(\lambda A) = D(A)$.

We denote by 1 the identity operator in X, and we often identify a scalar λ

with the operator $\lambda 1$. Thus $\gamma + \lambda A$ is the operator given by $(\gamma + \lambda A)u = \gamma u + \lambda A u$.

If $S \subset X$, AS is the union of the Au for all $u \in S$.

The product AB of two operators A, B is defined by $(AB)u = A(Bu)$, which is simply denoted by ABu. Thus ABu is the union of the Av for all $v \in Bu$, and $D(AB) \subset D(B)$.

3. Accretive operators.

DEFINITION 3.1. An operator A is said to be *accretive* if for each $\lambda > 0$ and u, $v \in D(A)$

$$|x - y| \geq |u - v| \quad \text{whenever} \quad x \in (1 + \lambda A)u,\ y \in (1 + \lambda A)v. \tag{3.1}$$

If A is accretive, the $(1 + \lambda A)u$ for different u are disjoint. Thus one can define for each $\lambda > 0$ a *single-valued* operator $J_\lambda = (1 + \lambda A)^{-1}$, with $D(J_\lambda) = R(1 + \lambda A)$ and $R(J_\lambda) = D(A)$; by definition

$$J_\lambda x = u \quad \text{if and only if} \quad x \in (1 + \lambda A)u = u + \lambda A u. \tag{3.2}$$

It follows from (3.1) that J_λ is *nonexpansive:*

$$|J_\lambda x - J_\lambda y| \leq |x - y|, \quad x, y \in D(J_\lambda). \tag{3.3}$$

Let F be the *duality map* of X into X^*; it is by definition the unique multiple-valued operator from X to X^* with $D(F) = X$ such that $f \in Fx$ if and only if $(x, f) = |x|^2 = |f|^2$.

LEMMA 3.2. *A is accretive if and only if for each u, $v \in D(A)$ and each $x \in Au$, $y \in Av$, there exists $f \in F(u - v)$ such that $(x - y, f) \geq 0$.*

PROOF. (3.1) can be written $|u - v + \lambda(x - y)| \geq |u - v|$ for each $\lambda > 0$, $x \in Au$, and $y \in Av$. Thus the result follows from [**5**], Lemma 1.1.

DEFINITION 3.3. We say A is *maximal accretive on* $S \subset X$ if A is accretive and if any accretive extension of A coincides on S with A.

LEMMA 3.4. *Let A be maximal accretive on $S \subset X$. Let $u \in S$, $x \in X$. If for each $v \in D(A)$ and $y \in Av$ there exists $f \in F(u - v)$ such that $(x - y, f) \geq 0$, then $u \in D(A)$ and $x \in Au$.*

PROOF. Define A' by $A'u = Au \cup \{x\}$ and $A'w = Aw$ for $w \neq u$. In view of Lemma 3.2, it follows from the assumption that A' is accretive. Hence A' coincides on S with A. Since $u \in S$, we have $A'u = Au$, hence $x \in Au$ (so that $u \in D(A)$).

LEMMA 3.5. *Let X^* be strictly convex. If A is maximal accretive on $S \subset X$, then Au is convex and closed for each $u \in D(A) \cap S$.*

PROOF. In this case F is single-valued. Let $u \in D(A) \cap S$, $x_1, x_2 \in Au$. We shall show that $x = \alpha_1 x_1 + \alpha_2 x_2 \in Au$ if α_1, $\alpha_2 \geq 0$ and $\alpha_1 + \alpha_2 = 1$. Since $(x_k - y, F(u - v)) \geq 0$, $k = 1, 2$, for each $v \in D(A)$ and $y \in Av$, we obtain $(x - y, F(u - v)) \geq 0$ by taking the weighted average with weights α_1, α_2 of those two inequalities. Since A is maximal accretive on S and $u \in S$, we obtain $x \in Au$ by Lemma 3.4.

To show that Au is closed, let $x_k \in Au$, $k = 1, 2, \ldots, x_k \to x \in X$. We have $(x_k - y, F(u - v)) \geq 0$ and going to the limit gives $(x - y, F(u - v)) \geq 0$. Hence $x \in Au$ as above.

DEFINITION 3.6. We say A is *demiclosed* if the following condition is satisfied: if $u_n \in D(A)$, $n = 1, 2, \ldots, u_n \to u \in X$ and if there are $x_n \in Au_n$ such that $x_n \rightharpoonup x$ (weak convergence), then $u \in D(A)$ and $x \in Au$. We say A is *almost demiclosed* if the above condition is satisfied except the assertion $x \in Au$.

LEMMA 3.7. *Let X^* be uniformly convex.* (a) *If A is maximal accretive on $[D(A)]$, A is demiclosed.* (b) *If A is maximal accretive on $D(A)$ and if A is almost demiclosed, then A is demiclosed.*

PROOF. In this case F is continuous (see [**5**]). Let u_n, u, x_n, x be as in Definition 3.6. For any $v \in D(A)$ and $y \in Av$, we have $(x_n - y, F(u_n - v)) \geq 0$. Since $x_n \rightharpoonup x$ and $u_n \to u$, we obtain $(x - y, F(u - v)) \geq 0$. (a) Since $u \in [D(A)]$ and A is maximal accretive on $[D(A)]$, we have $u \in D(A)$ and $x \in Au$ by Lemma 3.4. (b) We have $u \in D(A)$ because A is almost demiclosed. Then $x \in Au$ follows from Lemma 3.4 because A is maximal accretive on $D(A)$.

LEMMA 3.8. *Let X be reflexive. Let A be an operator in X, $u_n \in D(A)$, $n = 1, 2, \ldots, u_n \to u \in X$, and $\|Au_n\| \leq M$.* (a) *If A is almost demiclosed, then $u \in D(A)$.* (b) *If A is demiclosed, then $\|Au\| \leq M$ too.*

PROOF. $\|Au_n\| \leq M$ implies that there is $y_n \in Au_n$ such that $|y_n| \leq M + n^{-1}$. Since X is reflexive, there is a subsequence $\{y_{n_k}\}$ such that $y_{n_k} \rightharpoonup y \in X$, $k \to \infty$. If A is almost demiclosed, it follows that $u \in D(A)$. If A is demiclosed, we have $y \in Au$ in addition. Then $\|Au\| \leq |y| \leq \liminf |y_{n_k}| \leq M$.

DEFINITION 3.9. For each operator A, we define the *canonical restriction* A^0 of A by setting A^0u as equal to the set of all $x \in Au$ with $|x| = \|Au\|$.

LEMMA 3.10. *Let X and X^* be strictly convex and A maximal accretive on $D(A)$. Then the canonical restriction A^0 of A is single-valued. If in addition X is reflexive, then $D(A^0) = D(A)$.*

PROOF. By Lemma 3.5, Au is convex and closed for each $u \in D(A)$. Since X is strictly convex, $\|Au\| = \inf_{x \in Au} |x|$ is either attained by a unique element x or not attained. Thus $A^0u = \{x\}$ or empty, and A^0 is single-valued. If x is reflexive, the inf is always attained, hence $D(A^0) = D(A)$.

4. **Approximation of accretive operators.** Let A be an accretive operator. Let J_λ be given by (3.2), $\lambda > 0$, and set

$$A_\lambda = \lambda^{-1}(1 - J_\lambda). \tag{4.1}$$

Both J_λ and A_λ are single-valued with $D(J_\lambda) = D(A_\lambda) = R(1 + \lambda A) \equiv D_\lambda$. We recall that J_λ is nonexpansive (see (3.2)). A_λ is Lipschitz continuous too:

$$|A_\lambda x - A_\lambda y| \leq 2\lambda^{-1} |x - y|, \quad x, y \in D_\lambda. \tag{4.2}$$

LEMMA 4.1. *A_λ is accretive.*

PROOF. See [**5**, Lemma 2.3].

LEMMA 4.2. *$A_\lambda u \in AJ_\lambda u$, $\|AJ_\lambda u\| \le |A_\lambda u|$, for $u \in D_\lambda$.*

PROOF. Set $J_\lambda u = v$. Then $u = v + \lambda y$ with $y \in Av$, see (3.2). Hence $A_\lambda u = \lambda^{-1}(u - v) = y \in Av = AJ_\lambda u$, and $\|AJ_\lambda u\| \le |y| = |A_\lambda u|$.

LEMMA 4.3. *If $u \in D(A) \cap D_\lambda$, then $\|AJ_\lambda u\| \le |A_\lambda u| \le \|Au\|$.*

PROOF. If $u \in D(A)$, then $u = J_\lambda(u + \lambda x)$ for any $x \in Au$. If $u \in D_\lambda$ in addition, then

$$|A_\lambda u| = \lambda^{-1}\,|u - J_\lambda u| = \lambda^{-1}\,|J_\lambda(u + \lambda x) - J_\lambda u| \le \lambda^{-1}\,|u + \lambda x - u| = |x|.$$

Hence $|A_\lambda u| \le \inf_{x \in Au} |x| = \|Au\|$.

LEMMA 4.4. *Let $u \in D(A) \cap D_{\varepsilon_n}$ for all n, where $\varepsilon_n \downarrow 0$. Then $J_{\varepsilon_n} u \to u$.*

PROOF. $|u - J_{\varepsilon_n} u| = \varepsilon_n\,|A_{\varepsilon_n} u| \le \varepsilon_n\,\|Au\| \to 0$, by Lemma 4.3.

LEMMA 4.5. *Let X^* be uniformly convex and A almost demiclosed. Let $\varepsilon_n \downarrow 0$, $u_n \in D_{\varepsilon_n}$, $u_n \to u \in X$, and $A_{\varepsilon_n} u_n \rightharpoonup x \in X$. Then $u \in D(A)$. If A is demiclosed, then $x \in Au$* too.

PROOF. Set $v_n = J_{\varepsilon_n} u_n \in D(A)$. Then $u_n - v_n = \varepsilon_n A_{\varepsilon_n} u_n \to 0$ because $\{A_{\varepsilon_n} u_n\}$ is bounded. Hence $v_n \to u$. Since $A_{\varepsilon_n} u_n \in AJ_{\varepsilon_n} u_n = Av_n$ by Lemma 4.2 and $A_{\varepsilon_n} u_n \rightharpoonup x$, the results follow from Definition 3.6.

5. m-accretive and locally m-accretive operators.

DEFINITION 5.1. Let A be an accretive operator. We say A is *m-accretive* if $D_\lambda = X$ for every $\lambda > 0$. (It is known (see [**7**], [**10**]) that for this it suffices that $D_\lambda = X$ for some $\lambda > 0$.) We say A is *locally m-accretive* on $S \subset X$ if for each $x \in S$ there are a neighborhood U of x and a sequence $\varepsilon_n \downarrow 0$ such that $U \subset \bigcap D_{\varepsilon_n}$.

REMARK 5.2. Obviously an m-accretive operator is locally m-accretive on X. A deeper result is that, conversely, A is m-accretive if A is almost demiclosed and locally m-accretive on $D(A)$, provided X^* is uniformly convex. But we have been able to prove it only after solving the evolution equation associated with A (see Corollary 7.4). (We do not know whether it is true when X^* is not uniformly convex.) In spite of this equivalence, the notion of local m-accretiveness is useful because it is often easier to verify it than to verify the m-accretiveness directly.

LEMMA 5.3. *If A is locally m-accretive on S, A is maximal accretive on S.*

PROOF. Let $A' \supset A$ be accretive; we have to show that $A'u' = Au'$ for $u' \in S$. We may assume that $u' \in D(A')$, for otherwise both $A'u'$ and $Au' \subset A'u'$ are empty. Let $x' \in A'u'$. By hypothesis u' has a neighborhood U' contained in all D_{ε_n} for some sequence $\varepsilon_n \downarrow 0$. Now $u' + \varepsilon_n x' \in U'$ for sufficiently large n. Fix one such n. Then $J_{\varepsilon_n}(u' + \varepsilon_n x') = u \in D(A)$ is defined. The definition of

J_{ε_n} implies that $u' + \varepsilon_n x' = u + \varepsilon_n x$ for some $x \in Au \subset A'u$. Since A' is accretive, this is possible only if $u' = u$. Hence $x' = x$, $u' \in D(A)$ and $x' \in Au'$. This proves $A'u' = Au'$.

LEMMA 5.4. *Let A and A' be operators such that $A'u = \beta A(\alpha u + a) + b$, where $\alpha, \beta > 0$ and $a, b \in X$ are fixed and where $D(A') = \alpha^{-1}(D(A) - a)$. If A is demiclosed [almost demiclosed/accretive/maximal accretive on its domain/maximal accretive on the closure of its domain/m-accretive/locally m-accretive on its domain/locally m-accretive on the closure of its domain], the same is true of A', and vice versa.*

PROOF. Since A' is obtained from A by translations and dilatations in the domain and range spaces, the proof is trivial except for the cases of local m-accretiveness. To deal with these cases, we note that

$$(1 + \lambda A')u = \alpha^{-1}\{(1 + \lambda\alpha\beta A)(\alpha u + a) - a\} + \lambda b$$

and hence

$$R(1 + \lambda A') = \alpha^{-1}\{R(1 + \lambda\alpha\beta A) - a\} + \lambda b. \tag{5.1}$$

Suppose A is locally m-accretive on $D(A)$; we have to show that A' is locally m-accretive on $D(A')$. Let $x' \in D(A')$ and let x be such that $x' = \alpha^{-1}(x - a)$ so that $x \in D(A)$. There is a neighborhood U of x and a sequence $\varepsilon_n \downarrow 0$ such that $R(1 + \varepsilon_n\alpha\beta A) = D_{\varepsilon_n\alpha\beta} \supset U$. (5.1) shows that $D'_{\varepsilon_n} = R(1 + \varepsilon_n A') \supset U' + \varepsilon_n b$, where U' is a neighborhood of x'. Let U'' be a bounded neighborhood of x' with $[U''] \subset U'$. Since $\varepsilon_n \downarrow 0$, $D'_{\varepsilon_n} \supset U''$ for sufficiently large n. This proves the required result. The case of A being locally m-accretive on $[D(A)]$ is the same; it suffices to note that $[D(A')] = \alpha^{-1}([D(A)] - a)$.

LEMMA 5.5. *If A is demiclosed [almost demiclosed/m-accretive/locally m-accretive on $S \subset X$], the same is true of $A + \gamma$ for $\gamma \geq 0$.*

PROOF. Again the proof is trivial except for the case of local m-accretiveness. In this case we note that

$$(1 + \lambda(A + \gamma))u = (1 + \lambda\gamma)(1 + \lambda(1 + \lambda\gamma)^{-1}A)u$$

so that $R(1 + \lambda(A + \gamma)) = (1 + \lambda\gamma)R(1 + \lambda(1 + \lambda\gamma)^{-1}A)$. The rest of the proof is similar to that for Lemma 5.4.

LEMMA 5.6. *Let X^* be uniformly convex. Let A be locally m-accretive on $[D(A)]$. If A^0 is the canonical restriction of A, then $D(A^0) = D(A)$ and A^0u is a nonempty closed convex set for each $u \in D(A)$. All weak limits of sequences of the form $\{A_{\varepsilon_n}u\}$ with $\varepsilon_n \downarrow 0$ and $u \in D_{\varepsilon_n}$ belong to A^0u.*

PROOF. We first note that A is demiclosed by Lemmas 3.7 and 5.3. Let $u \in D(A)$. By hypothesis there is a sequence $\varepsilon_n \downarrow 0$ such that $u \in D_{\varepsilon_n}$ for all n. Since $|A_{\varepsilon_n}u| \leq \|Au\|$ by Lemma 4.3, $\{A_{\varepsilon_n}u\}$ has a subsequence with a weak limit $x \in X$ (note that X is reflexive with X^*). Thus $x \in Au$ by Lemma 4.5. Since $|x| \leq \limsup |A_{\varepsilon_n}u| \leq \|Au\|$, we have $x \in A^0u$ and A^0u is not empty. All weak limits of similar sequences $\{A_{\varepsilon_n} u\}$ belong to $A^0 u$ for the same reason. Since Au is closed convex by Lemmas 3.5 and 5.3, it is clear that $A^0 u$ has the same properties.

6. **The evolution equation and a priori estimates.** We now consider the differential equation

$$du(t)/dt \in -A(t)u(t), \tag{6.1}$$

where $A(t)$ is a (multiple-valued) operator depending on a real variable t. We may assume without loss of generality that $A(t)$ is defined for $-\infty < t < \infty$.

For any interval I of real numbers, we denote by $W_1^1(I; X)$ the set of all X-valued functions u on I such that u is an indefinite integral of a strongly integrable function v on I. Then $u'(t) = du(t)/dt$ exists and equals $v(t)$ almost everywhere (see [**11**, p. 134]), and u is absolutely continuous on the closure of I.

DEFINITION 6.1. We say u is a *strong solution* of (6.1) on an interval $I = I_u$ if u is continuous on I, if $u \in W_1^1(I'; X)$ for any finite closed interval contained in the interior of I, and if $u'(t) \in -A(t)u(t)$ for almost all $t \in I$.

LEMMA 6.2. *Let $C(t)$ be a single-valued operator, depending on $t \in (-\infty, \infty)$, with $D(C(t)) = X$, such that $|C(t)x - C(t)y| \leq \lambda(t)\,|x - y|$, where $\lambda(t)$ is a locally integrable function. Let $\gamma(t)$ be a locally integrable real function, and let $\beta(t)$ be an indefinite integral of $\gamma(t) - \lambda(t)$. Assume that $A(t) - \gamma(t) - C(t)$ is accretive for each t and that $\|A(t)x\|$ is measurable in t for each $x \in X$, where we set $\|A(t)x\| = \infty$ if $x \notin D(A(t))$. Then*

(a) *For any strong solution u of* (6.1) *we have*

$$|u(t) - u(r)| \leq \int_r^t e^{\beta(s)-\beta(t)}\,\|A(s)u(r)\|\,ds \tag{6.2}$$

for $r < t$ with $r, t \in I_u$.

(b) *If the right derivative $D^+u(r)$ exists at $r \in I_u$,*

$$|D^+u(r)| \leq \operatorname{ess\,lim\,sup}_{s\downarrow r} \|A(s)u(r)\|, \tag{6.3}$$

where "ess lim sup" *means that one may disregard a set of s of measure zero in taking the* lim sup.

(c) *If u, v are two strong solutions of* (6.1),

$$|u(t) - v(t)| \leq e^{\beta(r)-\beta(t)}\,|u(r) - v(r)| \tag{6.4}$$

for $r < t$ with $r, t \in I_u \cap I_v$. In particular $u(r) = v(r)$ implies $u(t) = v(t)$ for $t \geq r$ (forward uniqueness).

PROOF. (a) For each fixed $r \in I_u$ and for almost all $s \in I_u$ we have

$$\tfrac{1}{2}d\,|u(s) - u(r)|^2/ds = (u'(s), f(s)), \tag{6.5}$$

for any $f(s) \in F(u(s) - u(r))$; see [**5**, Lemma 1.3]. If $u(r) \in D(A(s))$, we can write

$$(u'(s), f(s)) = (u'(s) + \gamma(s)u(s) + C(s)u(s) - b(s) - \gamma(s)u(r) - C(s)u(r), f(s)) \tag{6.6}$$
$$- \gamma(s)(u(s) - u(r), f(s)) - (C(s)u(s) - C(s)u(r), f(s)) + (b(s), f(s)),$$

where $b(s)$ is an arbitrary element of $-A(s)u(r)$. The first term on the right of (6.6) is nonpositive for some $f(s)$ because $A(s) - \gamma(s) - C(s)$ is accretive, $u'(s) \in -A(s)u(s)$, and $b(s) \in -A(s)u(r)$. Using the relations

$$(u(s) - u(r), f(s)) = |u(s) - u(r)|^2,$$

$$|C(s)u(s) - C(s)u(r)| \leq \lambda(s)\, |u(s) - u(r)|,$$

and

$$|f(s)| = |u(s) - u(r)|,$$

and noting that $b(s) \in -A(s)u(r)$ is arbitrary so that $|b(s)|$ can be made arbitrarily close to $\|A(s)u(r)\|$, we obtain from (6.5) and (6.6) (cf. the proof of Lemma 4.2 of [**5**])

$$d\,|u(s) - u(r)|/ds \leq (\lambda(s) - \gamma(s))\,|u(s) - u(r)| + \|A(s)u(r)\|. \tag{6.7}$$

This is true even when $u(r) \notin D(A(s))$, for then $\|A(s)u(r)\| = \infty$ by convention. Thus (6.7) is true for almost all $s \in I_u$, and (6.2) follows on integrating (6.7).

(b) follows easily from (6.2) since $\beta(t)$ is continuous.

(c) As in (a) we have

$$\tfrac{1}{2}d\,|u(t) - v(t)|^2/dt \leq (u'(t) - v'(t), g(t))$$

almost everywhere, with any $g(t) \in F(u(t) - v(t))$. Using the facts that $A(t) - \gamma(t) - C(t)$ is accretive and that $u'(t) \in -A(t)u(t)$, $v'(t) \in -A(t)v(t)$, we obtain as above

$$d\,|u(t) - v(t)|/dt \leq (\lambda(t) - \gamma(t))\,|u(t) - v(t)|$$

almost everywhere. (6.4) then follows on integration.

LEMMA 6.3. *In Lemma 6.2 assume further that $D(A(t)) = D$ is independent of t and $\|A(t)x\|$ is continuous in t for each $x \in D$. Then $|u'(t)| = \|A(t)u(t)\|$ for almost all t. Thus any strong solution of* (6.1) *is a strong solution of the restricted equation*

$$du(t)/dt \in -A(t)^0u(t), \tag{6.8}$$

where $A(t)^0$ is the canonical restriction of $A(t)$.

PROOF. By definition $u'(r)$ exists and belongs to $-A(r)u(r)$ for almost all r. Since $\|A(s)u(r)\|$ is continuous in s, we see from (6.3) that $|u'(r)| \leq \|A(r)u(r)\|$ for almost all $r \in I_u$. Since the opposite inequality is true by $u'(r) \in -A(r)u(r)$, we have $|u'(r)| = \|A(r)u(r)\|$ and hence $u'(r) \in -A(r)^0u(r)$ almost everywhere.

REMARK 6.4. Lemma 6.3 is interesting since it shows that each strong solution u of (6.1) takes the smallest possible value $|u'(t)|$ compatible with (6.1). If in particular X and X^* are strictly convex and the $A(t) - \gamma(t) - C(t)$ are maximal accretive on D, then the $A(t)^0$ are single-valued (Lemma 3.9). In this case u must satisfy the proper differential equation

$$du(t)/dt + A(t)^0 u(t) = 0. \tag{6.9}$$

Next we shall estimate the derivatives of solutions of (6.1). For simplicity we consider a more restricted class of operators $A(t)$ than above.

LEMMA 6.5. *Let*

$$A(t)x = Ax + \gamma x - b(t), \qquad D(A(t)) = D(A), \tag{6.10}$$

where A is an accretive operator independent of t, γ is a real constant and $b \in W_1^1(I; X)$ on any finite interval I. If u is any strong solution of (6.1) *on a finite interval $I_u = [t', t'')$, then $|u'(t)|$ has an extension which is of bounded variation on any closed interval $[r, t'']$ with $t' < r < t''$ and which has no positive jumps. In particular the limits of $u(t)$ and $|u'(t)|$ exist as $t \uparrow t''$, and u is extended to a strong solution on $[t', t'']$.*

PROOF. Set $u_h(t) = u(t + h)$ for small $h > 0$. u_h is a strong solution on $I_u - h$ of a differential equation (6.1) with $b(t)$ in (6.10) replaced by $b_h(t) = b(t + h)$. A computation similar to the one given in the proof of Lemma 6.2, (c) gives $d\,|u_h(t) - u(t)|/dt \le -\gamma\,|u_h(t) - u(t)| + |b_h(t) - b(t)|$ a.e. Hence

$$|u_h(t) - u(t)| \le e^{-\gamma(t-r)}\,|u_h(r) - u(r)| + \int_r^t e^{-\gamma(t-s)}\,|b_h(s) - b(s)|\,ds \tag{6.11}$$

for sufficiently small $h > 0$, where $t' \le r < t < t''$. Dividing (6.11) by h and going to the limit $h \downarrow 0$, we obtain

$$|u'(t)| \le e^{-\gamma(t-r)}\,|u'(r)| + \int_r^t e^{-\gamma(t-s)}\,|b'(s)|\,ds \tag{6.12}$$

for $t' \le r < t < t''$, provided $u'(t)$ and $u'(r)$ exist. The difficulty in taking the limit under the integral sign may be avoided by using the inequality $|b_h(s) - b(s)| \le \int_s^{s+h} |b'(p)|\,dp$ and changing the order of integration before going to the limit.

(6.12) implies that

$$e^{\gamma t}\,|u'(t)| - \int^t e^{\gamma s}\,|b'(s)|\,ds \tag{6.13}$$

is monotone nonincreasing in t. Although $|u'(t)|$ need not be defined everywhere on I_u, it follows that it is of bounded variation on its domain of definition. Thus for each t in the interior of I_u, the limits of $|u'(s)|$ exist as $s \to t$ from above and from below, the two limits being equal except for at most countably many points t. But the jumps of $|u'(t)|$ at the exceptional t are always negative, for (6.13) is nonincreasing and the second term of (6.13) is continuous in t. Similarly $\lim |u'(t)| < \infty$ exists as $t \uparrow t''$. Thus $|u'(t)|$ may be extended to a function which is of bounded variation on $[r, t'']$ for any $r > t'$ and which has no positive jumps. This implies, in particular, that $|u'(t)|$ is bounded near t''. Hence $\lim u(t) = v$ exists as $t \uparrow t''$, and setting $u(t'') = v$ makes u a strong solution on $[t', t'']$.

LEMMA 6.6. *In Lemma* 6.5 *assume further that* X^* *is uniformly convex and* A *is demiclosed. Then* $u(t) \in D(A)$ *for all* $t \in (t', t'']$, *where* $u(t'') = \lim_{t\uparrow t''} u(t)$. $\|A(t)u(t)\|$ *is of bounded variation on* $[r, t'']$ *for any* $r > t'$, *with no positive jumps, and* $\|A(t)u(t)\| = |u'(t)|$ *almost everywhere.*

PROOF. We know already that $\|A(t)u(t)\| = |u'(t)|$ almost everywhere (Lemma 6.3) and that $|u'(t)|$ is of bounded variation (Lemma 6.5). What is important in this lemma is that $\|A(t)u(t)\|$ is defined *everywhere* for $t' < t \le t''$ and is of bounded variation.

We note that $v = \lim_{t\uparrow t''} u(t)$ exists by Lemma 6.5. Since $u'(t) \in -A(t)u(t) = -Au(t) - \gamma u(t) + b(t)$ almost everywhere, there is a sequence $t_k \uparrow t''$ such that $u'(t_k) + \gamma u(t_k) - b(t_k) \in -Au(t_k)$, $k = 1, 2, \ldots$. Since $\{|u'(t_k)|\}$ is bounded by Lemma 6.5, $\{u'(t_k)\}$ contains a subsequence with a weak limit $x \in X$. Since $u(t_k) \to v$ and $b(t_k) \to b(t'')$, and since A is demiclosed, it follows that $v \in D(A)$ with $x + \gamma v - b(t'') \in -Av$ or $x \in -A(t'')v$. Hence $\|A(t'')v\| \le |x| \le \lim_{t\uparrow t''} |u'(t)|$.

We can apply the same argument with t'' replaced by any $t \in (t', t'')$, with $t_k \uparrow t$ as well as $t_k \downarrow t$. In this way we obtain (note that $|u'(t)|$ has no positive jumps)

$$\|A(t)u(t)\| \le \lim_{s\downarrow t} |u'(s)| \le \lim_{s\uparrow t} |u'(s)|, \quad u(t) \in D(A). \tag{6.14}$$

To complete the proof, we return to (6.11). Using (6.2) to estimate $|u_h(r) - u(r)|$ on the right, we obtain

$$|u_h(t) - u(t)| \le e^{-\gamma(t-r)} \int_0^h e^{-\gamma(h-p)} \|A(r+p)u(r)\| \, dp + \int_r^t e^{-\gamma(t-s)} |b_h(s) - b(s)| \, ds, \quad t' \le r < t < t''. \tag{6.15}$$

Now we note that $\|A(t)x\|$ is continuous in t for each $x \in D(A)$. In fact, $A(t)x = Ax + \gamma x - b(t)$ is the set Ax shifted by the amount $\gamma x - b(t)$. Hence $\|A(t)x\|$, which is the distance of $A(t)x$ from the origin, is equal to the distance of the set Ax from the continuously varying point $b(t) - \gamma x$ and therefore depends on t continuously.

Since $u(r) \in D(A)$ for $r > t'$ as proved above, it follows that $\|A(r+p)u(r)\| \to \|A(r)u(r)\|$ as $p \downarrow 0$. Dividing (6.15) by $h > 0$ and going to the limit $h \downarrow 0$, we thus obtain

$$|u'(t)| \le e^{-\gamma(t-r)} \|A(r)u(r)\| + \int_r^t e^{-\gamma(t-s)} |b'(s)| \, ds, \quad t' < r < t < t'', \tag{6.16}$$

provided $u'(t)$ exists (note the remark after (6.12)).

For an arbitrary $t > r$, choose a sequence $t_k \to t$ such that $u'(t_k)$ exists. Applying (6.16) with t replaced by t_k and going to the limit $k \to \infty$, we obtain by (6.14)

$$\|A(t)u(t)\| \le e^{-\gamma(t-r)} \|A(r)u(r)\| + \int_r^t e^{-\gamma(t-s)} |b'(s)| \, ds \tag{6.17}$$

for any r, t such that $t' < r < t \leq t''$. The assertion of the lemma on $A(t)u(t)$ now follows from (6.17) exactly as for the corresponding one on $|u'(t)|$ proved in Lemma 6.5.

REMARK 6.7. If in Lemma 6.6 we further *assume* that $u(t') \in D(A)$, then (6.16) and (6.17) are obviously true even for $r = t'$. In particular it follows that $\|A(t)u(t)\|$ is of bounded variation on $[t', t'']$ with no positive jumps.

REMARK 6.8. Suppose we have two strong solutions u, v of (6.1) with $I_u = [t_0, t_1]$, $I_v = [t_1, t_2]$ and with $u(t_1) = v(t_1)$. According to Definition 6.1, the function w on $I_w = [t_0, t_2]$ obtained by combining u and v is not necessarily a strong solution, for w may not be in $W_1^1(I'; X)$ when $t_1 \in I'$. Under the assumptions of Lemma 6.6, however, w is certainly a strong solution. To see this it suffices to note that both $u'(t)$ and $v'(t)$ are bounded near $t = t_1$. For u' this follows from Lemma 6.5. For v' it follows from Remark 6.7, for $v(t_1) = u(t_1) \in D(A)$ by Lemma 6.6.

7. Existence theorems and consequences.

Throughout this section X^* is assumed to be uniformly convex. We consider the existence of solutions for (6.1).

THEOREM 7.1. *Let A be almost demiclosed and locally m-accretive on $D(A)$. Let $A(t)x = Ax + \gamma x - b(t)$, $0 \leq t < \infty$, with $D(A(t)) = D(A)$, where γ is a real constant and $b \in W_1^1(I; X)$ for each finite interval $I \subset [0, \infty)$. Then for each $a \in D(A)$, (6.1) has a unique strong solution u on $[0, \infty)$ with $u(0) = a$. u has the following additional properties.* (a) *$u(t) \in D(A)$ for all $t \geq 0$.* (b) *$\|A(t)u(t)\|$ is of bounded variation on any finite subinterval of $[0, \infty)$, with no positive jumps.* (c) *$u'(t) \in -A(t)^0u(t)$ for almost all $t \geq 0$, where $A(t)^0$ is the canonical restriction of $A(t)$.*

REMARK 7.2. 1. The assumptions on A imply that A is demiclosed. They are satisfied if A is locally m-accretive on $[D(A)]$ (see Lemmas 3.7 and 5.3). 2. (c) implies that u is actually a strong solution of (6.8). It should be noted, however, that $A(t)^0x$ is in general *not* equal to $A^0x + \gamma x - b(t)$, where A^0 is the canonical restriction of A. Thus one would need the whole of A, not merely A^0, to define $A(t)^0$.

The proof of Theorem 7.1 will be given in the following section. In this section we shall deduce some simple consequences of the theorem.

THEOREM 7.3. *Let A be almost demiclosed and locally m-accretive on $D(A)$. If $\gamma > 0$, the differential equation $du(t)/dt \in -(A + \gamma)u(t)$ has strong solutions u on $[0, \infty)$. $v = \lim_{t\to\infty} u(t)$ exists for each of these solutions and satisfies $0 \in (A + \gamma)v$.*

PROOF. The existence of solutions follows from Theorem 7.1. Since $\gamma > 0$, (6.12) with $b = 0$ shows that $|u'(t)|$ tends to zero exponentially as $t \to \infty$. Hence $v = \lim u(t)$ exists. Since $u'(t) \in -(A + \gamma)u(t)$ and $u'(t) \to 0$ at least along a sequence of values t tending to ∞ and since $A + \gamma$ is demiclosed (see Lemma 5.5 and Remark 7.2), it follows that $0 \in (A + \gamma)v$.

COROLLARY 7.4. *The following conditions are equivalent.* (i) *A is m-accretive.* (ii) *A is locally m-accretive on* $[D(A)]$. (iii) *A is almost demiclosed and locally m-accretive on $D(A)$.*

PROOF. Assume (iii) and set $A'u = Au - b$ with any fixed $b \in X$. Then A' is also almost demiclosed and locally m-accretive on $D(A')$ (see Lemma 5.4). According to Theorem 7.3, there is $v \in X$ such that $0 \in (A' + 1)v$ or $b \in (A + 1)v$. Thus $R(A + 1) = X$ and A is m-accretive. This shows that (iii) implies (i). But it is obvious that (i) implies (ii), and (ii) implies (iii) by Remark 7.2.

THEOREM 7.5. *In Theorem* 7.1 *assume further that X is uniformly convex too. Then $A(t)^0$ is single-valued with $D(A(t)^0) = D(A)$. For each $a \in D(A)$, there is a unique strong solution u on $[0, \infty)$ of the differential equation* (6.9): *$du(t)/dt + A(t)^0u(t) = 0$ with $u(0) = a$. In addition to properties* (a) *to* (c) *given in Theorem* 7.1, *u has the further properties:* (d) *$A(t)^0u(t)$ is right-continuous for all $t \in [0, \infty)$.* (e) *The right derivative $D^+u(t)$ exists for all $t \in [0, \infty)$ and equals $-A(t)^0u(t)$.* (f) *Except for countably many values of t, $du(t)/dt$ exists, equals $-A(t)^0u(t)$, and is continuous in t.*

PROOF. $A(t)^0$ is single-valued with $D(A(t)^0) = D(A(t)) = D(A)$ by Lemma 3.9. We note once for all that $|A(t)^0x| = \|A(t)x\|$ by definition. The solution u given by Theorem 7.1 clearly satisfies (6.9).

Let $t \geq 0$ and take any sequence $t_k \downarrow t$, so that $u(t_k) \to u(t)$. Since $|A(t)^0u(t_k)| = \|A(t_k)u(t_k)\|$ is bounded by Theorem 7.1, (b), there is a subsequence $\{s_k\}$ of $\{t_k\}$ such that $A(s_k)^0u(s_k) \rightharpoonup x \in X$. But $A(s_k)^0u(s_k) - \gamma u(s_k) + b(s_k) \in Au(s_k)$ and $u(s_k) \to u(t)$, $b(s_k) \to b(t)$. Since A is demiclosed, it follows that $x - \gamma u(t) + b(t) \in Au(t)$ or $x \in A(t)u(t)$. This implies that

$$(7.1) \qquad |x| \geq \|A(t)u(t)\| = |A(t)^0u(t)|.$$

On the other hand, $|x| \leq \lim |A(s_k)^0u(s_k)| \leq |A(t)^0u(t)|$; note that $|A(s)^0u(s)| = \|A(s)u(s)\|$ is of bounded variation in s and has no positive jumps by Theorem 7.1, (b). Combined with (7.1), this shows that $|x| = |A(t)^0u(t)|$ and so $x = A(t)^0u(t)$. Also $|A(s_k)^0u(s_k)| \to |x|$ and so $A(s_k)^0u(s_k) \to x$ strongly, for X is uniformly convex. We have thus proved that for any $t_k \downarrow t$, $\{A(t_k)^0u(t_k)\}$ contains a subsequence converging strongly to $A(t)^0u(t)$. This proves (d).

Since $u(t)$ is an indefinite integral of $-A(t)^0u(t)$, (e) follows immediately.

We know already that $|A(t)^0u(t)| = \|A(t)u(t)\|$ is of bounded variation and so continuous except for a countable number of points t. We shall show that $A(t)^0u(t)$ is continuous except for those points. For the proof it suffices to repeat the above argument with $t_k \uparrow t$; the continuity at t of $|A(t)^0u(t)|$ ensures that $\lim |A(s_k)^0u(s_k)| = |A(t)^0u(t)|$ and the argument goes through. Then (f) follows easily.

8. **Proof of Theorem** 7.1. We note once for all that A and the $A(t)$ are demiclosed and maximal accretive on $D(A)$; see Remark 7.2, Lemmas 5.3 to 5.5.

Since $a \in D(A)$ and A is locally m-accretive on $D(A)$, there are an open ball S with center a and radius R and a sequence $\varepsilon_n \downarrow 0$ such that $S \subset D_{\varepsilon_n}$ for all n, where $D_\lambda = D(J_\lambda) = D(A_\lambda)$, see §4. With this sequence ε_n, we consider the differential equations

$$du_n(t)/dt + (A_{\varepsilon_n} + \gamma)u_n(t) = b(t), \quad n = 1, 2, \ldots \tag{8.1}$$

where A_{ε_n} is given by (4.1) and is single-valued.

Since A_{ε_n} is Lipschitz continuous on S (see (4.2)), (8.1) has a unique continuously differentiable solution u_n on an interval $[0, T_n)$ with $u_n(0) = a$ and $u_n(t) \in S$. We may assume that $[0, T_n)$ is the maximal interval with this property.

Since A_{ε_n} is accretive (Lemma 4.1), we have by (6.12)

$$|u_n'(t)| \le e^{-\gamma t} |u_n'(0)| + \int_0^t e^{-\gamma(t-s)} |b'(s)| \, ds \le K(t), \tag{8.2}$$

$K(t)$ being continuous and independent of n; note that

$$|u_n'(0)| = |b(0) - (A_{\varepsilon_n} + \gamma)u_n(0)| \le |b(0)| + |A_{\varepsilon_n} a| + |\gamma| \, |a| \le |b(0)| + \|Aa\| + |\gamma| \, |a|$$

by Lemma 4.3. It follows that there is $\tau > 0$ such that $T_n \ge \tau$. In order words, all u_n exist on $I = [0, \tau)$. In what follows we consider the u_n only on I.

Now it can be proved that $\{u_n(t)\}$ is a uniform Cauchy sequence on I. The proof is almost the same as in [**5**] and need not be repeated (it is much simpler since the main part of $A(t)$ does not depend on t here); the only points to be noted are that the inequality corresponding to (4.10) of [**5**] is $0 \le (A_{\varepsilon_m} u_m(t) - A_{\varepsilon_n} u_n(t), Fy_{mn}(t))$ where $A_{\varepsilon_n} u_n(t) \in AJ_{\varepsilon_n} u_n(t)$, $y_{mn}(t) = J_{\varepsilon_m} u_m(t) - J_{\varepsilon_n} u_n(t)$, etc. (see Lemma 4.3), and that the appearance of the constant γ modifies (4.11) of [**5**] into $|x_{mn}(t)|^2 \le 4K' \int_0^t e^{-2\gamma(t-s)} |Fy_{mn}(s) - Fx_{mn}(s)| \, ds$.

It follows that $u(t) = \lim u_n(t)$ exists and is strongly continuous on I with $u(0) = a$.

LEMMA 8.1. *$u \in W_1^1(I; X)$, and $u_n' = du_n/dt$ converges weakly to $u' \in L^\infty(I; X)$ in any $L^p(I; X)$ with $1 < p < \infty$.*

PROOF. We have

$$-\int_0^\tau \phi(t) u_n'(t) \, dt = \int_0^\tau \phi'(t) u_n(t) \, dt \to \int_0^\tau \phi'(t) u(t) \, dt \tag{8.3}$$

for any $\phi \in C_0^1(0, \tau)$ (real smooth function with compact support in the open interval $(0, \tau)$). It follows that for each function f of the form

$$f(t) = \sum \phi_k(t) g_k, \quad \phi_k \in C_0^1(0, \tau), \quad g_k \in X^*,$$

$(u_n', f) = \int_0^\tau (u_n'(t), f(t)) \, dt$ forms a Cauchy sequence of real numbers. But $\{u_n'\}$ is a bounded sequence in $L^\infty(I; X)$ by (8.2). Since the f of the above form are dense in $L^1(I; X^*)$, $\{(u_n', f)\}$ forms a Cauchy sequence for each $f \in L^1(I; X^*)$.

In particular $\{(u_n', f)\}$ is Cauchy for each $f \in L^q(I; X^*)$, $1 < q < \infty$. But $L^q(I; X^*) = L^p(I; X)^*$ for $p^{-1} + q^{-1} = 1$ because X is reflexive (see e.g. [**4**], p. 89).

Hence $\{u'_n\}$ is weakly convergent in $L^p(I; X)$, which is reflexive by $L^p(I; X)^{**} = L^q(I; X^*)^* = L^p(I; X^{**}) = L^p(I; X)$. Consequently, $\{u'_n\}$ has a weak limit v in $L^p(I; X)$.

It is now easy to see that the left member of (8.3) tends weakly to $-\int_0^\tau \phi(t)v(t)\,dt$ in X. Hence $\int_0^\tau \phi(t)v(t)\,dt = -\int_0^\tau \phi'(t)u(t)\,dt$, from which it follows easily that $u(t) = a + \int_0^t v(s)\,ds$ for $t \in I$. Thus $u \in W_1^1(I; X)$, and $u' = v$ almost everywhere. (This shows that v is independent of the p used in its construction. Also $v \in L^\infty(I; X)$ follows from the fact that $|u(t) - u(s)| \leq$ const. $|t - s|$ for $s, t \in I$.)

Lemma 8.2. *For each $t \in I$ let $V(t)$ be the set of all weak cluster points in X of the sequence $\{u'_n(t)\}$, and let $\hat{V}(t)$ be the closed convex hull of $V(t)$. Then $u'(t) \in \hat{V}(t)$ for almost all $t \in I$.*

Proof. By Lemma 8.1, u' is the weak limit in $L^p(I; X)$ of $\{u'_n\}$. Hence u' is the strong limit of a certain sequence $\{v_k\}$ consisting of convex combinations of the u'_n with $n \geq k$ (see [**4**, p. 36]).

Since $v_k \to u'$ strongly in $L^p(I; X)$, there is a subsequence of $\{v_k\}$ converging strongly to u' pointwise for almost all $t \in I$. Changing the notation if necessary, we may assume that

$$v_k(t) \to u'(t), \quad k \to \infty, \tag{8.4}$$

strongly in X for almost every $t \in I$.

Fix any t for which (8.4) is true. Let M be any open half-space of X containing $V(t)$. Then $u'_n(t) \in M$ for sufficiently large n; otherwise there would exist a weak cluster point of the bounded sequence $\{u'_n(t)\}$ outside of M, contradicting the assumption $V(t) \subset M$. Hence $v_k(t) \in M$ for sufficiently large k, and $u'(t) = \lim v_k(t) \in [M]$. Since this is true for any half-space $M \supset V(t)$, $u'(t)$ must belong to $\hat{V}(t)$.

Lemma 8.3. *$u(t) \in D(A)$ and $\hat{V}(t) \subset -A(t)u(t)$ for all $t \in I$.*

Proof. Since $A(t)u(t)$ is convex and closed (Lemma 3.5), it suffices to show that $V(t) \subset -A(t)u(t)$; note that $V(t)$ is not empty since $\{u'_n(t)\}$ is bounded. Since each $x \in V(t)$ is the weak limit of some subsequence of $\{u'_n(t)\}$, where $u'_n(t) + \gamma u_n(t) - b(t) = -A_{\varepsilon_n} u_n(t)$, and since $u_n(t) \to u(t)$, it follows from Lemma 4.5 that $x + \gamma u(t) - b(t) \in -Au(t)$ or $x \in -A(t)u(t)$.

Lemmas 8.2 and 8.3 show that $u'(t) \in -A(t)u(t)$ for almost all $t \in I$. In view of Lemma 8.1, we have thus solved (6.1) on I. In other words, (6.1) has a local strong solution u for any initial value $u(0) = a \in D(A)$. We note that $u' \in L^\infty(I; X)$.

It is now easy to prove the existence of a global solution. In view of the local existence theorem just proved and the forward uniqueness of the solution given by Lemma 6.2, it is clear that there is a strong solution with $u(0) = a$ and with a maximal interval $[0, T)$ of existence. We have only to show that $T < \infty$ leads to a contradiction.

If $T < \infty$, $\lim_{t \uparrow T} u(t) = v$ exists and belongs to $D(A)$ by Lemma 6.6. If we apply the local existence theorem with the initial time T and initial condition

$u(T) = v$, we obtain a continuation of the solution u beyond T, contradicting the definition of T (see Remark 6.8).

We shall complete the proof of Theorem 7.1 by verifying properties (a) to (c). (a) and (b) follow directly from Lemma 6.6 and Remark 6.7. (c) follows from Lemma 6.3; note that $\|A(t)x\|$ is continuous in t for each $x \in D(A)$, as was shown in the proof of Lemma 6.6.

9. **Remarks on semigroups determined by m-accretive operators.** Let X^* be uniformly convex and A an m-accretive operator. (In view of Corollary 7.4, there is no need to use the notion of local m-accretiveness any longer!)

According to Theorem 5.1, the evolution equation

$$du(t)/dt \in -Au(t) \tag{9.1}$$

has a unique solution on $[0, \infty)$ for each initial value $u(0) = a \in D(A)$. On setting $u(t) = U(t)a$, we define a family of single-valued operators $U(t)$, $0 \leq t < \infty$, with domain $D(A)$ and ranges in $D(A)$. Lemma 6.2 and Theorem 7.1 show that $\{U(t)\}$ has the following properties. It is a semigroup: $U(t)U(s) = U(t + s)$. The $U(t)$ are nonexpansive: $|U(t)a - U(t)b| \leq |a - b|$. $U(t)a$ is strongly continuous in t. By continuity the $U(t)$ can be extended to single-valued operators, again denoted by $U(t)$, with domain $[D(A)]$ and ranges in $[D(A)]$. The properties stated above are preserved for the extended $U(t)$.

We might say that $\{U(t)\}$ is generated by $-A$. But the meaning of "generate" is somewhat vague in this case. In fact we might as well say that it is generated by $-A^0$, the canonical restriction of $-A$, for $u(t) = U(t)a$ with $a \in D(A)$ is already a solution of

$$du(t)/dt \in -A^0u(t) \tag{9.2}$$

(see Theorem 7.1).

In any case there are the following questions.

QUESTION 9.1. Does $\{U(t)\}$ determine A or A^0? More precisely, let A and B be m-accretive and let $\{U(t)\}$, $\{V(t)\}$ be the semigroups determined from A, B respectively. Suppose $U(t) = V(t)$ for all $t \geq 0$ (this implies, in particular, that $[D(A)] = [D(B)]$). Does it follow that $A = B$ or $A^0 = B^0$?

QUESTION 9.2. Does A^0 determine A? More precisely, let A, B be m-accretive. Then does $A^0 = B^0$ imply $A = B$?

10. **Criteria for m-accretiveness.** The foregoing results show that m-accretive operators are rather important, and it is useful to know when a given operator is m-accretive. We shall give several conditions of this kind in this and following sections.

In this connection the following should be remarked. We are interested in m-accretive operators mainly because the evolution equation of the form (6.1) is well-posed if the main part of $A(t)$ is m-accretive. However, it happens frequently that one can prove the m-accretiveness of A only by showing that (6.1) with $A(t) = A$ is well-posed. The proof given above for the equivalence of m-accretiveness and local m-accretiveness (Corollary 7.4) is a typical example of the situation, but we shall encounter other examples below. Such an interrelationship was already remarked by Browder [**1**].

Throughout this and the following sections we assume that X^* is uniformly convex.

THEOREM 10.1. *A single-valued hemicontinuous accretive operator A with $D(A) = X$ is m-accretive.* (*This is a slight generalization of a theorem given in* [**1**].)

PROOF. A is hemicontinuous if it is continuous from each line segment to the weak topology. It is known (see [**6**]) that under the assumptions of the theorem, A is actually demicontinuous, i.e., continuous from the strong topology to the weak topology.

We shall prove the theorem by showing that (9.1), which reduces in this case to the proper differential equation $du(t)/dt + Au(t) = 0$, is well-posed. Let $a \in X$ be arbitrary. Since A is demicontinuous, Ax is bounded, say $|Ax| \le K$, when x varies over a ball S with center a and radius R sufficiently small. In S we apply a variant of the Cauchy polygonal method to construct approximate solutions $u_n(t)$, $n = 1, 2, \ldots$, for (9.1) such that

$$du_n(t)/dt + Au_n(t - \varepsilon_n) = 0, \qquad 0 \le t < R/K, \tag{10.1}$$

where $\varepsilon_n \downarrow 0$ and we set $u_n(t) = a$ for $t < 0$ for convenience. Note that the u_n exist for $t < R/K$ and belong to S because $|u_n'(t)| = |Au_n(t - \varepsilon_n)| \le K$.

We shall show that $x_{mn}(t) = u_m(t) - u_n(t) \to 0$ uniformly for $t < R/K$. As in [**5**], we have

$$\begin{aligned} d\,|x_{mn}(t)|^2/dt &= -2(Au_m(t - \varepsilon_m) - Au_n(t - \varepsilon_n), Fx_{mn}) \\ &\le -2(Au_m(t - \varepsilon_m) - Au_n(t - \varepsilon_n), Fx_{mn}(t) - Fy_{mn}(t)), \end{aligned} \tag{10.2}$$

where $y_{mn}(t) = u_m(t - \varepsilon_m) - u_n(t - \varepsilon_n)$. Since $|Au_n(t - \varepsilon_n)| \le K$, the argument used in [**5**] is directly applicable to (10.2) to prove the desired convergence.

Thus $\lim u_n(t) = u(t)$ exists and is continuous for $t < R/K$. We note also that $u_n(t - \varepsilon_n) \to u(t)$ uniformly.

For each $f \in X^*$ we have $(u_n(t), f) = (a, f) - \int_0^t (Au_n(s - \varepsilon_n), f)\, ds$. Since $Au_n(s - \varepsilon_n) \rightharpoonup Au(s)$ boundedly for $s < R/K$ by the demicontinuity of A and by $|Au_n(s - \varepsilon_n)| \le K$, we obtain $(u(t), f) = (a, f) - \int_0^t (Au(s), f)\, ds$ and so $u(t) = a + \int_0^t Au(s)\, ds$, where $Au(t)$ is weakly continuous and hence strongly integrable. Thus u is a strong solution of (9.1) on $[0, R/K)$.

Now it is easy to show that u can be continued to a global solution as in §8. It suffices to note that A is demiclosed because it is demicontinuous with domain X.

Let $A'x = (A + 1)x - b$ with any fixed $b \in X$. A' is also single-valued, accretive and demicontinuous. Thus we can apply the above result to the differential equation (9.1) with A replaced by A'. If u is any solution of it on $[0, \infty)$, $|A'u(t)| = |u'(t)| \to 0$ exponentially as in the proof of Theorem 7.3. Thus $u(t) \to x \in X$ as $t \to \infty$, and $A'x = \text{weak} \lim A'u(t) = 0$. Hence $(A + 1)x = b$. Thus $R(A + 1) = X$ and A is m-accretive.

THEOREM 10.2. *Let A and B be (multiple-valued and) m-accretive. Let B be locally A-bounded with A-bound < 1 in the following sense: $D(A) \subset D(B)$, and for each $x \in X$ there are a neighborhood U of x and constants $L < 1$ and K such that*

$$\|Bu\| \leq K + L\,\|Au\| \quad \text{for } u \in D(A) \cap U. \tag{10.3}$$

Then $A + B$ is m-accretive. (This is a slight generalization of a result due to Crandall and Pazy, Cf. [**2**].)

PROOF. I. Define B_ε as we defined A_ε, namely $B_\varepsilon = \varepsilon^{-1}(1 - (1 + \varepsilon B)^{-1})$, see (4.1). For $\varepsilon > 0$, B_ε is single-valued and accretive and has domain X because B is m-accretive.

First we shall show that $A + B_\varepsilon$ is m-accretive for any $\varepsilon > 0$. Since it is obviously accretive, it suffices to show that for some $\lambda > 0$, the equation

$$b \in u + \lambda(A + B_\varepsilon)u \tag{10.4}$$

has a solution u for each $b \in X$. (10.4) is equivalent to $b - \lambda B_\varepsilon u \in u + \lambda A u$, which in turn reduces to

$$u = J_\lambda(b - \lambda B_\varepsilon u) \tag{10.5}$$

where J_λ is given by (3.2); note that $D(J_\lambda) = X$ because A is m-accretive. But (10.5) has a solution u for any b if $2\lambda < \varepsilon$, for the single-valued function of u on the right of (10.5) is a contraction map of X into itself:

$$\begin{aligned} |J_\lambda(b - \lambda B_\varepsilon u) - J_\lambda(b - \lambda B_\varepsilon v)| &\leq |(b - \lambda B_\varepsilon u) - (b - \lambda B_\varepsilon v)| \\ &= \lambda\,|B_\varepsilon u - B_\varepsilon v| \leq 2\lambda\varepsilon^{-1}\,|u - v| \end{aligned}$$

(see (3.3) and (4.2)).

II. Since $A + B$ is accretive, the theorem will be proved if we show that $b \in (1 + A + B)v$ has a solution v for each $b \in X$. To this end we consider the solutions v_n of the approximate equations

$$b \in (1 + A + B_{\varepsilon_n})v_n, \quad n = 1, 2, \ldots, \tag{10.6}$$

where $\varepsilon_n \downarrow 0$; the v_n exist because $A + B_{\varepsilon_n}$ are m-accretive as shown above.

We shall prove below that $\lim v_n = v$ exists. Anticipating this, we shall show that v is the required solution.

Let U be a neighborhood of v for which (10.3) is true. We note that (10.3) is true even if A is replaced by $1 + A$ if K is changed slightly, for $\|Au\| \leq \|(1 + A)\,u\| + |u|$ (see (2.3)). Since $|B_\varepsilon u| \leq \|Bu\|$ for $u \in D(B)$ by Lemma 4.3, we have

$$|B_\varepsilon u| \leq K + L\,\|(1 + A)u\|, \quad u \in D(A) \cap U, \quad \varepsilon > 0, \tag{10.7}$$

with the modified K.

Since $v_n \to v$, (10.7) is true for $u = v_n$ for sufficiently large n. Since (10.6) implies $|b| \geq \|(1 + A + B_{\varepsilon_n})v_n\|$, we have by (2.3)

$$\begin{aligned} \|(1 + A)v_n\| &\leq \|(1 + A + B_{\varepsilon_n})v_n\| + |B_{\varepsilon_n}v_n| \\ &\leq |b| + K + L\,\|(1 + A)v_n\|. \end{aligned}$$

Hence

$$(10.8)\qquad \begin{aligned} \|(1+A)v_n\| &\le (1-L)^{-1}(|b|+K),\\ |B_{\varepsilon_n}v_n| \le K + L\,\|(1+A)v_n\| &\le (1-L)^{-1}(L\,|b|+K), \end{aligned}$$

and these expressions are uniformly bounded. By replacing $\{v_n\}$ by a suitable subsequence if necessary, we may assume that $B_{\varepsilon_n}v_n \rightharpoonup y \in X$. Then $v \in D(B)$ and $y \in Bv$ by Lemma 4.5. But $b - B_{\varepsilon_n}v_n \in (1+A)v_n$ by (10.6). Since $b - B_{\varepsilon_n}v_n \rightharpoonup b - y$ and $v_n \to v$ and since $1 + A$ is demiclosed, we have $v \in D(A)$ and $b - y \in (1+A)v$. Hence $b = b - y + y \in (1+A)v + Bv = (1+A+B)v$, as we wished to show.

III. We shall now give the proof that $\lim v_n$ exists. We do this somewhat indirectly. According to Theorem 7.3, $v_n = \lim u_n(t)$ as $t \to \infty$ where u_n is the solution of

$$(10.9)\qquad du_n(t)/dt \in -(1+A+B_{\varepsilon_n})u_n(t) + b, \quad u_n(0) = a,$$

with $a \in D(A)$ arbitrary but fixed. The existence of $\lim v_n = \lim u_n(\infty)$ will be proved by "continuous induction" by establishing the following propositions.

(i) If $s < \infty$ and $\lim_n u_n(s)$ exists, there is $\delta > 0$ such that $\lim_n u_n(t)$ exists for $s \le t < s + \delta$.

(ii) If $s_k \uparrow s \le \infty$ and $\lim_n u_n(s_k)$ exists for all k, then $\lim_n u_n(s)$ exists.

Since $\lim u_n(0) = a$ exists, it is clear that (i) and (ii) imply the existence of $\lim u_n(\infty)$.

The proof of (ii) is simple and uses the fact that the u_n are equicontinuous on the compactified interval $[0, \infty]$. In fact we have $|u_n'(t)| \le e^{-t}\,\|(1+A+B_{\varepsilon_n})a - b\|$ by (6.16), in which

$$\|(1+A+B_{\varepsilon_n})a - b\| \le \|(1+A)a - b\| + |B_{\varepsilon_n}a| \le \|(1+A)a - b\| + \|Ba\|.$$

IV. It remains to prove (i). Let $\lim u_n(s) = z$. There is a neighborhood U of z for which (10.7) is true. Since the u_n are equicontinuous, there is $\delta > 0$ such that $u_n(t) \in U$ for $s \le t < s + \delta$ for almost all n. Thus

$$(10.10)\qquad |B_{\varepsilon_n}u_n(t)| \le K + L\,\|(1+A)u_n(t)\|, \quad L < 1,$$

for $s \le t < s + \delta$ and almost all n. On the other hand, (6.17) gives

$$\|(1+A+B_{\varepsilon_n})u_n(t) - b\| \le e^{-t}\,\|(1+A+B_{\varepsilon_n})a - b\| \le \|(1+A)a - b\| + \|Ba\|$$

as in III. Hence $\|(1+A+B_{\varepsilon_n})u_n(t)\|$ is uniformly bounded. In view of (10.10), the argument used in deducing (10.8) then shows that $\|(1+A)u_n(t)\|$ and $|B_{\varepsilon_n}u_n(t)|$ are bounded uniformly in such t and n.

Consider now $x_{mn}(t) = u_m(t) - u_n(t)$. We have

$$\begin{aligned} d\,|x_{mn}(t)|^2/dt &= 2(u_m'(t) - u_n'(t), Fx_{mn}(t))\\ &\le -2(B_{\varepsilon_m}u_m(t) - B_{\varepsilon_n}u_n(t), Fx_{mn}(t)) \end{aligned}$$

almost everywhere, for $u_n'(t) + B_{\varepsilon_n}u_n(t) - b \in -(1+A)u_n(t)$ etc. and $1 + A$ is accretive. The argument used in [**5**] then shows that $|x_{mn}(t)|^2 - |x_{mn}(s)|^2 \to 0$ for

$t \leq s < t + \delta$ as $m, n \to \infty$; here it is essential that the $B_{\varepsilon_n} u_n(t)$ are uniformly bounded. Since $x_{mn}(s) \to 0$ by hypothesis, we have $x_{mn}(t) \to 0$ and $\lim u_n(t)$ exists for $s \leq t < t + \delta$.

This completes the proof of Theorem 10.2.

COROLLARY 10.3. *Let A be m-accretive and let B be single-valued, accretive, and hemicontinuous with $D(B) = X$. Then $A + B$ is m-accretive* (*cf.* [**1**]).

PROOF. B is m-accretive by Theorem 10.1. Thus the result follows from Theorem 10.2.

11. **Criteria for m-accretiveness** (continued).

THEOREM 11.1. *Let A be m-accretive and B single-valued and accretive (but not necessarily m-accretive). Assume that B is locally A-bounded with A-bound < 1 (as in Theorem* 10.2). *Furthermore, assume that for each $x \in D(A)$ there are a neighborhood U of x and constants $L < 1$, K such that*

$$(11.1) \quad |Bu - Bv| \leq K\,|u - v| + L\,\|Au - Av\| \quad \text{for } u, v \in D(A) \cap U.$$

Then $A + B$ is m-accretive. (*This is a generalization of a result in* [**9**].)

PROOF. According to Corollary 7.4, it suffices to prove the following two propositions.

(i) $A + B$ is almost demiclosed.

(ii) $A + B$ is locally m-accretive on $D(A + B) = D(A)$.

The proof of (i) is simple. Suppose $v_n \in D(A)$, $x_n \in (A + B)v_n$, $v_n \to v \in X$ and $x_n \rightharpoonup x \in X$; we have to show that $v \in D(A)$. By hypothesis there is a neighborhood U of v for which (10.3) is true. Since $v_n \to v$, $|Bv_n| \leq K + L\,\|Av_n\|$ for sufficiently large n. Since $\|Av_n\| \leq \|(A + B)v_n\| + |Bv_n| \leq \|(A + B)v_n\| + K + L\,\|Av_n\|$ and since $\|(A + B)v_n\| \leq |x_n|$ is bounded in n, it follows by $L < 1$ that $\|Av_n\|$ is bounded. Thus there are $y_n \in Av_n$ with $y_n \rightharpoonup y \in X$ along a subsequence of $\{n\}$. Since A is m-accretive and hence demiclosed (Lemmas 3.7 and 5.3), it follows that $v \in D(A)$.

The proof of (ii) is more complicated. Let $a \in D(A)$. It suffices to show that a has a neighborhood contained in $R(1 + \varepsilon(A + B))$ for all sufficiently small $\varepsilon > 0$. Thus we shall show that there are $\delta, \rho > 0$ such that if $0 < \varepsilon \leq \delta$ and $|b - a| \leq \rho$, the equation

$$(11.2) \qquad b \in z + \varepsilon(A + B)z, \quad \text{i.e. } b - \varepsilon Bz \in z + \varepsilon Az$$

has a solution $z \in D(A)$.

(11.2) is equivalent to $z = J_\varepsilon(b - \varepsilon Bz)$. Setting $y = b - \varepsilon Bz$, we have $z = J_\varepsilon y$ and hence

$$(11.3) \qquad y = b - \varepsilon BJ_\varepsilon y \equiv Gy.$$

It is easy to see that, conversely, (11.3) implies (11.2). Thus it suffices to solve (11.3) for y. To this end we shall apply the fixed point theorem for a contraction map. We note that G is a single-valued operator with $D(G) = X$, for $D(J_\varepsilon) = X$ by the m-accretiveness of A and $R(J_\varepsilon) = D(A) \subset D(B)$.

Let us denote by S_r the closed ball with center a and radius r. By hypothesis there is $r > 0$ such that (11.1) is true with $U = S_{2r}$. For the moment we assume that $L < \frac{1}{2}$.

LEMMA 11.2. *If* $\varepsilon \leq r/\|Aa\|$, $x \in S_r$ *implies* $J_\varepsilon x \in S_{2r}$.

PROOF. $|J_\varepsilon x - a| \leq |J_\varepsilon x - J_\varepsilon a| + |a - J_\varepsilon a| \leq |x - a| + \varepsilon \|Aa\| \leq r + r \leq 2r$ (see Lemma 4.3).

LEMMA 11.3. *Let* $2L < \sigma < 1$. *If* $\varepsilon \leq \min\{r/\|Aa\|, (\sigma - 2L)/K\}$, *then* $|Gx - Gy| \leq \sigma |x - y|$ *for* $x, y \in S_r$.

PROOF. If $x, y \in S_r$, then $J_\varepsilon x, J_\varepsilon y \in S_{2r}$ by Lemma 11.2, so that

$$|Gx - Gy| = \varepsilon |BJ_\varepsilon x - BJ_\varepsilon y| \leq \varepsilon K |J_\varepsilon x - J_\varepsilon y| + \varepsilon L \|AJ_\varepsilon x - AJ_\varepsilon y\|$$

by (11.1). But $|J_\varepsilon x - J_\varepsilon y| \leq |x - y|$ and $A_\varepsilon x \in AJ_\varepsilon x$, $A_\varepsilon y \in AJ_\varepsilon y$ so that $\|AJ_\varepsilon x - AJ_\varepsilon y\| \leq |A_\varepsilon x - A_\varepsilon y| \leq 2\varepsilon^{-1} |x - y|$ (see Lemma 4.2 and (4.2)). This leads immediately to the desired result.

LEMMA 11.4. *Let* $\rho = (1 - \sigma)r/2$, $\delta = \min\{\rho(\|Aa\| + |Ba|)^{-1}, (\sigma - 2L)/K\}$. *If* $|b - a| \leq \rho$ *and* $\varepsilon \leq \delta$, *then* G *maps* S_r *into itself.*

PROOF. If $\varepsilon \leq \delta$, ε satisfies the assumptions of Lemmas 11.2 and 11.3. Let $x \in S_r$. We have by Lemma 11.3

$$|Gx - a| \leq |Gx - Ga| + |Ga - a| \leq \sigma |x - a| + |b - a| + \varepsilon |BJ_\varepsilon a|.$$

But $|BJ_\varepsilon a| \leq |Ba| + |BJ_\varepsilon a - Ba| \leq |Ba| + K |J_\varepsilon a - a| + L \|AJ_\varepsilon a - Aa\| \leq |Ba| + (\varepsilon K + 2L) \|Aa\| \leq |Ba| + \sigma \|Aa\| \leq |Ba| + \|Aa\|$; note that $\|AJ_\varepsilon a\| \leq \|Aa\|$ by Lemma 4.3. Hence

$$|Gx - a| \leq \sigma r + |b - a| + \varepsilon(|Ba| + \|Aa\|) \leq \sigma r + \rho + \rho \leq r.$$

Lemmas 11.3 and 11.4 show that G has a fixed point in S_r under the assumptions that $|b - a| \leq \rho$ and $\varepsilon \leq \delta$. This proves the theorem under the restriction that $L < \frac{1}{2}$ in (11.1).

To remove the restriction we may use the continuity argument. Consider the family of operators $A + tB$ with $0 \leq t \leq 1$. We have proved that $A + tB$ is m-accretive for $t \leq \frac{1}{2}$. On the other hand, (11.1) implies

$$|Bu - Bv| \leq K |u - v| + L(\|(A + tB)u - (A + tB)v\| + t |Bu - Bv|),$$

$$|Bu - Bv| \leq (1 - tL)^{-1}\{K |u - v| + L \|(A + tB)u - (A + tB)v\|\}.$$

If $A + tB$ is known to be m-accretive, application of the above result shows that $A + t'B = (A + tB) + (t' - t)B$ is m-accretive if $(t' - t)(1 - t)^{-1} \leq \frac{1}{2}$. In this way we see that $A + tB$ is accretive for all $t < 1$.

We can even go to $t = 1$. To this end it suffices to prove (ii) for $t = 1$. Let $a \in D(A)$ be fixed and again choose r so that (11.1) is true with $U = S_{2r}$. We repeat the above argument for $A + B = (A + tB) + (1 - t)B$ with $1 - t$ sufficiently small. Here we must replace L by $(1 - t)(1 - tL)^{-1}$, which should not be larger than $\frac{1}{2}$ if the previous argument should go through. This condition is satisfied if $1 - t \leq (1 - L)/2$. In this way we see that (11.2) is solvable for sufficiently small $|b - a|$ and $\varepsilon > 0$, completing the proof of (ii) and hence of the theorem.

THEOREM 11.2. *Let A be m-accretive. Let B be single-valued and accretive with $D(B) \supset D(A)$. Assume that for each $x \in D(A)$ and $M > 0$, there exist a neighborhood U of x and $K > 0$ such that if $u, v \in D(A) \cap U$ and $\|Au\| \leq M$, $\|Av\| \leq M$, then*

$$|Bu - Bv| \leq K\,|u - v|. \tag{11.4}$$

Assume, furthermore, that $A + B$ is demiclosed. Then $A + B$ is m-accretive.

PROOF. I. Let $a \in D(A)$ and choose an $M > \|Aa\| + 2\,|Ba|$. Choose a neighborhood U of a and $K > 0$ as stated in the assumption, with $x = a$. We may assume that U contains a closed ball with center a and radius r. Let $0 < T \leq r/M$, and define the set Σ of X-valued functions $t \to v(t)$ on $[0, T]$ with the following properties.

$$v(0) = a, \quad v(t) \in D(A), \qquad \|Av(t)\| \leq M, \quad t \in [0, T]. \tag{11.5}$$

$$|v(t) - v(s)| \leq M\,|t - s|, \quad s, t \in [0, T]. \tag{11.6}$$

We note that $v(t) \in U$ since $|v(t) - a| \leq Mt \leq MT \leq r$.

For each given $v \in \Sigma$, we shall solve the evolution equation

$$du(t)/dt \in -A(t)u(t), \quad 0 \leq t \leq T, \tag{11.7}$$

where

$$A(t)u = Au - b(t), \qquad b(t) = -Bv(t). \tag{11.8}$$

Since $v(t) \in D(A) \cap U$ and $\|Av(t)\| \leq M$, we have

$$|b(t) - b(s)| \leq K\,|v(t) - v(s)| \leq KM\,|t - s|. \tag{11.9}$$

Thus $b \in W_1^1([0, T]; X)$ with $|b'(t)| \leq KM$ almost everywhere (see e.g. [7]). Since A is m-accretive, (11.7) has a unique solution u on $[0, T]$ such that $u(0) = a$ and $u(t) \in D(A)$ for all $t \in [0, T]$ (Theorem 7.1 and Lemma 6.6). We shall show that $u \in \Sigma$ if T is chosen sufficiently small.

We have by (6.16) (see Remark 6.7)

$$|u'(t)| \leq \|A(0)u(0)\| + \int_0^t |b'(s)|\,ds \leq \|A(0)a\| + KMT \qquad \text{a.e.}$$

and similarly by (6.17)

$$\|A(t)u(t)\| \leq \|A(0)a\| + KMT, \quad t \in [0, T].$$

But $\|A(0)a\| = \|Aa - b(0)\| \leq \|Aa\| + |Ba|$, so that

$$|u'(t)| \leq \|Aa\| + |Ba| + KMT \quad \text{a.e.},$$

$$\begin{aligned} \|Au(t)\| &\leq \|A(t)u(t)\| + |b(t)| \\ &\leq \|A(0)a\| + KMT + |b(t) - b(0)| + |b(0)| \\ &\leq \|Aa\| + 2\,|Ba| + 2KMT, \quad t \in [0, T]. \end{aligned}$$

Since $|u'(t)| \leq L$ a.e. implies that $|u(t) - u(s)| \leq L\,|s - t|$, it follows that $u \in \Sigma$ if T is so small that

$$\|Aa\| + 2\,|Ba| + 2KMT \leq M, \tag{11.10}$$

which is possible since $\|Aa\| + 2\,|Ba| < M$.

II. On setting $u = \Phi v$, we have thus defined a map Φ of Σ into itself.

We shall now show that Φ is a strict contraction if we make Σ into a metric space with the distance function

$$d(v_1, v_2) = \sup_{t \in [0,T]} |v_1(t) - v_2(t)|.$$

In fact, let $u_k = \Phi v_k$, $k = 1, 2$. It is easy to show that

$$|u_1(t) - u_2(t)| \leq \int_0^t |b_1(s) - b_2(s)|\, ds$$

where $b_k(t) = -Bv_k(t)$; the proof is similar to the proof of (6.11). Since

$$|b_1(s) - b_2(s)| \leq K\,|v_1(s) - v_2(s)| \leq Kd(v_1, v_2),$$

it follows that $d(u_1, u_2) \leq KTd(v_1, v_2)$, where $KT < 1$ by (11.10).

Finally we note that Σ is a complete metric space. The only nontrivial part in this proof is that if $v_n(t) \in D(A)$ and $v_n(t) \to v(t)$ with $\|Av_n(t)\| \leq M$, then $v(t) \in D(A)$ with $\|Av(t)\| \leq M$. But this is a direct consequence of Lemma 3.8.

Thus the fixed point theorem for a contraction map is applicable to Φ on Σ, with the result that there exists $u \in \Sigma$ with $\Phi u = u$. This means that u is a strong solution on $[0, T]$ of the evolution equation

$$du(t)/dt \in -(A + B)u(t). \tag{11.11}$$

Thus we have proved that (11.11) has a local strong solution u with u' bounded, for any initial value $u(0) = a \in D(A) = D(A + B)$. Now it can be proved that u can be continued to a global solution on $[0, \infty)$, by the argument given at the end of §8. In this proof it is essential that if u is a strong solution on $[0, T')$ then $u(T') = \lim_{t \uparrow T'} u(t) \in D(A + B)$; this follows from Lemmas 6.5 and 3.8 in virtue of the assumption that $A + B$ is demiclosed.

III. All the results deduced above are true if A is replaced by A' given by $A'u = (1 + A)u - c$ with $c \in X$. In particular, for each $c \in X$

$$du(t)/dt \in -(1 + A + B)u(t) + c \tag{11.12}$$

has global solutions. If $u(t)$ is one of them, $u'(t) \to 0$ and $u(t) \to v \in X$ as $t \to \infty$ (see the proof of Theorem 7.3). Then $u'(t) + u(t) - c \in -(A + B)u(t)$ and $u'(t) + u(t) - c \to v - c$. Since $A + B$ is demiclosed, it follows that $v \in D(A + B)$ and $v - c \in -(A + B)v$ or $c \in (1 + A + B)v$. Since $c \in X$ was arbitrary, we have $R(1 + A + B) = X$ and $A + B$ is m-accretive.

Bibliography

1. F. E. Browder, *Nonlinear equations of evolution and nonlinear accretive operators in Banach spaces*, Bull. Amer. Math. Soc. **73** (1967), 867–874.

2. M. G. Crandall and A. Pazy, *Nonlinear semigroups of contractions and dissipative sets*, (to appear).

3. J. R. Dorroh, *A nonlinear Hille-Yosida-Phillips theorem*, J. Functional Anal. (to appear).

4. E. Hille and R. S. Phillips, *Functional analysis and semigroups*, Amer. Math. Soc. Colloq. Publ., vol. 31, 1957, Amer. Math. Soc., Providence, R.I.

5. T. Kato, *Nonlinear semigroups and evolution equations*, J. Math. Soc. Japan **19** (1967), 508–520.

6. ———, *Demicontinuity, hemicontinuity, and monotonicity*. II, Bull. Amer. Math. Soc. **73** (1967), 886–889.

7. Y. Kōmura, *Nonlinear semigroups in Hilbert space*, J. Math. Soc. Japan **19** (1967), 493–507.

8. ———, *Differentiability of nonlinear semigroups*, J. Math. Soc. Japan (to appear).

9. J. Mermin, *Accretive operators and nonlinear semigroups*, Thesis, University of California, Berkeley, 1968.

10. S. Oharu, *Note on the representation of semigroups of nonlinear operators*, Proc. Japan Acad. **42** (1967), 1149–1154.

11. K. Yosida, *Functional analysis*, Springer-Verlag, Berlin, 1965.

University of California, Berkeley

FIXED POINT THEOREMS FOR NONEXPANSIVE MAPPINGS

William A. Kirk

Let K be a subset of a Banach space X. A mapping T of K into itself is called *nonexpansive* if $\|Tx - Ty\| \leq \|x - y\|$ for all x, y in K. Here we discuss the problem of determining conditions on K and on T (T is not assumed to be linear) such that T will have a fixed point in K.

The first fixed point theorems of a general type for nonlinear nonexpansive mappings in noncompact settings were those obtained independently by Browder [**8**] and by Göhde [**16**] who proved that a nonexpansive self-mapping of a bounded closed convex subset K of a uniformly convex Banach space X has a fixed point, and by the author [**18**] who obtained the same result under the slightly weaker assumptions that X be reflexive and K be a bounded closed convex subset of X which has "normal structure." Other recent work includes papers of Browder (e.g. [**7**], [**9**], and [**11**]) treating the relationship of nonexpansive mappings to the theory of monotone operators in Hilbert space and, in more general spaces, to the theory of J-monotone and accretive mappings. Theorems for calculating or approximating fixed points using iterative techniques for general nonexpansive mappings are given by Browder and Petryshyn [**13**] and weak convergence of successive approximations is discussed by Opial [**22**]. We make no attempt to review all of the results on fixed point theorems for nonexpansive mappings here, but we do call attention to the applications of this theory to the existence of periodic solutions of nonlinear equations of evolution by Browder [**6**], [**12**], and to recent comprehensive bibliographies on this subject, e.g. [**10**] and [**14**].

In §1 of this paper we discuss conditions on K sufficient to yield fixed point theorems for nonexpansive mappings. The emphasis in §2 is on conditions on the mapping T, and the main result of that section is a fixed point theorem for a certain class of nonexpansive mappings which holds in any weakly compact set.

1. In considering conditions on K under which mappings $T: K \to K$ which are nonexpansive always have a fixed point, much of our previous work has dealt with the concept of normal structure and with certain strengthenings or weakenings of that concept. In this section we give a brief review of results along this line.

A bounded convex set K in a Banach space X has *normal structure* (Brodskii-Milman [**5**]) if for every convex subset H of K which contains more than one point there is a point x in H such that x is not a diametral point of H, that is, for which

$$\sup \{\|x - y\| : y \in H\} < \delta(H) = \sup \{\|u - v\| : u, v \in H\}.$$

This is a property which all bounded convex subsets of uniformly convex spaces possess, as do all compact convex sets.

Our original theorem [**18**] may be stated as follows:

THEOREM 1. *Let K be a nonempty, weakly compact, convex subset of a Banach space X, and suppose K has normal structure. Then every nonexpansive mapping $T:K \to K$ has a fixed point.*

In [**1**] Belluce and the author extend this theorem by showing that any finite family of commuting nonexpansive self-mappings of such a set K always has a common fixed point. If the norm of X is strictly convex this result holds for infinite families, but our attempts to prove a common fixed point theorem for arbitrary families *without* assuming strict convexity of the norm resulted in the need for a strengthening of normal structure called "complete normal structure" (see [**2**]).

Efforts to generalize Theorem 1 by weakening the assumption of normal structure have been unsuccessful, although an apparent slight weakening does yield a theorem for strictly contractive mappings ($T:K \to K$ is *strictly contractive* if $\|Tx - Ty\| < \|x - y\|$ for all x, y in K, $x \neq y$).

DEFINITION [**20**]. A bounded convex subset K of X is said to have *normal structure relative to F*, $F \subseteq X$, if for each bounded convex subset H of K which contains more than one point, there is a point x in F such that

$$\sup \{\|x - y\| : y \in H\} < \delta(H).$$

THEOREM 2. *Let K be a nonempty, weakly compact, convex subset of X and suppose K has normal structure relative to K. If $T:K \to K$ is strictly contractive, then T has a fixed point in K.*

The investigations discussed above have given rise to many questions of geometric type in Banach spaces. For example, it is not clear whether wider application might result from relative normal structure, nor do we know of an example of a convex set which possesses normal structure but not complete normal structure. A more thorough study of the concept of normal structure has been initiated in Belluce-Kirk-Steiner [**4**], where examples of noncompact convex subsets of nonstrictly convex spaces which possess normal structure are obtained. Spaces shown by M. M. Day to be strictly convex, reflexive, and not isomorphic to any uniformly convex space, are shown in [**4**] to have the property that each of their bounded convex subsets has normal structure, and in addition an example of R. C. James is included which shows that normal structure is not implied by reflexivity.

An immediate consequence of Theorem 3 of [**3**] is the following generalization of Theorem 1.

THEOREM 3. *Suppose $K \subset X$ is nonempty weakly compact and convex, and let $T:K \to K$ be nonexpansive. If for each x in K it is the case that* conv $\{x, Tx, T^2x, \ldots\}$ *has normal structure, then T has a fixed point in K.*

Thus with K as in Theorem 3 and T nonexpansive, if it is the case that the sequence $\{T^n x\}$ is precompact for each x in K, then T will have a fixed point in K. In particular, if $\{T^n x\}$ is *finite*—that is, if T is pointwise periodic, then T will have a fixed point in K. This latter fact is, of course, trivial for affine T since

$$1/n(x + Tx + \cdots + T^{n-1}x)$$

is a point of K which is fixed under T if $T^n x = x$ and T is affine. However, nonexpansive mappings $T: K \to K$ exist which have periodic points, yet are *not* affine. Also if T is *not* nonexpansive then periodicity of T need not imply the existence of a fixed point, even if T is a homeomorphism of the unit ball of Hilbert space into itself (Klee [**21**]).

2. An alternate approach is offered in this section in that here we discuss some fixed point theorems for nonexpansive T with no normal structure or uniform convexity assumption on the space. The theorems given here will apply in any bounded closed convex subset of a reflexive Banach space.

We begin with an inspection of the classical Contraction Mapping Principle for complete metric spaces. Suppose M is a metric space and suppose $f: M \to M$ is a strict contraction with contractive constant $\alpha < 1$. (Thus $d(f(x), f(y)) \leq \alpha\, d(x, y)$.) Let

$$Q(x) = \{x, f(x), f^2(x), \ldots\}.$$

For $A \subset M$ let $\delta(A) = \sup\{d(x, y): x, y \in A\}$, and define $\rho(f^n(x)) = \delta(Q(f^n(x))$. It follows that $\rho(f^n(x)) \leq \alpha^n \rho(x)$. The closures of the sets $Q(f^n(x))$ thus form a descending sequence with the property that

$$\lim_{n \to \infty} \delta(Q(f^n(x)) = \lim_{n \to \infty} \rho(f^n(x)) = 0.$$

If M is complete there is exactly one point in $\bigcap_{n=1}^{\infty} \overline{Q(f^n(x))}$, which is fixed under f.

For an arbitrary mapping $f: M \to M$ the sequence $\rho(f^n(x))$ is nonincreasing and has limit $r(x) \geq 0$. We call $r(x)$ (which may be infinite) the *limiting orbital diameter of f at x*. If for each $x \in M$ it is the case that $\rho(x) < \infty$ and $r(x) < \rho(x)$ when $\rho(x) > 0$, then f is said to have *diminishing orbital diameters* on M.

The following theorem illustrates the force of this condition in compact settings. In this theorem (which appears in [**19**]) the mapping f need not be nonexpansive; it need only satisfy:

(i) there is a constant C such that for each positive integer k and for each x, y in M, $d(f^k(x), f^k(y)) \leq C\, d(x, y)$.

THEOREM 4. *Let f be a mapping of a metric space M into itself which satisfies* (i) *and which has diminishing orbital diameters. If for some $x \in M$ a subsequence of the sequence $\{f^n(x)\}$ has limit z, then $\lim_{n\to\infty} f^n(x) = z$ and $f(z) = z$.*

All of the remaining theorems discussed here are for nonexpansive mappings T in a Banach space X. Such a mapping T has diminishing orbital diameter at

$x \in X$ if, for example, here exists a number $\alpha(x)$, $0 \leq \alpha(x) < 1$, such that for some integer N (depending on x),

$$\|T^N x - T^{N+i}x\| \leq \alpha(x) \sup_j \|x - T^j x\|, \quad i = 1, 2, \ldots .$$

An interesting feature of the next theorem is that convexity of the set K is not required. This theorem generalizes a corollary in Belluce-Kirk [**3**], where it is assumed K is convex.

THEOREM 5. *Suppose K is a nonempty weakly compact subset of a Banach space X. If T is a nonexpansive mapping of K into itself which has diminishing orbital diameters, then there is a point x in K such that $Tx = x$.*

PROOF. Because this is essentially the same as the central argument of [**3**] we abbreviate the proof here. Suppose K_1 is a subset of K which is minimal with respect to being nonempty, weakly compact, and invariant under T, and suppose $\delta(K_1) > 0$. If $x \in K_1$ then $Tx \neq x$, and there exists $x_0 \in K_1$ such that for some $r < \delta(K_1)$, $\|T^i x - x_0\| \leq r$ if i is sufficiently large. (This is because T has diminishing orbital diameters at x; x_0 can be taken to be $T^n x$ for n sufficiently large.) For each $\epsilon > 0$, let

$$S_\epsilon = \{y \in X : \|y - T^i x\| \leq r + \epsilon \text{ for almost all } i\}$$

and let

$$S = \bigcap_{\epsilon > 0} S_\epsilon.$$

Suppose $z_n \in S$ and $z_n \to z$. Let $\epsilon > 0$. Then there exists n such that $\|z_n - z\| < \epsilon/2$ and there exists i_0 such that if $i \geq i_0$, $\|z_n - T^i x\| \leq r + \epsilon/2$. Thus for $i \geq i_0$,

$$\|z - T^i x\| \leq \|z - z_n\| + \|z_n - T^i x\| \leq r + \epsilon.$$

We conclude $z \in S_\epsilon$ for each $\epsilon > 0$ and thus $z \in S$; so S is closed. Since S is also convex, $M = S \cap K_1$ is a weakly compact nonempty ($x_0 \in M$) subset of K_1. Furthermore, $T(M) \subset M$ so by minimality of K_1, $M = K_1$.

Let $\epsilon > 0$ and let $w \in K_1$. Since $w \in S_\epsilon$, for n sufficiently large, $\|w - T^n x\| \leq r + \epsilon$. It follows that spherical balls $\bar{U}(w; r + \epsilon)$ centered at points w of K_1 with radius $r + \epsilon$ have the finite intersection property, so

$$C_\epsilon = \left\{ \bigcap_{w \in K_1} \bar{U}(w; r + \epsilon) \right\} \cap K_1 \neq \varnothing .$$

Thus $C = \bigcap_{\epsilon > 0} C_\epsilon \neq \varnothing$, and if $u \in C$, $K_1 \subset \bar{U}(u; r)$. This means that K_1 has a nondiametral point and the argument of [**18**] may be used to show that $R \cap K_1$ is a nonempty, weakly compact, proper subset of K_1 invariant under T, where

$$R = \{y \in X : K_1 \subset \bar{U}(y; r)\}.$$

Therefore the assumption $\delta(K_1) > 0$ leads to a contradiction, so K_1 consists of a single point which is fixed under T.

The hypothesis of weak compactness cannot be removed in the above theorem.

For example, in the space $C[0, 1]$ of continuous functions with supremum norm, let g_i be the function whose graph is the x-axis for $x \leq 1 - 1/2^i$ and the line joining $(1 - 1/2^i, 0)$ to $(1, 1/2 + 1/2^{i+1})$ for $x > 1 - 1/2^i$. The mapping $T: g_i \to g_{i+1}$ is nonexpansive, has diminishing orbital diameters, and it has no fixed point. The sequence $\{g_i\}$ is closed but not weakly compact, having no weak limit points. (A trivial example shows that diminishing orbital diameters cannot be removed.)

A result more general than Theorem 5 can be obtained if K is convex.

THEOREM 6. *Let K be a nonempty, weakly compact convex subset of X and let $T: K \to K$ be nonexpansive. Suppose there is an integer N such that T^N has diminishing orbital diameters on K. Then T has a fixed point in K.*

PROOF. Let K_1 be a subset of K minimal with respect to being nonempty, closed, convex, and mapped into itself by T. By Theorem 2 there exists a point $x \in K_1$ such that $T^N x = x$. We show that $Tx = x$. Let

$$M = \{x, Tx, \dots, T^{N-1}x\}.$$

Suppose $\delta(M) > 0$. Then the closed convex hull $\overline{\text{co}}M$ of M is compact so there is a point $x_0 \in \overline{\text{co}}M$ which is not a diametral point of $\overline{\text{co}}M$ (cf. [**15**, Lemma 1]); hence there is a number $r < \delta(K_1)$ such that $M \subset \bar{U}(x_0; r)$. It can be shown that the set

$$R = \{y \in K_1 : K_1 \subset \bar{U}(y; r)\}$$

is a proper subset of K_1 which is closed, convex, and invariant under T, and a contradiction is obtained.

A simple example of a mapping T which satisfies the hypothesis of Theorem 6 but not of Theorem 5 is a nonexpansive mapping of period $N > 1$. (Such a mapping is necessarily an isometry.) It might also be noted that if Theorem 5 is applied directly to T^N in Theorem 6 then the existence of a point $x \in K$ such that $T^N x = x$ is immediate, but this alone is not sufficient to ensure the existence of a fixed point for T if T is not affine.

The fact that a nonexpansive mapping of K into itself (K as in Theorem 6) which is pointwise periodic must have a fixed point leads one to suspect that Theorem 6 remains true if it is only required that for each x in K there exists an integer N (depending on x) such that T^N has diminishing orbital diameter at x; for example, if for each x in K there exists $\alpha(x)$, $0 \leq \alpha(x) < 1$, and integers $K(x)$, $N(x)$ such that for all i:

$$\|T^{NK}x - T^{N(K+i)}x\| \leq \alpha(x) \sup_j \|x - T^{Nj}x\|.$$

By requiring T to be strictly contractive we are able to make the suggested weakening.

THEOREM 7. *Let K be a nonempty, weakly compact subset of X and let $T : K \to K$ be strictly contractive. Suppose for each $x \in K$ there is an integer N (depending on x) such that T^N has diminishing orbital diameter at x. Then T has a fixed point in K.*

PROOF. Again let K_1 be a subset of K which is minimal with respect to being nonempty, weakly compact, and mapped into itself by T. For each n let K_n be a subset of K_1 which is minimal with respect to being nonempty, weakly compact, and mapped into itself by T^n.

CASE 1. Suppose $\bigcap_{n=1}^{\infty} K_n = \varnothing$. Then, since each of these sets is weakly compact, there is an integer m such that

$$\bigcap_{n=1}^{m} K_n = \varnothing \quad \text{and} \quad \bigcap_{n=1}^{m-1} K_n \neq \varnothing .$$

Let $p = m!$, and let $H = \bigcap_1^{m-1} K_n$. Because H and K_m are nonempty disjoint weakly compact sets it follows that [**17**]

$$\inf \{\|x - y\| : x \in H, y \in K_m\} = d > 0.$$

Furthermore there exist points $x \in H$, $y \in K_m$ such that $\|x - y\| = d$. But $\|T^p x - T^p y\| < \|x - y\|$ and $T^p x \in H$, $T^p y \in K_m$, so we have a contradiction.

CASE 2. Suppose $\bigcap_{n=1}^{\infty} K_n \neq \varnothing$. Let z be a point which is in each of the K_n. Then there exists an integer N such that T^N has diminishing orbital diameter at z *and* $z \in K_N$. The argument of Theorem 5 yields a point $x \in K_N$ such that $T^N x = x$ (in fact, $K_N = \{x\}$). But this can only happen for strictly contractive T if $Tx = x$.

The approach of this section is designed to obtain existence of fixed points for nonexpansive T which hold in as wide a class of spaces as possible. Except for Theorem 4, which has the quite stringent hypothesis of a converging subsequence of iterates, weak compactness is fundamental in the theorems here. While the hypothesis of diminishing orbital diameters is not sufficient to ensure the existence of a fixed point in general in nonweakly compact settings, the question remains as to what additional assumptions are required on T (short of those of Theorem 4) in order to obtain such a fixed point theorem.

REFERENCES

1. L. P. Belluce and W. A. Kirk, *Fixed-point theorems for families of contraction mappings*, Pacific J. Math. **18** (1966), 213–218.

2. ———, *Nonexpansive mappings and fixed-points in Banach spaces*, Illinois J. Math **11** (1967), 474–479.

3. ———, *Fixed-point theorems for certain classes of nonexpansive mappings*, Proc. Amer. Math. Soc. **20** (1969), 141–146.

4. L. P. Belluce, W. A. Kirk, and E. F. Steiner, *Normal structure in Banach spaces*, Pacific J. Math. **26** (1968), 433–440.

5. M. S. Brodskiĭ and D. P. Milman, *On the center of a convex set*, Dokl. Akad. Nauk SSSR (N.S.), **59** (1948), 837–840.

6. F. E. Browder, *Existence of periodic solutions for nonlinear equations of evolution*, Proc. Nat. Acad. Sci. **53** (1965), 1100–1103.

7. ———, *Fixed point theorems for noncompact mappings in Hilbert space*, Proc. Nat. Acad. Sci. **54** (1965), 1272–1276.

8. ———, *Nonexpansive nonlinear operators in a Banach space*, Proc. Nat. Acad. Sci. **54** (1965), 1041–1044.

9. ———, *Fixed point theorems for nonlinear semicontractive mappings in Banach spaces*, Arch. Rational Mech. Anal. **21** (1966), 259–269.

10. ———, *Convergence theorems for sequences of nonlinear operators in Banach spaces*, Math. Z. **100** (1967), 201–225.

11. ———, *Nonlinear mappings of nonexpansive and accretive type in Banach spaces*, Bull. Amer. Math. Soc. **73** (1967), 875–881.

12. ———, *Periodic solutions of nonlinear equations of evolution in infinite dimensional spaces*, Lectures on Differential Equations, Van Nostrand (to appear).

13. F. E. Browder and W. V. Petryshyn, *The solution by iteration of nonlinear functional equations in Banach spaces*, Bull. Amer. Math. Soc. **72** (1966), 571–575.

14. ———, *Construction of fixed points of nonlinear mappings in Hilbert spaces*, J. Math. Anal. Appl. **20** (1967), 197–228.

15. R. DeMarr, *Common fixed-points for commuting contraction mappings*, Pacific J. Math. **13** (1963), 1139–1141.

16. D. Göhde, *Zum Prinzip der kontraktiven Abbildung*, Math. Nachr. **30** (1965), 251–258.

17. R. C. James, *Weak compactness and separation*, Canad. J. Math. **16** (1964), 204–206.

18. W. A. Kirk, *A fixed point theorem for mappings which do not increase distances*, Amer. Math. Monthly **72** (1965), 1004–1006.

19. ———, *On mappings with diminishing orbital diameters*, J. London Math. Soc. **44** (1969), 107–111.

20. ———, *Nonexpansive mappings and the weak closure of sequences of iterates* (submitted).

21. V. L. Klee, *Convex bodies and periodic homeomorphisms in Hilbert space*, Trans. Amer. Math. Soc. **74** (1953), 10–43.

22. Z. Opial, *Weak convergence of the sequence of successive approximations for nonexpansive mappings*, Bull. Amer. Math. Soc. **73** (1967), 591–597.

UNIVERSITY of IOWA

ON SOME NONLINEAR PARTIAL DIFFERENTIAL EQUATIONS RELATED TO OPTIMAL CONTROL THEORY

J. L. Lions

Introduction. We study in this paper, by a rather indirect method, a number of nonlinear partial differential equations, stationary or of evolution type, with nonlinearities of quadratic type.

Problems of this type arise in optimal control theory (feedback, Bucy-Kalman filters etc.) for systems governed by *ordinary* differential equations. We extend this theory to systems governed by *partial* differential equations. The usual arguments do not seem to extend to our situation. We follow here a rather different approach (which seems also to give the classical results in a simpler way). This is done in §1, an example being given in §2 and some variants being given in §3. We obtain in this way *global* existence and uniqueness theorems for nonlinear partial differential equations for which a direct solution does not seem to be obvious.

The analysis of the proof of §1 leads also to another situation, with no control theory involved, where one can get global existence and uniqueness theorems for "some other" nonlinear partial differential equations. It would be interesting to know "up to where" one can go in this direction, but this is unsettled here.

The plan is as follows:

1. Distributed controls—Functional equation of Riccati type.

1.1. *Setting of the problem.* We consider a system whose state y is given through the solution of an *evolution equation;* more precisely, let V and H be two *real* hilbert spaces, $V \subset H$, V dense in H, the identity mapping $V \to H$ being continuous. We denote by $(\; , \;)$ the scalar product in H and by $| \; |$ the associate norm. We consider a family of operators

(1.1) $$A(t) \in \mathscr{L}(V; V'), t \in [0, T], T > 0 \text{ given},$$

where V' denotes the dual of V, when H is identified with its dual; therefore $V \subset H \subset V'$; if $f \in V'$, $v \in V$, we denote by (f, v) their scalar product. On the family $A(t)$ we assume that

(1.2) $t \to (A(t)\varphi, \psi)$ is bounded and measurable on $[0, T]$, $\forall \varphi, \psi \in V$,

(1.3) there exists $\lambda \in \boldsymbol{R}$ and $\alpha > 0$ such that

$$(A(t)\varphi, \varphi) + \lambda |\varphi|^2 \geqslant \alpha \|\varphi\|_V^2, \quad \alpha > 0, \quad \forall \varphi \in V.$$

We denote by u, v, the *controls;* to simplify the exposition (see more general situations in [**10**], Chapter 4—and also another situation in §3 below) we assume that

1.4) the space $\mathscr{U}$ of controls (which is always assumed to be a real hilbert space) coincides with $L^2(0, T; H)$[1].

Let $B(t)$ be a family of operators satisfying

(1.5) $B(t) \in \mathscr{L}(H; H)$,

$t \to (B(t)\varphi, \psi)$ is bounded and measurable in $[0, T]\forall \varphi, \psi \in H$.

If v is a control, Bv denotes the function $t \to B(t)v(t)$.

With these notations and hypotheses, we now define *the state* $y(t; v) = y(v)$ *of the system as the unique solution of*

(1.6) $dy(v)/dt + A(t)y(v) = f + Bv, f$ given in $L^2(0, T; H)$,

(1.7) $y(0; v) = (y(t; v)|_{t=0}) = y_0$, y_0 given in H,

where

(1.8) $$y(v) \in L^2(0, T; V)$$

(and then, necessarily,

(1.9) $$dy(v)/dt \in L^2(0, T; V')).$$

We refer to [**9**] for the proof that (1.6), (1.7), (1.8) *uniquely defines* $y(v)$. We now consider *the cost functional*[2]

(1.10) $$J(v) = \int_0^T |y(t; v) - z_d(t)|^2 \, dt + \int_0^T (Nv, v) \, dt$$

[1] $L^2(0, T; \mathscr{H})$ denotes the space of (classes of) functions f measurable from $[0, T] \to \mathscr{H}$ such that $t \to \|f(t)\|_{\mathscr{H}}$ belongs to $L^2(0, T)$.

[2] Other functionals are considered in [**10**].

where z_d is given in $L^2(0, T; H)$, and

$$(1.11)\quad N \in \mathscr{L}(H; H), \qquad N^* = N, \qquad (N\varphi, \varphi) \geqslant \gamma\,|\varphi|^2, \qquad \gamma > 0, \qquad \forall \varphi \in H$$

(we could as well consider N to depend on t). Due to (1.11) and due to the fact that the mapping $v \to y(v)$ is affine *continuous* (see e.g. [**9**]) from $L^2(0, T; H) \to L^2(0, T; V)$,[3] the functional $v \to J(v)$ is quadratic *continuous* on $L^2(0, T; H)$ and *positive definite*. Therefore there exists a unique element u (the optimal control) such that $J(u) \leqslant J(v) \forall v \in L^2(0, T; H)$. This element is characterized by (Euler's equation)

$$(1.12)\quad \int_0^T (y(t; u) - z_d, y(t; v) - y(t; 0))\,dt + \int_0^T (Nu, v)\,dt = 0 \qquad \forall v \in L^2(0, T; H).$$

We are now going to transform (1.12) by introducing the *adjoint state*.

1.2. *The adjoint state. The coupled system of partial differential equations.* Let $A^*(t) \in \mathscr{L}(V; V')$ be the adjoint of $A(t)$ (i.e. $(A^*(t)\varphi, \psi) = (\varphi, A(t)\psi) \forall \varphi, \psi \in V$). We define the *adjoint state* $p(u) = p$ as the *unique solution*[4] of

$$(1.13)\quad -dp/dt + A^*(t)p = y(t; u) - z_d, \quad t \in]0, T[,$$

$$(1.14)\quad p(T) = 0,$$

$$(1.15)\quad p \in L^2(0, T; V).$$

It follows from (1.13) that

$$\int_0^T (y(t; u) - z_d, y(t; v) - y(t; 0))\,dt = \int_0^T \left(-\frac{dp}{dt} + A^*(t)p, y(t; v) - y(t; 0)\right) dt$$

and it is known [**9**] that under these circumstances one can integrate by parts to obtain:

$$(1.16)\quad \begin{aligned} &\int_0^T (y(t; u) - z_d, y(t; v) - y(t; 0))\,dt \\ &= -(p(T), y(T; v) - y(T; 0)) + (p(0), y(0, v) - y(0; 0)) \\ &+ \int_0^T \left(p, \left(\frac{d}{dt} + A(t)\right)\left(y(t; v) - y(t; 0)\right)\right) dt. \end{aligned}$$

Using (1.14), (1.6) and (1.7), the right-hand side of (1.16) equals

$$\int_0^T (p, Bv)\,dt = \int_0^T (B^*p, v)\,dt$$

[3] Hence in particular from $L^2(0, T; H)$ into itself.

[4] For the existence and uniqueness of p, we apply the same result [**9**] as for the existence and uniqueness of y.

so that (1.12) reduces to

$$\int_0^T (B^*p + Nu, v)\, dt = 0 \quad \forall v \in L^2(0, T; H)$$

i.e.

$$B^*p + Nu = 0, \tag{1.17}$$

or

$$u = -N^{-1}B^*p. \tag{1.17 bis}$$

We can sum up: *the optimal control u is given by* (1.17 bis) *where p is given through the solution of the following coupled system of equations in* $\{y, p\}$,

$$dy/dt + A(t)y + BN^{-1}B^*p = f \quad \text{in }]0, T[, \tag{1.18}$$

$$-dp/dt + A^*(t)p - y = -z_d \quad \text{in }]0, T[, \tag{1.19}$$

$$y(0) = y_0, \tag{1.20}$$

$$p(T) = 0, \tag{1.21}$$

$$y, p \in L^2(0, T; V). \tag{1.22}$$

The system (1.18) $\cdots$ (1.22) admits a unique solution (being equivalent to the optimal control problem).

Our aim is now to "uncouple" system (1.18) $\cdots$ (1.22).

1.3. *Uncoupling. The functional equation of Riccati type.* Let us consider a system analogous to (1.18) $\cdots$ (1.22) but with the interval $[0, T]$ replaced by $[S, T]$. More precisely, let S be arbitrarily chosen in $[0, T]$ and let h be given in H; we consider the system of equations in $\{\varphi, \psi\}$:

$$\begin{gathered} d\varphi/dt + A(t)\varphi + BN^{-1}B^*\psi = f \quad \text{in }]S, T[, \\ -d\varphi/dt + A^*(t)\psi - y = -z_d \quad \text{in }]S, T[, \\ \varphi(S) = h, \qquad \psi(T) = 0, \qquad \varphi, \psi \in L^2(S, T; V). \end{gathered} \tag{1.23}$$

Problem (1.23) admits a unique solution, since it corresponds to an optimal control problem analogous to the first one, with $[0, T]$ replaced by $[S, T]$ (and y_0 by h).

Therefore $\psi(S)$ is uniquely defined in H, hence a mapping

$$h \to \psi(S) \text{ from } H \text{ into itself.} \tag{1.24}$$

One can easily check that this mapping is *continuous;* since it is affine, we obtain *the existence of* $P(S) \in \mathscr{L}(H; H)$ such that

$$\psi(S) = P(S)h + r(S), \quad r(S) \in H. \tag{1.25}$$

If in (1.23) we take $h = y(S)$ then (due to the *uniqueness* of the solution)

$$\varphi = y, \qquad \psi = p \quad \text{in } [S, T]$$

and in particular $\psi(S) = p(S)$ and (1.25) becomes

$$p(S) = P(S)y(S) + r(S). \tag{1.26}$$

But since S is *arbitrary,* we have finally proved that

(1.27) *there exists a family $P(t)$ of operators is $\mathscr{L}(H; H)$ such that*

$$p(t) = P(t)y(t) + r(t).$$

The problem is now to obtain (if possible) equations characterizing P and r.

We proceed by a formal identification calculation (which can be justified—[**10**]). We take the t-derivative of the identity (1.27):

$$dp/dt = P(dy/dt) + dP/dt \cdot y + dr/dt. \tag{1.28}$$

Using (1.18) and bringing (1.28) into (1.19) we obtain

$$\begin{aligned} -dP/dt \cdot y - P(t)(-A(t)y - BN^{-1}B^*p + f) - dr/dt \\ + A^*(t)(P(t)y + r) - y = -z_d \end{aligned} \tag{1.29}$$

and after replacing p by its value (1.27) in (1.29) we get

$$\begin{aligned} (-dP(t)/dt + P(t)A(t) + A^*(t)P(t) + P(t)BN^{-1}B^*P - I)y(t) \\ + (-dr/dt + A^*(t)r + P(t)BN^{-1}B^*r(t) - P(t)f + z_d) = 0. \end{aligned} \tag{1.30}$$

But (1.30) is true for *every* $y(t) \in H$ (t fixed) (for the justification of this point, see [**10**, Chapter 4]) so that (1.30) is equivalent to

$$-dP(t)/dt + P(t)A(t) + A^*(t)P(t) + P(t)BN^{-1}B^*P(t) = I \tag{1.31}$$

and to

$$-dr/dt + A^*(t)r + P(t)BN^{-1}B^*r(t) = P(t)f - z_d. \tag{1.32}$$

Observe that (1.31) is an (abstract) *evolution equation with a nonlinearity of Riccati type.*

We have now used (1.18), (1.19) and (implicitly) (1.22). Condition (1.21) is equivalent to

$$P(T) = 0, \tag{1.33}$$

$$r(T) = 0. \tag{1.34}$$

One can also prove (see [**10**]) that $t \to P(t)h$ is $\forall h \in H$ once continuously differentiable from $[0, T] \to H$ and that

$$P^*(t) = P(t). \tag{1.35}$$

The final result is:

(1.36) *the Riccati evolution equation* (1.31), *with "initial" condition* (1.32), *admits a unique solution such that* $t \to P(t)h$ *is once continuously differentiable from* $[0, T] \to H$ *and such that* (1.35) *holds true.*

Once $P(t)$ is known, $r(t)$ is given by (1.32), (1.34).

The optimal control $u(t)$ is given by (see (1.17 bis))

$$u(t) = -N^{-1}B^*(P(t)y(t) + r(t)), \tag{1.37}$$

which is the "feedback formula".

REMARK 1.1. *We want to emphasize the fact that the preceding reasoning proves the existence and uniqueness of a global (i.e. in* $[0, T]$*) solution* $P(t)$ *of the nonlinear evolution equation* (1.31).

A *direct* proof of this fact (i.e. without the use of the control problem) does not seem to be known (see also Remark 1.2 below).

REMARK 1.2. The above procedure is classical in *finite dimension* [**1**]. It is the "field method" in the calculus of variations, see [**2**], [**3**], [**6**], [**7**], [**8**] but with different approaches: one "tries" a representation of type (1.27); a formal calculation leads to (1.31), (1.33) and one proves the *global* existence of $P(t)$ using the a-priori estimate

$$|P(t)|_{\mathscr{L}(H;H)} \leqslant \text{constant}.$$

But this estimate is not enough here.

2. Example.

2.1. *The partial differential equation giving the state.* We consider the simplest possible example of application of §1. Let Ω be an open set in $\boldsymbol{R}^n$, say bounded with smooth boundary Γ. We consider the situation of 1.1 with

$$V = H_0^1(\Omega) = \left\{\varphi \mid \varphi, \frac{\partial\varphi}{\partial x_1}, \ldots, \frac{\partial\varphi}{\partial x_n} \in L^2(\Omega), \quad \varphi = 0 \text{ on } \Gamma\right\},$$

$$A(t) = -\Delta, \quad H = L^2(\Omega),$$

$$B(t) = \text{identity in } L^2(\Omega).$$

Therefore the state $y(v) = y(x, t; v)$ is given by the solution of the heat equation

$$\begin{aligned} &\partial y/\partial t - \Delta y = f + v \quad \text{in } Q = \Omega \times]0, T[, \\ &y = 0 \quad \text{if } x \in \Gamma,\ t \in [0, T], \\ &y(x, 0) = y_0(x), \quad y_0 \text{ given in } L^2(\Omega). \end{aligned} \tag{2.1}$$

The cost functional (1.10) becomes, assuming to simplify a step further that $N = \nu I$, $\nu > 0$:

$$J(v) = \int_Q |y(x, t; v) - z_d(x, t)|^2\, dx\, dt + \nu \int_Q v^2\, dx\, dt. \tag{2.2}$$

We can apply the results of §1. In what follows, we want to make explicit the Riccati equation satisfied by P.

2.2. *The partial differential equation of Riccati type.* By the Schwartz's kernel theorem [**14**] we know that

$$P(t)\varphi = \int_\Omega P(x, \xi, t)\varphi(\xi)\, d\xi, \qquad \varphi \in \mathscr{D}(\Omega_\xi) \tag{2.3}$$

where

$$\begin{aligned} &P(x, \xi, t) \text{ is, for every fixed } t, \text{ a distribution over } \Omega_x \times \Omega_\xi; \\ &\text{the } kernel\ P(x, \xi, t) \text{ is such that} \end{aligned} \tag{2.4}$$

$$\forall h \in H = L^2(\Omega), \quad P(t)h \in H_0^1(\Omega) \quad \text{for } t < T. \tag{2.5}$$

Condition (1.35) becomes

$$P(x, \xi, t) = P(\xi, x, t). \tag{2.6}$$

The equation (1.31) gives:

$$-\partial P/\partial t - \Delta_x P - \Delta_\xi P + \frac{1}{\nu}\int_\Omega P(x, \hat{x}, t)P(\hat{x}, \xi, t)\, d\hat{x} = \delta(x - \xi) \tag{2.7}$$ [5]

together with the boundary conditions[6]

$$\begin{aligned} P(x, \xi, t) &= 0 \quad \text{if } x \in \Gamma,\ \xi \in \Omega,\ t \in]0, T[, \text{ and} \\ P(x, \xi, t) &= 0 \quad \text{if } x \in \Omega,\ \xi \in \Gamma,\ t \in]0, T[, \end{aligned} \tag{2.8}$$

and the "initial" condition

$$P(x, \xi, T) = 0. \tag{2.9}$$

We know therefore *the existence and uniqueness of a solution of* (2.7), (2.8), (2.9) (and P automatically satisfies (2.6)) *in the class of "functions"* $P(x, \xi, t)$ *which are kernels of mappings from* $L^2(\Omega)$ *into* $H_0^1(\Omega)$ *and such that* $t \to P(t)h$ *is once continuously differentiable from* $[0, T] \to L^2(\Omega)$ $\forall h \in L^2(\Omega)$.

REMARK 2.1. It would be interesting to solve directly (2.7), (2.8), (2.9) and to obtain regularity results on P—in particular at the boundary.

REMARK 2.2. The above situation is of course the simplest possible. One can extend the above considerations to the case when A is an elliptic operator of order $2m$ subject to m boundary conditions of "general type" (such that the corresponding boundary value problem for $\partial/\partial t + A$ is well set).

REMARK 2.3. For the case of boundary control, see §3 below.

REMARK 2.4. One can find similar examples (but treated in a formal way) in [**5**], [**15**].

---

[5] Which is the kernel of I.

[6] Because of (2.5).

3. Some remarks.

3.1. *Boundary controls.* We confine ourselves to an example. See more general situations in [**10**]. We consider a system whose state is given as the solution of

$$\partial y(v)/\partial t + Ay(v) = f \quad \text{in } Q = \Omega \times]0, T[, \tag{3.1}$$

where A is a second order linear elliptic operator given by

$$A\varphi = -\sum_{i,j=1}^{n} \frac{\partial}{\partial x_i}\left(a_{ij}(x, t)\frac{\partial \varphi}{\partial x_j}\right), \tag{3.2}$$

subject to the boundary and initial conditions

$$\partial y(v)/\partial \nu_A = v \quad \text{on } \Sigma = \Gamma \times]0, T[, \tag{3.3}$$

$$y(x, 0; v) = y_0(x), \quad x \in \Omega. \tag{3.4}$$

The control v appears as a *boundary control*; we assume that

$$v \in L^2(\Sigma). \tag{3.5}$$

We assume next that the *cost functional* is given by

$$J(v) = \int_Q (y(v) - z_d)^2 \, dx \, dt + \nu \int_\Sigma v^2 \, d\Sigma \tag{3.6}$$

where ν is given > 0 and z_d is given in $L^2(Q)$.

One can check (see details in [**10**, Chapter 4]) that the optimal control u (which exists and is unique) is given by

$$u = -\frac{1}{\nu} p\Big|_\Sigma, \tag{3.7}$$

where p is given through the solution of the following system in $\{y, p\}$:

$$\partial y/\partial t + Ay = f \quad \text{in } Q, \tag{3.8}$$

$$-\partial p/\partial t + A^*p - y = -z_d \quad \text{in } Q, \tag{3.9}$$

$$\partial y/\partial \nu_A + p/\nu = 0 \quad \text{on } \Sigma, \tag{3.10}$$

$$\partial p/\partial \nu_{A^*} = 0 \quad \text{on } \Sigma, \tag{3.11}$$

$$y(x, 0) = y_0(x) \quad \text{on } \Omega, \tag{3.12}$$

$$p(x, T) = 0 \quad \text{on } \Omega. \tag{3.13}$$

The boundary value problem (3.8) $\cdots$ (3.13) admits a unique solution. This problem can be "uncoupled" by the same method as in §1. We obtain (see details in [**10**, Chapter 4])

$$p(x, t) = \int_\Omega P(x, \xi, t) y(\xi, t) \, d\xi + r(x, t) \tag{3.14}$$

where the kernels $P(x, \xi, t)$ satisfy:

$$(3.15)\quad -\frac{\partial P}{\partial t} + (A_x^* + A_\xi^*)P(x, \xi, t) + \frac{1}{\nu}\int_\Gamma P(x, \hat{x}, t)P(\hat{x}, \xi, t)\, d\Gamma_{\hat{x}} = \delta(x - \xi), \quad \text{in } \Omega_x \times \Omega_\xi \times]0, T[,$$

$$(3.16)\quad \begin{aligned} &\frac{\partial P}{\partial \nu_{A_x^*}}(x, \xi, t) = 0 \quad \text{if } x \in \Gamma,\ \xi \in \Omega,\ t \in]0, T[, \\ &\frac{\partial P}{\partial \nu_{A_\xi^*}}(x, \xi, t) = 0 \quad \text{if } x \in \Omega,\ \xi \in \Gamma,\ t \in]0, T[, \end{aligned}$$

and

$$(3.17)\quad P(x, \xi, T) = 0 \quad \text{on } \Omega_x \times \Omega_\xi.$$

(We have also $P(x, \xi, t) = P(\xi, x, t)$). One can make here a remark entirely similar to Remark 2.1.

3.2. *Stationary case.* Assume that in the situation of §1 one has $A(t) = A$ independent of t and assume that (1.3) holds true with $\lambda = 0$. We can take $T = +\infty$. We also assume that $B(t)$ does not depend on t.

Then the system (1.18) $\cdots$ (1.22) is replaced by

$$(3.18)\quad \begin{gathered} dy/dt + Ay + BN^{-1}B^*p = f \quad \text{in }]0, +\infty[, \ (N^{-1} = 1/\nu), \\ -dp/dt + A^*p - y = -z_d \quad \text{in }]0, +\infty[, \\ y(0) = y_0, \quad y, p \in L^2(0, \infty; V). \end{gathered}$$

We have the identity (same method as in §1)

$$(3.19)\quad p(t) = P_\infty y(t) + r(t)$$

where P_∞ *does not depend on* t.

One finds that

$$(3.20)\quad P_\infty A + A^*P_\infty + (1/\nu)P_\infty BB^*P_\infty = I.$$

This is an "abstract" nonlinear elliptic equation.

EXAMPLE. We can apply the above remark to the situation of §2. We obtain for the *kernel* $P_\infty(x, \xi)$ of the operator P the *nonlinear partial differential equation*

$$(3.21)\quad -\Delta_x P_\infty - \Delta_\xi P_\infty + \frac{1}{\nu}\int_\Omega P_\infty(x, \hat{x})P_\infty(\hat{x}, \xi)\, d\hat{x} = \delta(x - \xi),$$

together with the boundary conditions

$$(3.22)\quad P_\infty(x, \xi) = 0 \quad \text{if } x \in \Gamma,\ \xi \in \Omega \text{ or if } x \in \Omega,\ \xi \in \Gamma.$$

REMARK 3.1. (See [**10**]).

Let $P_T(x, \xi, t)$ denote the kernel solution of (3.15), (3.16), (3.17). Then $P_T(x, \xi, t) \to P_\infty(x, \xi)$ as $T \to \infty$, in the sense:

$$\int_\Omega P_T(x, \xi, t)\varphi(\xi)\, d\xi \to \int_\Omega P_\infty(x, \xi)\varphi(\xi)\, d\xi \quad \text{in } L^2(\Omega) \tag{3.23}$$
$$\forall \varphi \in L^2(\Omega), \quad \text{for } t \text{ fixed and } T \to +\infty.$$

3.3. *Other remarks.*

REMARK 3.2. In case one has "convex constraints" on the control (i.e. the control v is subject to stay in a closed convex set different from the whole space of controls) systems of equations of the form (1.18)–(1.22) are replaced by *variational inequalities* in the sense of [**13**]—See [**12**] and [**10**].

REMARK 3.3. For systems whose state is given by *nonlinear* equations or in cases the cost functional is *not quadratic*, the Riccati equation is replaced (see [**10**]) by a functional equation involving ordinary partial derivatives and Fréchet derivatives (as in [**4**]); one can also (in some cases, see [**11**]) develop in series the (now nonlinear) mapping $h \to p(S)$; one obtains a chain of equations, the 1st one being of Riccati's type, the following ones being linear. See [**11**].

REMARK 3.4. All that has been said extends to the case of systems whose state is given through similar equations but where $d/dt + A$ is replaced by $d^2/dt^2 + A$, with $A^* = A$. See [**10**, Chapter 5].

REMARK 3.5. For extensions to stochastic cases, noisy observations etc., see (for the infinite dimensional case with *unbounded* operators) [**1**].

4. Other examples of uncoupling of partial differential equations.

4.1. *A general remark.* Let V and H be given as in §1,[7] and a family of operators $A_j(t)$, $j = 1, \dots, m$, satisfying

$$\begin{aligned} &A_j(t) \in \mathcal{L}(V; V'), \quad t \in [0, T], \forall j, \\ &t \to (A_j(t)\varphi, \psi) \text{ bounded and measurable } \forall \varphi, \psi \in V \text{ and } \forall j, \\ &(A_j(t)\varphi, \varphi) \geqslant \alpha_j \|\varphi\|_V^2, \quad \alpha_j > 0, \forall \varphi \in V. \end{aligned} \tag{4.1}$$

We introduce

$$\mathcal{H} = \underbrace{H \times \cdots \times H}_{(m \text{ factors})} = (H)^m, \quad \mathcal{V} = (V)^m, \tag{4.2}$$

and we set

$$\mathcal{A}(t) = \begin{pmatrix} A_1(t) & & & & \\ & A_2(t) & & \bigcirc & \\ & & \cdot & & \\ & & & \cdot & \\ & \bigcirc & & & \cdot \\ & & & & A_m(t) \end{pmatrix} \in \mathcal{L}(\mathcal{V}; \mathcal{V}'). \tag{4.3}$$

[7] We could also consider spaces V_j, H_j, $j = 1, \dots, m$.

Let us consider an operator $\mathscr{C}$ satisfying:

$$\mathscr{C} \in \mathscr{L}(\mathscr{H}\,;\mathscr{H}), \quad (\mathscr{C}\vec{\varphi}, \vec{\varphi})_{\mathscr{H}} \geqslant 0 \quad \forall \vec{\varphi} \in \mathscr{H}. \tag{4.4}$$

We now consider the following (abstract) boundary value problem: we look for a function $\vec{y} = \{y_1, \ldots, y_m\}$ which satisfies

$$\Lambda\vec{y} + \mathscr{A}\vec{y} + \mathscr{C}\vec{y} = \vec{f} \tag{4.5}$$

where

$$\Lambda = \begin{pmatrix} \frac{d}{dt} & & & & \bigcirc \\ & -\frac{d}{dt} & & & \\ & & -\frac{d}{dt} & & \\ & \bigcirc & & \ddots & \\ & & & & -\frac{d}{dt} \end{pmatrix}, \tag{4.6}$$

$\vec{f} = \{f_1, \ldots, f_m\}$ is given in $L^2(0, T; \mathscr{H})$, and where $\vec{y}$ satisfies also

$$\vec{y} \in L^2(0, T; \mathscr{V}) \quad (\text{i.e. } y_j \in L^2(0, T; V)\forall j), \tag{4.7}$$

and

$$\begin{aligned} &y_1(0) = y_0, \quad y_0 \text{ given in } H, \\ &y_2(T) = y_3(T) = \cdots = y_m(T) = 0. \end{aligned} \tag{4.8}$$

This problem admits a unique solution, applying [**12**, Theorem 1.1. of Chapter 3]; observe that Λ considered as an unbounded operator in $L^2(0, T; \mathscr{H})$, with domain $D(\Lambda) = \{\vec{\varphi} \mid \varphi, \Lambda\varphi \in L^2(0, T; \mathscr{H}), \varphi_1(0) = 0, \varphi_2(T) = \cdots = \varphi_m(T) = 0\}$, is the infinitesimal generator of a contraction semigroup, which also acts in $L^2(0, T; \mathscr{V})$ and its dual.

We can now "split" this problem in the following manner, which parallels the method of §1; let h be given in H and let S be arbitrarily chosen in $]0, T[$; we consider (compare to (1.23)) the system of equations

$$\begin{aligned} &\Lambda\vec{\varphi} + \mathscr{A}\vec{\varphi} + \mathscr{C}\vec{\varphi} = \vec{f} \quad \text{in }]S, T[, \\ &\varphi_1(S) = h, \varphi_2(T) = \cdots = \varphi_m(T) = 0, \\ &\vec{\varphi} \in L^2(0, T; \mathscr{V}). \end{aligned} \tag{4.9}$$

This problem admits a unique solution, so that

$$\vec{\psi}(S) = \{\varphi_2(S), \ldots, \varphi_m(S)\} \in (H)^{m-1}$$

is uniquely defined—hence a *continuous* affine mapping

$$h \rightarrow \vec{\psi}(S) \quad \text{from } H \rightarrow (H)^{m-1}.$$

Consequently

$$\vec{\psi}(S) = P(S)h + r(S) \qquad P(S) \in \mathscr{L}(H; (H)^{m-1}),$$

and since S is arbitrary we have the *identity*

$$p(t) = P(t)y_1(t) + r(t) \tag{4.10}$$

if we set $\vec{p} = \{y_2, \ldots, y_m\}$.

One can then obtain an equation satisfied by $P(t)$ by a calculation of identification.

We will now consider a very simple example.

4.2. *An example.* We consider the system

$$\begin{aligned} dy_1/dt + A_1y_1 - y_2 &= 0, \\ -dy_2/dt + A_2y_2 + y_1 - y_3 &= 0, \\ -dy_3/dt + A_3y_3 + y_2 &= 0, \end{aligned} \tag{4.11}$$

together with

$$y_1(0) = y_0, \quad y_2(T) = 0, \quad y_3(T) = 0. \tag{4.12}$$

Equation (4.10) gives

$$y_2(t) = P_2(t)y_1(t), \quad y_3(t) = P_3(t)y_1(t)$$

which we rewrite:

$$y_2(t) = P(t)y_1(t), \quad y_3(t) = Q(t)y_1(t). \tag{4.13}$$

We obtain in this way:

$$-\partial P/\partial t + PA_1 + A_2P - P^2 + I = Q, \tag{4.14}$$

and

$$-\partial Q/\partial t + QA_1 + A_3Q - QP + P = 0. \tag{4.15}$$

The "initial" conditions are

$$P(T) = 0, \quad Q(T) = 0. \tag{4.16}$$

Due to the special form of (4.14) it is a simple matter to eliminate Q. We obtain, assuming to simplify that $A_1 = A_2 = A_3 = A$:

$$\begin{aligned} \frac{\partial^2 P}{\partial t^2} + 2\frac{\partial P}{\partial t}P + P\frac{\partial P}{\partial t} - 2\frac{\partial P}{\partial t}A - 2A\frac{\partial P}{\partial t} + P^3 - P^2A - 2AP^2 \\ - PAP + PA^2 + 2APA + A^2P + 2A = 0 \end{aligned} \tag{4.17}$$

and (4.16) gives

$$P(T) = 0 \frac{\partial P(T)}{\partial t} = I. \tag{4.18}$$

One can of course apply (in case A is a partial differential operator) the Schwartz kernel theorem; the kernel $P(x, \xi, t)$ is then a global solution of a nonlinear partial differential equation corresponding to (4.17).

Bibliography

1. A. Bensoussan, Cahiers IRIA, vol. 1, Thèse, Paris, 1969.

2. R. S. Bucy and P. D. Joseph, *Filtering for stochastic processes with applications to guidance*, Interscience, New York, 1968.

3. R. Bellman, P. Kalaba and G. M. Wing, *Invariant imbedding and mathematical physics*, J. Math. Phys. **1** (1960), 280–308.

4. M. D. Donsker and J. L. Lions, *Frechet Volterra variational equations, boundary value problems and function space integrals*, Acta Math. **108** (1962), 147–228.

5. H. Erzberger and M. Kim, *Optimum boundary control of distributed parameter systems*, Information and Control, **9** (1966), 265–278.

6. I. M. Gelfand and S. V. Fomin, *Calculus of variations*, Prentice-Hall, Englewood Cliffs, N.J., 1963.

7. R. E. Kalman, *Contributions to the theory of optimal control*, Bol. Soc. Mat. Mexicana (1960), 102–119.

8. R. E. Kalman and R. S. Bucy, *New results in linear filtering and prediction theory*, J. Basic Engineering, (1961), 95–107.

9. J. L. Lions, *Equations différentielles opérationelles et problèmes aux limites*. Vol. III, Springer-Verlag, Berlin, 1961; rev. ed, 1970.

10. ———, *Sur le contrôle optimal de systèmes gouvernés par des équations aux dérivées partielles*, Dunod, Paris, 1968.

11. ———, *Sur le feedback nonlinéaire*. (To appear.)

12. J. L. Lions and E. Magenes, *Problèmes aux limites nonhomogènes et applications*, Vol. 1 and 2, Dunod, Paris, 1968.

13. J. L. Lions and G. Stampacchia, *Variational inequalities*, Comm. Pure Appl. Math. (3) **20** (1967), 493–519.

14. L. Schwartz, *Théorie des noyaux*, Proc. of the Internat. Congress of Math., **1** (1950), 220–230.

15. P. K. C. Wang, "Control of distributed parameter systems" in *Advances in control systems*. Vol. 1, edited by C. T. Leondes, Academic Press, New York, 1964, pp. 75–172.

University of Paris

PERTURBATION OF VARIATIONAL INEQUALITIES

Umberto Mosco[1]

1. We shall describe in this lecture some results concerning the continuous dependence of the solutions of a variational inequality such as

$$(*) \qquad u \in K: \langle Tu, v - u \rangle \geq 0 \quad \text{for all } v \in K$$

on the map T and the set K. We are assuming here that T is a monotone map from a real reflexive Banach space X to its dual X^* and K is a closed convex subset of the domain $D(T)$ of T. Moreover, $\langle . , . \rangle$ denotes the duality pairing between X and X^*.

Many examples of inequalities of this form have been discussed by Professor G. Stampacchia in his lecture at this Symposium. We only recall here the following basic

EXISTENCE THEOREM (F. E. BROWDER [1], P. HARTMAN—G. STAMPACCHIA [8]). *Let T be a monotone, hemicontinuous map from X to X^*, K a closed convex subset of $D(T)$. If in addition, either K is bounded or T is coercive on K, then there exists at least one solution of $(*)$. Moreover, the set of all solutions of $(*)$ is a bounded closed convex subset of K, which consists of a single vector if T is strictly monotone.*

[A map T from X to X^* is said to be *monotone* if $\langle Tu - Tv, u - v \rangle \geq 0$ for all u, $v \in D(T)$ and *strictly monotone* if the sign $>$ holds above whenever $u \neq v$. T is said to be *hemicontinuous* if $D(T)$ is convex and for any u, $v \in D(T)$ the map $t \to T(tu + (1 - t)v)$ of $[0, 1]$ to X^* endowed with the weak topology is continuous. Finally, T is said to be *coercive on a subset* K of $D(T)$ if

$$\langle Tv, v \rangle / \|v\| \to +\infty \quad \text{as } \|v\| \to +\infty, v \in K$$

and simply *coercive* if it is coercive on $D(T)$.]

We shall consider a real-parametrized perturbation of the inequality $(*)$, that is, a family of inequalities

$$(*_\varepsilon) \qquad u_\varepsilon \in K_\varepsilon: \langle T_\varepsilon u_\varepsilon, v - u_\varepsilon \rangle \geq 0 \quad \text{for all } v \in K_\varepsilon$$

where, for every $\varepsilon > 0$, T_ε is a monotone map from X to X^* which is supposed to converge to T in some sense as $\varepsilon \to 0$, and K_ε is a subset of the domain $D(T_\varepsilon)$ of T_ε, that also converges to K as $\varepsilon \to 0$ in a sense to be specified. Our main problem

[1] Gruppo CNR n. 23.

is, indeed, to find out what these convergences should be in order that the solutions u_ε of $(*_\varepsilon)$ converge, weakly or strongly in X, to a solution u of $(*)$ as $\varepsilon \to 0$.

2. **The general theorems.** When the solution of $(*)$ is unique, which is always true if T is strictly monotone, then we can prove the convergence of any bounded family of perturbed solutions under fairly general hypotheses on the perturbations allowed.

The basic assumptions on the map T and its perturbation T_ε are the following:

I *T is a strictly monotone, hemicontinuous map from X to X^*; $(T_\varepsilon)_{\varepsilon>0}$ is a family of uniformly bounded, monotone mappings from X to X^*, which converges to T as $\varepsilon \to 0$ in the following sense: for every $v \in D(T)$ there exists for each ε a vector $v_\varepsilon \in D(T_\varepsilon)$, such that v_ε converges strongly to v in X and $T_\varepsilon v_\varepsilon$ converges strongly to Tv in X^* as $\varepsilon \to 0$.*

[A map T from X to X^* is said to be *bounded* if it carries bounded subsets of $D(T)$ into bounded subsets of X^*; a family $(T_\varepsilon)_{\varepsilon>0}$ of mappings from X to X^* is said to be *uniformly bounded* if for any bounded subset B of X, there exists a bounded subset B' of X^*, such that $T_\varepsilon(B \cap D(T_\varepsilon)) \subset B'$ for every ε.]

To describe the convergence of K_ε to K, we shall use the following notation, in which *strong lim* denotes convergence in the strong topology of X, while *weak lim* denotes convergence in the weak topology of X:

$$s\text{-Lim inf } K_\varepsilon = \{v \in X : v = \text{strong lim } v_\varepsilon \text{ as } \varepsilon \to 0,\ v_\varepsilon \in K_\varepsilon \text{ for every } \varepsilon > 0\}$$

$$w\text{-Lim sup } K_\varepsilon = \{v \in X : v = \text{weak lim } v_j \text{ as } j \to +\infty,$$
$$v_j \in K_{\varepsilon_j} \text{ for every integer } j,\ \varepsilon_j \to 0 \text{ as } j \to +\infty\}.$$

Clearly, s-Lim inf $K_\varepsilon \subset w$-Lim sup K_ε.

We say that $(K_\varepsilon)_{\varepsilon>0}$ *converges* in X if the two sets above coincide and we write

$$K = \text{Lim } K_\varepsilon \quad \text{in } X$$

if K is a subset of X such that

$$s\text{-Lim inf } K_\varepsilon = w\text{-Lim sup } K_\varepsilon = K.$$ [2]

Thus, our hypotheses on K and its perturbation K_ε are the following:

II *K is a nonempty closed convex subset of $D(T)$; $(K_\varepsilon)_{\varepsilon>0}$ is a family of subsets of X, with $K_\varepsilon \subset D(T_\varepsilon)$ for every ε, such that $K =$ Lim K_ε in X in the sense stated above.*

---

[2] It can be proved that by this definition of Lim we introduce a convergence in the family of all *closed convex* subsets of X, which generalizes the Hausdorff metric convergence for uniformly bounded sets and the opening convergence for linear subspaces of X, and makes that family an $\mathscr{L}^*$-space in the sense of Kuratowski (see [**9**, p. 83]). We refer to [**15**] and [**16**] for more details on this point.

Our main results can be stated as follows:

THEOREM 1. (*Weak convergence of perturbed solutions*): *Suppose there exists for each* $\varepsilon > 0$ *a solution* u_ε *of* $(*_\varepsilon)$ *and that the family* $(u_\varepsilon)_{\varepsilon>0}$ *is bounded in* X. *Then, under the assumptions* I *and* II *above, there exists a unique solution* u *of* $(*)$ *and* u_ε *converges weakly to* u *in* X *as* $\varepsilon \to 0$. *Furthermore,* $\langle T_\varepsilon u_\varepsilon - Tu, u_\varepsilon - u\rangle \to 0$ *as* $\varepsilon \to 0$.

If, in addition, the following assumption is satisfied

III *For any* $w \in K$, *there exists a continuous, strictly increasing function* $\beta:[0, +\infty) \mapsto [0, +\infty)$, *with* $\beta(0) = 0$, *such that*

$$\beta(\|v - w\|) \leq \liminf |\langle T_\varepsilon v - Tw, v - w\rangle| \text{ as } \varepsilon \to 0, \quad v \in D(T_\varepsilon),$$

uniformly as v *varies in a bounded subset of* X

then the conclusion of Theorem 1 can be improved to assert the convergence of the perturbed solutions in the *strong* topology of X. We have in fact

THEOREM 2. (*Strong convergence of perturbed solutions*): *Suppose that there exists for each* $\varepsilon > 0$ *a solution* u_ε *of* $(*_\varepsilon)$ *and that the family* $(u_\varepsilon)_{\varepsilon>0}$ *is bounded in* X. *Then, under the Assumptions* I, II *and* III *above, there exists a unique solution* u *of* $(*)$ *and* u_ε *converges strongly to* u *in* X *as* $\varepsilon \to 0$.

Theorem 1 and Theorem 2 rely on the assumption that there exists a bounded family of perturbed solutions. Now, a sufficient condition for the existence for every ε of a solution of $(*_\varepsilon)$ is furnished by the general existence theorem stated at the beginning: namely, such a solution exists if, in addition to what is assumed in I and II, we suppose that T_ε is hemicontinuous, K_ε is closed and convex and, moreover, either K_ε is bounded or T_ε is coercive on K_ε. On the other hand, a sufficient condition which implies that any given family $(u_\varepsilon)_{\varepsilon>0}$ of solutions u_ε of $(*_\varepsilon)$ is bounded in X, is given by the following

THEOREM 3. (*"A priori" bound for perturbed solutions*): *Let* $(T_\varepsilon)_{\varepsilon>0}$ *be a family of uniformly bounded mappings from* X *to* X^*, $(K_\varepsilon)_{\varepsilon>0}$ *a family of subsets of* X, $K_\varepsilon \subset D(T_\varepsilon)$ *for every* ε. *Suppose that the following condition holds: There exists a function* $\gamma:[0, +\infty) \mapsto (-\infty, +\infty]$, *with* $\gamma(r) \to +\infty$ *as* $r \to +\infty$, *such that for any family* $(v_\varepsilon)_{\varepsilon>0}$ *in* X, *with* $v_\varepsilon \in K_\varepsilon$ *for each* ε, *there exists a bounded family* $(z_\varepsilon)_{\varepsilon>0}$ *in* X, $z_\varepsilon \in K_\varepsilon$ *for every* ε, *such that*

$$\|v_\varepsilon - z_\varepsilon\| \, \gamma(\|v_\varepsilon - z_\varepsilon\|) \leq \langle T_\varepsilon v_\varepsilon - T_\varepsilon z_\varepsilon, v_\varepsilon - z_\varepsilon\rangle \quad \textit{for all } \varepsilon.$$

Then, any family $(u_\varepsilon)_{\varepsilon>0}$, *with* u_ε *a solution of* $(*_\varepsilon)$ *for each* ε, *is bounded in* X.

REMARK. The condition stated in Theorem 3 is trivially satisfied if $0 \in \bigcap_{\varepsilon>0} K_\varepsilon$ and T_ε is *uniformly coercive on* K_ε, in the following sense: There exists a function γ as above, such that

$$\|v\| \, \gamma(\|v\|) \leq \langle T_\varepsilon v, v\rangle \quad \text{for every } \varepsilon \text{ and all } v \in K_\varepsilon.$$

By combining the Existence Theorem with Theorem 1 and Theorem 3 and taking the preceding remark into account, we obtain the following special result.

THEOREM 4. *Let T be a coercive, bounded, strictly monotone and hemicontinuous map from X to X^*. Let f and f_ε, $\varepsilon > 0$, be elements of X^*, with f_ε converging strongly to f as $\varepsilon \to 0$. Let K be a closed convex subset of $D(T)$, $(K_\varepsilon)_{\varepsilon>0}$ a family of closed convex subsets of $D(T)$, with $0 \in \bigcap_{\varepsilon>0} K_\varepsilon$, such that $K = \operatorname{Lim} K_\varepsilon$ in X. Then, there exists for each ε a unique solution u_ε of the inequality*

$$(*_\varepsilon') \qquad u_\varepsilon \in K_\varepsilon\colon \langle Tu_\varepsilon, v - u_\varepsilon\rangle \geq \langle f_\varepsilon, v - u_\varepsilon\rangle \quad \textit{for all } v \in K_\varepsilon$$

and u_ε converges weakly as $\varepsilon \to 0$ to the unique solution u of

$$(*') \qquad u \in K\colon \langle Tu, v - u\rangle \geq \langle f, v - u\rangle \quad \textit{for all } v \in K.$$

Furthermore, $\langle Tu_\varepsilon - Tu, u_\varepsilon - u\rangle \to 0$ as $\varepsilon \to 0$.

A trivial consequence of Theorem 4 is that u_ε converges *strongly* to u in X, provided T has the following additional property: if v_ε converges weakly to v in X as $\varepsilon \to 0$, v, $v_\varepsilon \in D(T)$ and $\langle Tv_\varepsilon - Tv, v_\varepsilon - v\rangle \to 0$ as $\varepsilon \to 0$, then v_ε converges strongly to v in X as $\varepsilon \to 0$.

A more special result is the following theorem, that can be deduced from the Existence Theorem and Theorems 2 and 3 above:

THEOREM 5. *Let T be a bounded, hemicontinuous map from X to X^*, such that*

$$\|v - w\|\, \alpha(\|v - w\|) \leq \langle Tv - Tw, v - w\rangle \quad \textit{for all } v, w \in D(T),$$

where α is a continuous, strictly increasing function from $[0, +\infty)$ to $[0, +\infty]$, with $\alpha(0) = 0$ and $\alpha(r) \to +\infty$ as $r \to +\infty$. Let f and f_ε, $\varepsilon > 0$, be elements of X^*, f_ε converging strongly to f as $\varepsilon \to 0$. Let K be a closed convex subset of $D(T)$, $(K_\varepsilon)_{\varepsilon>0}$ a family of closed convex subsets of $D(T)$, with $\bigcap_{\varepsilon>0} K_\varepsilon \neq \varnothing$, such that $K = \operatorname{Lim} K_\varepsilon$ in X. Then, there exists for each ε one and only one solution u_ε of $(*_\varepsilon')$ and u_ε converges strongly in X as $\varepsilon \to 0$ to the unique solution u of $(*')$.

The theorems above are a slightly different formulation of results proved in [**15**].

3. **Application to variational problems for differential operators.** We shall apply the results of the previous section to some "perturbed" boundary value problems and more general variational problems for a nonlinear (elliptic) partial differential operator A of order $2m$ in divergence form

$$(1) \qquad Av = \sum_{|\alpha| \leq m} (-1)^{|\alpha|} D^\alpha A_\alpha(x, Dv, \ldots, D^m v).$$

We are denoting by m a positive integer, by $x \equiv (x_1, \ldots, x_n)$ the general point in an n-dimensional Euclidean space $\boldsymbol{R}^n$ and for any n-tuple $\alpha \equiv (\alpha_1, \ldots, \alpha_n)$ of nonnegative integers we use the standard notation

$$D^\alpha = \prod_{i=1}^n (\partial/\partial x_i)^{\alpha_i}, \qquad |\alpha| = \sum_{i=1}^n \alpha_i.$$

We suppose that for each multi-index α, A_α is a real function of x in an open subset Ω of $\boldsymbol{R}^n$ and of the vector $\xi \equiv \{\xi_\alpha : \xi_\alpha \in \boldsymbol{R}, |\alpha| \leq m\}$ in the space $\boldsymbol{R}^l$, l being the number of derivatives of order $\leq m$ in the space $\boldsymbol{R}^n$, which satisfies the following

ASSUMPTION (a). *A_α is measurable in $x \in \Omega$ for fixed $\xi \in R^l$ and is continuous in $\xi \in R^l$ for almost all $x \in \Omega$; moreover, A_α is of polynomial growth in ξ, that is, we have*

$$|A_\alpha(x, \xi)| \leq k(x) + c\,|\xi|^{p-1}, \qquad x \in \Omega,\ \xi \in R^l,$$

with $1 < p < +\infty$, $c > 0$ and k a given function in $L^{p'}(\Omega)$, $p' = p/(p-1)$.

If $W^{m,p}(\Omega)$ denotes as usual the Sobolev space of all real functions $v \in L^p(\Omega)$ whose distribution derivatives of order $\leq m$ also belong to $L^p(\Omega)$, normed by

$$\|v\|_{m,p} = \left(\sum_{|\alpha| \leq m} \|v\|^p_{L^p(\Omega)} \right)^{1/p},$$

then we find, as a consequence of the hypothesis (a), that for every α and all functions $v \in W^{m,p}(\Omega)$,

$$A_\alpha(x, v, Dv, \ldots, D^m v) \in L^{p'}(\Omega),$$

where $A(x, v, Dv, \ldots, D^m v) \equiv A_\alpha(x, \xi(u)(x))$, and $\xi(u)(x) = \{D^\alpha u(x) : |\alpha| \leq m\}$ for $x \in \Omega$.

Thus, the generalized Dirichlet form

$$a(u, v) = \sum_{|\alpha| \leq m} \int_\Omega A_\alpha(x, u, Du, \ldots, D^m u) D^\alpha v \, dx,$$

associated to the given operator A, is well defined for every u and v in $W^{m,p}(\Omega)$ and satisfies an inequality of the form

$$|a(u, v)| \leq g(\|u\|_{m,p})\, \|v\|_{m,p} \tag{2}$$

with g some continuous function of $r \in \boldsymbol{R}$.

Now, let X be a given closed linear subspace of $W^{m,p}(\Omega)$; X is a real reflexive Banach space, for $W^{m,p}(\Omega)$ is such, and we shall denote as above by $\langle . , . \rangle$ the pairing between X and its dual X^*.

We shall consider the following problem

$$u \in K : a(u, v - u) \geq \langle f, v - u \rangle \quad \text{for all } v \in K, \tag{P}$$

involving the Dirichlet form a of the operator A, a closed convex subset K of X and an element f of X^*.

We can write (P) as a variational inequality of the form

$$u \in K : \langle Tu, v - u \rangle \geq \langle f, v - u \rangle \quad \text{for all } v \in K, \tag{Q}$$

by introducing the map T of X into X^*, defined by the identity

$$a(u, v) = \langle Tu, v \rangle, \quad u, v \in X.$$

By the estimate (2) above, this equality indeed defines, for each u of X, an element Tu of X^* and the resulting mapping T of X into X^* turns out to be bounded and continuous.

Let us suppose now that the operator A satisfies the following ellipticity condition:

ASSUMPTION (b). *There exists a constant $c_0 > 0$, such that*

$$\sum_{|\alpha| \leq m} (A_\alpha(x, \xi) - A_\alpha(x, \xi'))(\xi_\alpha - \xi'_\alpha) \geq c_0 |\xi - \xi'|^p$$

for all $x \in \Omega$ and every pair $\xi, \xi' \in \boldsymbol{R}^l$.

Then, the map T determined as above by the operator A and the subspace X satisfies the inequality

$$\langle Tu - Tv, u - v \rangle \geq \alpha(\|u - v\|_{m,p}) \, \|u - v\|_{m,p}, \quad u, v \in X,$$

where $\alpha(r) = c_0 r^{p-1}$, $r \in \boldsymbol{R}$. In particular, T is also coercive and strictly monotone.[3]

Therefore, by applying the Existence Theorem of the previous section to the map $T - f : v \mapsto Tv - f$, we find that there exists a unique solution u of (Q), which is to say, a unique solution u of the original problem (P).

Now, in order to describe how the solution u of (P) depends on f and K, let us consider a family $(f_\varepsilon)_{\varepsilon > 0}$ of vectors of X^* and a family $(K_\varepsilon)_{\varepsilon > 0}$ of closed convex subsets of X and, for each ε, the problem

$$(\mathrm{P}_\varepsilon) \qquad u_\varepsilon \in K_\varepsilon : a(u_\varepsilon, v - u_\varepsilon) \geq \langle f_\varepsilon, v - u_\varepsilon \rangle \quad \text{for all } v \in K_\varepsilon.$$

Again, this problem corresponds to the variational inequality

$$(\mathrm{Q}_\varepsilon) \qquad u_\varepsilon \in K_\varepsilon : \langle Tu_\varepsilon, v - u_\varepsilon \rangle \geq \langle f_\varepsilon, v - u_\varepsilon \rangle \quad \text{for all } v \in K_\varepsilon,$$

and therefore has a unique solution u_ε.

Our map T satisfies all the hypotheses of Theorem 5. Therefore, by that theorem, if f_ε converges strongly to f and K_ε converges to K in X in the sense of §2 as $\varepsilon \to 0$, then u_ε converges strongly to u as $\varepsilon \to 0$.

Summarizing, the following theorem holds:

THEOREM 6. *Let A be a partial differential operator of order $2m$ in divergence form* (1) *satisfying assumptions* (a) *and* (b) *above in an open subset Ω of $\boldsymbol{R}$, with $1 < p < +\infty$. Let X be a closed linear subspace of $W^{m,p}(\Omega)$, K a closed convex subset of X and f an element of X^*. Let $(f_\varepsilon)_{\varepsilon>0}$ be a family of elements of X^* converging strongly to f in X as $\varepsilon \to 0$, $(K_\varepsilon)_{\varepsilon>0}$ a family of closed convex subsets of X, with $\bigcap_{\varepsilon>0} K_\varepsilon \neq \varnothing$, such that*

$$(3) \qquad K = \operatorname{Lim} K_\varepsilon \quad \text{in } X,$$

---

[3] A detailed discussion, in connection with the theory of monotone and semimonotone operators, of the properties of the map T arising from a differential operator such as A can be found, for instance, in [**2**] and [**10**], see also [3].

in the sense of §2. *Then, there exists for each* ε *a unique solution* u_ε *of problem* (P_ε) *and* u_ε *converges strongly in* X *as* $\varepsilon \to 0$ *to the unique solution* u *of problem* (P).

More general results, taking, for instance, perturbations of the operator A also into account and under weaker assumptions than (a) and (b), could be obtained by applying the more general theorems of §2.

However, in the following section we shall confine ourselves to illustrate Theorem 6 by some examples and, in particular, to show what kind of concrete analytical conditions may be involved in the convergence (3).

4. **Two special problems.** The general problem (P) considered in the previous section includes as a special case any variational boundary value problem for a partial differential operator such as A.

In fact, let V be any closed *linear subspace* of $W^{m,p}(\Omega)$ and let us assume $K = V$ in (P). Then, by the linearity of $v \mapsto a(u, v) - \langle f, v\rangle$, (P) is easily seen to be equivalent to the problem

$$(\mathrm{P}') \qquad u \in V: a(u, v) = \langle f, v\rangle \quad \text{for all } v \in V.$$

Now, suppose that $W_0^{m,p}(\Omega) \subset V$, where $W_0^{m,p}(\Omega)$ denotes as usual the closure in $W^{m,p}(\Omega)$ of all (real) C^∞ functions with compact support in Ω. It is well known that (P′) is then the variational formulation of the boundary value problem for the equation $Au = f$ with boundary conditions corresponding to the given subspace V of $W^{m,p}(\Omega)$. A solution u of (P′) will be called a *variational solution* of the original boundary value problem.

For example, if

$$V = W_0^{m,p}(\Omega), \tag{4}$$

then (P′) is the variational formulation of the m-order Dirichlet problem

$$(\mathrm{D}) \qquad \begin{aligned} Au &= f \quad \text{in} \quad \Omega, \\ D^\beta u &= 0 \quad \text{on} \quad \partial\Omega,\ |\beta| \le m-1 \end{aligned}$$

where $\partial\Omega$ denotes the boundary of Ω.

Now, let us suppose that we have such a problem (D) and we want to consider the following perturbation of it:

$$(\mathrm{D}_\epsilon) \qquad \begin{aligned} Au_\epsilon &= f_\epsilon \quad \text{in} \quad \Omega_\epsilon \\ D^\beta u_\epsilon &= 0 \quad \text{on} \quad \partial\Omega_\epsilon, \quad |\beta| \le m-1 \end{aligned}$$

where for each ε, f_ε is a distribution in the dual $W^{-m,p'}(\Omega)$ of $W^{m,p}(\Omega)$[4] and $\Omega_\varepsilon = \Omega - E_\varepsilon$, E_ε being a compact subset of Ω.

It is natural to expect that if f_ε converges to f in a suitable sense and E becomes "small" as $\varepsilon \to 0$, then the solution u_ε of (D_ε) should converge to the solution u of

[4] It is well known that the dual of $W_0^{n,p}(\Omega)$ can be identified with the space $W^{-m,p'}(\Omega)$ of all distributions f in Ω, such that $f = \sum_{|\alpha| \le m} D^\alpha v_\alpha$, with $v_\alpha \in L^{p'}(\Omega)$, $p' = p/(p-1)$.

(D) as $\varepsilon \to 0$. We shall see, indeed, that a result of this type can be obtained by applying Theorem 6 to the variational problems associated to (D) and (D_ε).

We have already noted that the variational formulation of (D) is given by (P′), with the choice (4) of V. Similarly, if we identify the space $W_0^{m,p}(\Omega_\varepsilon)$ with a (closed linear) subspace V_ε of $W_0^{m,p}(\Omega)$, by simply defining any function $v \in W_0^{m,p}(\Omega_\varepsilon)$ to be $\equiv 0$ on E_ε, then the variational formulation of (D_ε) is given by

$$(\mathrm{P}'_\varepsilon) \qquad u_\varepsilon \in V_\varepsilon \colon a(u_\varepsilon, v) = \langle f_\varepsilon, v\rangle \quad \text{for all } v \in V_\varepsilon.$$

We are in position now to apply Theorem 6, with $X = K = V = W_0^{m,p}(\Omega)$ and $K_\varepsilon = V_\varepsilon = W_0^{m,p}(\Omega_\varepsilon)$. However, let us first describe in concrete analytical terms what the convergence of K_ε to K assumed in that theorem amounts to in the case at hand.

The notion of $M_{\alpha,s}$ *capacity* of a compact subset E of R^n turns out to be useful in this respect. If $\alpha \geq 1$ is an integer and $1 \leq s < +\infty$, $M_{\alpha,s}(E)$ is defined by

$$M_{\alpha,s}(E) = \inf \|\varphi\|_{\alpha,s}$$

where the infimum is taken over all C^∞ functions having compact support in R^n and ≥ 1 near E.

The $M_{\alpha,s}$ capacity has been used, when $\alpha = 1$, by W. Littman—G. Stampacchia—H. F. Weinberger [**14**] and J. Serrin [**18**] in investigating questions concerning regular points and removable singularities for second order partial differential operators, and it is closely related, when $n - \alpha s \geq 0$, to the classical α-capacity of the potential theory, see H. Wallin [**19**] and W. Littman [**13**].

LEMMA 1. *Let Ω be an open subset of R^n, $(E_\varepsilon)_{\varepsilon>0}$ a family of compact subsets of Ω, α a positive integer and $1 \leq s < +\infty$. Let us consider the space $W^{\alpha,s}(\Omega)$ and, for each ε, let us identify the space $W_0^{\alpha,s}(\Omega_\varepsilon)$, where $\Omega_\varepsilon = \Omega - E_\varepsilon$, with a (closed linear) subspace of $W_0^{\alpha,s}(\Omega)$. Then, we have*

$$W_0^{\alpha,s}(\Omega) = \operatorname{Lim} W_0^{\alpha,s}(\Omega_\varepsilon) \quad \text{in } W_0^{\alpha,s}(\Omega),$$

according to the definition of §2, if and only if for each compact subset Ω' of Ω,

$$M_{\alpha,s}(E_\varepsilon \cap \Omega') \to 0 \quad \text{as } \varepsilon \to 0.$$

When $\alpha = 1$ this lemma was proved in [**15**], Lemma 1.8, and that proof can be easily generalized to all α by using Lemma 2 of [**13**].

By applying Theorem 6 as we have described above and taking Lemma 1 into account, we can prove then the following

THEOREM 7. *Let A be a partial differential operator of order $2m$ in divergence form* (1) *satisfying assumptions* (a) *and* (b) *in an open subset Ω of R^n, $1 < p < +\infty$. Let f be a given distribution in the dual $W^{-m,p'}(\Omega)$ of $W_0^{m,p}(\Omega)$, $p' = p/(p-1)$, and $(f_\varepsilon)_{\varepsilon>0}$ a family of distributions in $W^{-m,p'}(\Omega)$ converging strongly to f as $\varepsilon \to 0$. Let $(E_\varepsilon)_{\varepsilon>0}$ be a family of compact subsets of Ω such that for each compact subset Ω' of Ω,*

$$M_{m,p}(E_\varepsilon \cap \Omega') \to 0 \quad \text{as } \varepsilon \to 0.$$

Then, there exists for every ε *one and only one variational solution* u_ε *of problem* (D_ε), *and* u_ε *converges strongly in* $W_0^{m,p}(\Omega)$ *as* $\varepsilon \to 0$ *to the unique variational solution* u *of problem* (D).

Let us remark at this point that in the variational boundary value problems that we have considered so far, only the special form (P′) of problem (P) is involved.

On the other hand, we know that more general variational problems for a differential operator of the form (1) naturally arise in the general form of a variational inequality such as (P); for instance, when A is the Fréchet differential of a m-order multiple integral in Ω of type

$$(5) \qquad F(v) = \int_\Omega \chi(x, v, Dv, \ldots, D^m v)\, dx,$$

and we have to minimize F on a closed convex subset K of the space $W^{m,p}(\Omega)$. Then, unless a minimizing vector u is an *interior* point of K, in which case u is a solution of the Euler-Lagrange equation $Au = 0$, hence a solution of a problem of the form (P′), in the general case such u simply is a solution of the inequality $\langle Au, v - u\rangle \geq 0$ for all $v \in K$, that is, a solution of a variational inequality like (P).

Many examples of problems (P) have been discussed by Professor Stampacchia in his lecture. One of these problems can be stated as follows.

Let Ω be again an open subset of R^n, E a compact subset of Ω. For any pair v_1, v_2 of functions of $W_0^{m,p}(\Omega)$, we shall write

$$v_1 \geq v_2 \quad \text{on } E \text{ in } W_0^{m,p}(\Omega),$$

if $v_1 - v_2 \in W_0^{m,p}(\Omega, E)$, where $W_0^{m,p}(\Omega, E)$ denotes the closure in $W_0^{m,p}(\Omega)$ of the convex cone $C_0^\infty(\Omega, E)$ of all real C^∞ functions having compact support in Ω and ≥ 0 near E.

Now, let A be a partial differential operator of order $2m$ in divergence form (1) and ψ a given function of $W_0^{m,p}(\Omega)$. Let us consider the closed convex cone

$$K(\psi, E) = \{v \in W_0^{m,p}(\Omega) : v \geq \psi \quad \text{on } E \text{ in } W_0^{m,p}(\Omega)\}$$

and the problem

$$(\mathrm{P}'') \qquad u \in K(\psi, E) : a(u, v - u) \geq \langle f, v - u\rangle \quad \text{for all } v \in K(\psi, E)$$

where a is the Dirichlet form of A and f is a given distribution in $W^{-m,p'}(\Omega)$.

As we know from §3, if the operator A satisfies the assumptions (a) and (b) of that section, then the problem above has one and only one solution u. The properties of this solution have been investigated, when A is a second order elliptic operator, by H. Lewy and G. Stampacchia [**11**].

We want here to describe the dependence of u on ψ, E and f. To this end, let us consider a family $(\psi_\varepsilon)_{\varepsilon>0}$ of functions of $W_0^{m,p}(\Omega)$, a family $(E_\varepsilon)_{\varepsilon>0}$ of compact subsets of Ω, a family $(f_\varepsilon)_{\varepsilon>0}$ of distributions of $W^{-m,p'}(\Omega)$, and, for each ε, the problem

$$(\mathrm{P}''_\varepsilon) \qquad u_\varepsilon \in K(\psi_\varepsilon, E_\varepsilon) : a(u_\varepsilon, v - u_\varepsilon) \geq \langle f_\varepsilon, v - u_\varepsilon\rangle \quad \text{for all } v \in K(\psi_\varepsilon, E_\varepsilon),$$

where $K(\psi_\varepsilon, E_\varepsilon)$ is defined as above. Again, this problem has one and only one solution u_ε.

The following lemma provides a sufficient condition for the convergence of $K(\psi_\varepsilon, E_\varepsilon)$ to $K(\psi, E)$ in $W_0^{m,p}(\Omega)$ in the sense of §2. For sake of simplicity we shall confine ourselves to the case $E \subset E_\varepsilon$.

LEMMA 2. *Let ψ and ψ_ε, $\epsilon > 0$, be functions of $W_0^{m,p}(\Omega)$ with ψ_ε converging strongly to ψ as $\varepsilon \to 0$. Let E and E_ε, $\varepsilon > 0$, be compact subsets of Ω, with $E \subset E_\varepsilon$ such that for each compact subset Ω' of Ω,*

$$M_{m,p}(E_\varepsilon - E \cap \Omega') \to 0 \quad as\ \varepsilon \to 0.$$

Then,

$$K(\psi, E) = \operatorname{Lim} K(\psi_\varepsilon, E_\varepsilon) \quad in\ W_0^{m,p}(\Omega).$$

When $E_\varepsilon = E$ for every ε, Lemma 2 is a consequence of Lemma 1.7 of [**15**]. The general case will be considered elsewhere.

Taking Lemma 2 into account, we then deduce from Theorem 6 the following

THEOREM 8. *Let A, f, and $(f_\varepsilon)_{\varepsilon>0}$ be as in Theorem 7, ψ, $(\psi_\varepsilon)_{\varepsilon>0}$, and E, $(E_\varepsilon)_{\varepsilon>0}$ as in Lemma 2. Then, there exists for each ε a unique solution u_ε of* (P$''_\varepsilon$) *and such u_ε converges strongly in $W_0^{m,p}(\Omega)$ to the unique solution u of* (P$''$) *as $\varepsilon \to 0$.*

5. **The "degenerate" case.** A key role, for instance in Theorem 5, for the existence of a bounded family $(u_\varepsilon)_{\varepsilon>0}$ of perturbed solutions is played by the *coerciveness* of the map T, while the basic assumption yielding the convergence of the whole family $(u_\varepsilon)_{\varepsilon>0}$ to a solution u of the initial inequality (*) is the *uniqueness* of such u, actually implied by the strict monotonicity of T.

Therefore, when T is not necessarily coercive and a solution of (*) may not be unique, considerable changes in generalizing the perturbation theory of §2 should be expected. As we shall briefly describe below, the basic device to deal with this "degenerate" case is the so called "elliptic regularization," that has been used in connection with variational inequalities by J. L. Lions and G. Stampacchia, [**12**] and F. E. Browder [**6**].

This device consists in adding to the given map T a coercive, strictly monotone term εM, $\varepsilon > 0$, solving the variational inequality ($*_\varepsilon$) for the map $T_\varepsilon = T + \varepsilon M$ and the given K, and then letting $\varepsilon \to 0$.

As in §2, we shall assume below that a simultaneous perturbation K_ε of K is also allowed. The set K_ε, however, is now required to converge rapidly enough to K, to keep the correction εM acting as $\varepsilon \to 0$.

If K and K_ε, $\varepsilon > 0$, are subsets of X and α is a nonnegative real number, we say that K_ε *converges of order* $\geq \alpha$ in X as $\varepsilon \to 0$, if the following two conditions are satisfied (notation from §2):

(j) For each $v \in K$, we have

$$0 \in s = \operatorname{Lim\,inf} \varepsilon^\alpha (K_\varepsilon - v)^5.$$

[5] If v is a vector and S a subset of X, then for any $r \in \boldsymbol{R}$ we put

$$r(S - v) = \{z \in X : z = r(w - v),\ w \in S\},$$

and similarly for $r(v - S)$.

(jj) For any weakly convergent sequence (v_j), with $v_j \in K_{\varepsilon_j}$ for every j and $\varepsilon_j \to 0$ as $j \to +\infty$, we have

$$0 \in w\text{-Lim sup } \varepsilon_j^{\alpha}(v_j - K).$$

Note that K_ε converges to a closed convex K in the sense of §2, i.e. $K = \text{Lim } K_\varepsilon$, if and only if K_ε converges to K of order ≥ 0, according to the definition above. Thus, if K_ε converges to K of order $\geq \alpha$, for some $\alpha \geq 0$, then, in particular, $K = \text{Lim } K_\varepsilon$.

The following analogue of Theorem 1 holds:

THEOREM 9. *Let T and M be bounded, monotone and hemicontinuous mappings from X to X^*, with $D(T) = D(M)$ and M, in addition, strictly monotone. Let T_ε be for each $\varepsilon > 0$, the map $T + \varepsilon M$ and K_ε a subset of $D(T)$ converging of order $\geq \alpha > 0$ in X as $\varepsilon \to 0$ to a closed convex subset K of $D(T)$. Suppose there exists for each ε a solution u_ε of $(*_\varepsilon)$, the family $(u_\varepsilon)_{\varepsilon>0}$ being bounded in X. Then, u_ε converges weakly to a solution u of $(*)$ as $\varepsilon \to 0$, and such u is uniquely determined in the set S of all solutions of $(*)$ as the solution of the variational inequality*

$$u \in S: \langle Mu, w - u\rangle \geq 0 \quad \textit{for all } w \in S.$$

Furthermore,

$$\langle Mu_\varepsilon - Mu, u_\varepsilon - u\rangle \to 0 \quad \textit{as } \varepsilon \to 0.$$

It follows trivially from the last conclusion of the theorem that u_ε converges *strongly* to u in X, whenever M has the property stated for the map T after Theorem 4 of §2. In particular, when M is a *duality mapping* in X (see F. E. Browder [**4**], [**5**]).

The proof of Theorem 9 can be found in [**15**]. The theorem is false if we only assume that K_ε converges to K in the sense of §2, but not necessarily of order $\geq \alpha$, with $\alpha > 0$, see [**16**].

Finally, let us notice that an analogue of Theorem 3 could be also proved. However, to obtain an "a priori" bound for the solutions of $(*_\varepsilon)$, we now need to *assume*, in addition to a suitable coerciveness of M and a certain "uniform closeness at ∞" of K_ε to K (conditions IV_0 and IV_2 of §5 of [**15**], respectively), that some solution of the original inequality $(*)$ actually exists.

6. **Final remarks.** Let F be a convex functional on the space X, K a closed convex subset of X and suppose that we want to describe how the solutions of the minimum problem

$$u \in K: F(u) \leq F(v) \quad \text{for all } v \in K$$

depend on F and K.

If F has a Fréchet differential ∇F, which is then a monotone map of X into X^*, then we can apply the perturbation theory of §2 to the variational inequality

$$u \in K: \langle \nabla Fu, v - u\rangle \geq 0 \quad \text{for all } v \in K,$$

which characterizes the solution u of the minimum problem above.

Perturbation results of this type, involving a multiple integral F of the form (5), can be obtained, for example, from Theorem 6 of §3 and the more special Theorems 7 and 8 of §4.

Let us briefly describe now how the theory of §2 can be also applied in the framework of the "direct methods" of the calculus of variations, in connection with a not necessarily differentiable F.

To this end, let us consider in the product space $X \times \boldsymbol{R}$ the subset $\mathscr{K}$ of all vectors $[v, \beta]$, with $v \in K$ and $\beta \geq F(v)$. Let Φ be the projection

$$\Phi : [v, \beta] \mapsto \beta$$

of $X \times \boldsymbol{R}$ into $\boldsymbol{R}$. If u is a vector of K that minimizes F on K, then, clearly, the functional Φ achieves its minimum over $\mathscr{K}$ at the point $[u, F(u)]$.

This means that $[u, F(u)]$ is a solution of the variational inequality

$$[u, \alpha] \in \mathscr{K} : \langle \nabla\Phi[u, \alpha], [v, \beta] - [u, \alpha] \rangle \geq 0 \quad \text{for all } [v, \beta] \in \mathscr{K} \tag{6}$$

where $\nabla\Phi$ is the gradient of Φ, that is, $\nabla\Phi = 0 \oplus 1$ with $0 \oplus 1$ the map $[v, \beta] \mapsto [0, 1]$ of $X \times \boldsymbol{R}$ into $(X \times \boldsymbol{R})^* \simeq X^* \times \boldsymbol{R}$, and $\langle [v_1, \beta_1], [v_2, \beta_2] \rangle = \langle v_1, v_2 \rangle + \beta_1\beta_2$ for every pair $[v_1, \beta_1], [v_2, \beta_2] \in X \times \boldsymbol{R}$.

Thus, the original minimum problem has been reduced to a special variational inequality involving the (monotone) map $0 \oplus 1$ and the set $\mathscr{K}$, which is closed and convex provided K is such and F is lower-semicontinuous and convex, and then the theory of §2 can be applied.

Existence and perturbation results obtained along these lines can be found in [**17**] and [**15**]. In these papers a generalization of (6) is treated, with the map $\nabla\Phi \equiv 0 \oplus 1$ replaced by the map

$$T \oplus 1 : [v, \beta] \mapsto [Tv, 1],$$

T being any monotone map of X into X^*. It is easy to see that (6) is then equivalent to the inequality

$$u \in K : \langle Tu, v - u \rangle \geq F(u) - F(v) \quad \text{for all } v \in K,$$

that has been already studied by F. E. Browder and others, see [**7**].

Finally, let us remark that the results of §2 and §5 can be also used to prove the convergence of certain finite-dimensional approximations of solutions of variational inequalities. The results obtained, for suitable finite-dimensional choices of T_ε and K_ε, are related both to the "projection methods" of F. E. Browder and Petryshyn and to the "injective methods" of Cea, Aubin, and others, see [**15**] for further references.

References

1. F. E. Browder, *Nonlinear monotone operators and convex sets in Banach spaces*, Bull. Amer. Math. Soc. **71** (1965), 780–785.

2. ———, *Nonlinear elliptic boundary value problems*, II, Trans. Amer. Math. Soc. **117** (1965), 530–550.

3. ———, *Existence and uniqueness theorems for solutions of nonlinear boundary value problems*, Proc. Sympos. Appl. Math., Vol. 17, 1965, 24–29.

4. ———, *On a theorem of Beurling and Livingston*, Canad. J. Math. **17** (1965).

5. ———, *Les problèmes nonlinéaires*, Les Presses de l'Univ. de Montréal, (1965).

6. ———, *Existence and approximation of solutions of nonlinear variational inequalities*, Proc. Nat. Acad. Sci. U.S.A. **56** (1966), 1080–1086.

7. ———, *On the unification of the calculus of variations and the theory of monotone nonlinear operators in Banach spaces*, Proc. Nat. Acad. Sci. U.S.A. **56** (1966), 419–425.

8. P. Hartman and G. Stampacchia, *On some nonlinear elliptic differential functional equations*, Acta Math. **115** (1966), 271–310.

9. C. Kuratowski, *Topologie*, I, Warszawa-Wroclaw, 1948.

10. J. Leray and J. L. Lions, *Quelques resultats de Visik sur les problèmes elliptic nonlinéaires par les methodes de Minty-Browder*, Bull. Soc. Math. France **93** (1965), 97–107.

11. H. Lewy and G. Stampacchia, *On the regularity of the solution of a variational inequality*, Comm. Pure Appl. Math. (to appear).

12. J. L. Lions and G. Stampacchia, *Variational inequalities*, Comm. Pure Appl. Math. **20** (1967), 493–519.

13. W. Littman, *A connection between* α*-capacity and* m-p *polarity*, Bull. Amer. Math. Soc. **73** (1967), 862–866.

14. W. Littman, G. Stampacchia and H. F. Weinberger, *Regular points for elliptic equations with discontinuous coefficients*, Ann. Scuola Norm Sup. Pisa, **17** (1963), 45–79.

15. U. Mosco, *Convergence of convex sets and of solutions of variational inequalities*, Advances in Math. (to appear).

16. ———, *Convergence of solutions of variational inequalities*, Lectures Nato Summer School on Monotone operators, Venezia 1968.

17. ———, *A remark on a theorem of F. E. Browder*, J. Math. Anal. Appl. **20** (1967), 90–93.

18. J. Serrin, *Local behavior of solutions of quasilinear equations*, Acta Math. **3** (1964), 247–302.

19. H. Wallin, *A connection between* α*-capacity and* L^p*-classes of differentiable functions*, Ark. Mat. **5** (1964), 331–341.

Istituto Matematico
dell'Università e
Scuola Normale Superiore
Pisa

MANIFOLDS OF SECTIONS OF FIBER BUNDLES AND THE CALCULUS OF VARIATIONS

Richard S. Palais[1]

Let M be a compact, n-dimensional, C^∞ manifold, possibly with boundary. Denote by $FB(M)$ the category of C^∞ fiber bundles over M. A morphism $f:E \to F$ of $FB(M)$ is a C^∞ map such that $f_x = f \mid E_x$ maps E_x into F_x for each $x \in M$, where E_x is the fiber of E at x. If E and F are C^∞ vector bundles and each f_x is linear we call f a vector bundle morphism. We denote by $VB(M)$ the category of C^∞ vector bundles over M and vector bundle morphisms and by $FVB(M)$ the mongrel category of C^∞ vector bundles over M and fiber bundle morphisms.

Our goals in this talk are the following:

(1) To describe the Sobolev Functors L_k^p $(k > n/p)$ from $FB(M)$ to the category of C^∞ Banach manifolds and C^∞ maps. (As a set $L_k^p(E)$ is a certain subset of the continuous sections of E, roughly speaking those which in local coordinates and local trivializations have derivatives of order $\leq k$ which are pth power summable. If $f:E \to F$ is a morphism then $L_k^p(f):L_k^p(E) \to L_k^p(F)$ is given by $s \mapsto f \circ s$.)

(2) To interpret the Calculus of Variations as the study of the critical points of a certain type of differentiable real valued functions on the manifolds $L_k^p(E)$.

(3) To describe how using the latter point of view the strong global existence theorems provided by Morse Theory and Lusternik-Schnirelman theory, as well as certain classical smoothness theorems, extend naturally from one independent variable problems to certain problems with several independent variables in the calculus of variations.

A considerably more detailed account of the ideas and results presented here will be found in the authors *Foundations of global non-linear analysis* published by W. A. Benjamin Inc. (1968) and in the doctoral dissertation of Mrs. Karen Uhlenbeck (Brandeis, 1968).

1. **The L_k^p spaces of a vector bundle.** Let ξ be a C^∞ vector bundle over M and $1 \leq p < \infty$. Given a strictly positive smooth measure μ on M and a Riemannian structure $\langle\ ,\ \rangle$ for ξ define $L^p(\xi)$ to be the Banach space of measurable sections s of ξ such that $\|s\|_{L^p}^p = \int \langle s(x), s(x)\rangle^{p/2}\, d\mu(x)$ is finite. If we change μ or the Riemannian structure we get the same topological vector space with an equivalent norm, i.e., $L^p(\xi)$ is a well-defined topological vector space independent of any

[1] Research supported in part by Sloan Foundation Fellowship and Air Force Grant No. AFOSR-68-1403.

particular choice of μ or $\langle\ ,\ \rangle$. Let $J^k(\xi)$ denote the k-jet bundle of ξ and $j_k : C^\infty(\xi) \to C^\infty(J^k(\xi))$ the k-jet extension map. We denote by $L^p_k(\xi)$ (k a nonnegative integer) the completion of $C^\infty(\xi)$ under the topology induced from the injection

$$C^\infty(\xi) \xrightarrow{j_k} C^\infty(J^k(\xi)) \xrightarrow{\subseteq} L^p(J^k(\xi)).$$

We note that we have a continuous inclusion $L^p_{k+1}(\xi) \subseteq L^p_k(\xi)$ which by Rellich's theorem is even completely continuous. This allows us to define $L^p_k(\xi)$ for nonintegral positive k by using the "complex method of interpolation" between the integral values. If $f : \xi \to \eta$ is a vector bundle morphism over M then $s \mapsto f \circ s$ defines a continuous linear map $L^p_k(f) : L^p_k(\xi) \to L^p_k(\eta)$. This establishes each L^p_k as a functor from $VB(M)$ to the category of Banach spaces and continuous linear maps.

If $f : \xi \to \eta$ is a *fiber* bundle morphism of vector bundles (i.e. a morphism of $FVB(M)$) then for $pk > n$, $s \mapsto f \circ s$ still defines a continuous (but of course nonlinear) map $L^p_k(f) : L^p_k(\xi) \to L^p_k(\eta)$ and it follows easily that $L^p_k(f)$ must indeed be C^∞ (if $pk \leq n$ then in general $f \circ s$ will not even be in $L^p_k(\eta)$ for all $s \in L^p_k(\xi)$). Thus we have

THEOREM. *For $pk > n$, L^p_k extends to a functor from $FVB(M)$ to the category of Banach spaces and C^∞ maps.*

The above is a fundamental result for our approach to nonlinear analysis in general and the calculus of variations in particular and we sketch the basic reasons for it.

First of all we have the classical results:

SOBOLEV EMBEDDING THEOREMS. *If $k - n/p \geq l - n/q$ and $k \geq l$ then $L^p_k(\xi) \subseteq L^q_l(\eta)$ and the inclusion map is continuous (and even completely continuous if the inequalities are strict). Also if $k - n/p = l + \alpha$, $0 < \alpha < 1$ then $L^p_k(\xi) \subseteq C^{l+\alpha}(\xi)$ and the inclusion is continuous.*

Given vector bundles $\xi_1, \ldots, \xi_r, \eta$ over M let $L^r(\xi_1, \ldots, \xi_r, \eta)$ denote the bundle of r-linear maps of $\xi_1 \oplus \cdots \oplus \xi_r$ into η. Given a section T of the latter and sections s_i of ξ_i we get a section $T(s_1, \ldots, s_r)$ of η whose value at x is $T(x)(s_1(x), \ldots, s_r(x))$. The following is an easy generalization (and consequence) of Hölder's inequality.

THEOREM. *The function $(T, s_1, \ldots, s_r) \mapsto T(s_1, \ldots, s_r)$ defines a continuous multilinear map of $C^0(L^r(\xi_1, \ldots, \xi_r; \eta)) \oplus L^{q_1}(\xi_1) \oplus \cdots \oplus L^{q_r}(\xi_r)$ into $L^q(\eta)$ provided $1/q \geq \sum_{i=1}^r 1/q_i$.*

Putting the latter together with the Sobolev embedding theorem we obtain the following basic result.

MULTIPLICATION THEOREM. *The function $(T, s_1, \ldots, s_r) \mapsto T(s_1, \ldots, s_r)$ is a continuous multilinear map of $C^0(L^r(\xi_1, \ldots, \xi_r; \eta)) \oplus L^{p_1}_{k_1}(\xi_1) \oplus \cdots \oplus L^{p_r}_{k_r}(\xi_r)$ into $L^q_l(\eta)$ provided $n/q - l \geq \sum_{i \in A} (n/p_i - k_i)$, where A is the set of indices $i = 1, 2, \ldots,$*

r such that $n/p_i - k_i > 0$ and where the inequality must be strict if, for some i, $n/p_i = k_i$.

It is from the latter result that one derives, rather straightforwardly, that for $pk > n$ a fiber bundle morphism $f: \xi \to \eta$ defines a C^∞ map $s \mapsto f \circ s$ of $L^p_k(\xi)$ to $L^p_k(\eta)$. The details will be found in §9 of *Foundations of global non-linear analysis*.

2. Vector bundle neighborhoods and the differentiable structure of $L^p_k(E)$. If E and F are C^∞ fiber bundles over M, the F is called a sub-bundle of E if $F \subseteq E$ and the inclusion map is a fiber bundle morphism. If in addition F is open (closed) in E it is called an open (closed) sub-bundle of E. By a vector bundle neighborhood (VBN) in E we shall mean a vector bundle ξ over M which, considered as a C^∞ fiber bundle, is an open sub-bundle of E. We have the following fundamental existence theorem, whose proof is analogous to that of the tubular neighborhood theorem.

EXISTENCE THEOREM FOR VBN. *If E is a C^∞ fiber bundle over M, $s \in C^0(\xi)$, and $\mathcal{O}$ any neighborhood of $s(M)$ in E then there exists a VBN ξ in E with $s(M) \subseteq \xi \subseteq \mathcal{O}$ (and in particular $s \in C^0(\xi)$). If $s \in C^\infty(E)$ we can choose ξ so that s is the zero section of ξ.*

Now let E be a C^∞ fiber bundle over M, $s \in C^0(E)$, and ξ a VBN of s in E (i.e. a VBN in E such that $s \in C^0(\xi)$), the existence of which is assured by the above theorem. If $pk > n$ we say $s \in L^p_k(E)$ if $s \in L^p_k(\xi)$. It is easily seen that this condition is independent of the choice of the VBN of s, i.e. that if $s \in L^p_k(\xi)$ then $s \in L^p_k(\eta)$ for every VBN η of s in E. Thus $L^p_k(E)$ is the union of the $L^p_k(\xi)$ over all VBN ξ of E. Moreover the theorem of the §1, that L^p_k is a functor from $FVB(M)$ to Banach manifolds and C^∞ maps leads easily to the following

L^p_k MANIFOLD STRUCTURE THEOREM. *If $pk > n$ then for each C^∞ fiber bundle E over M there is a unique C^∞ Banach manifold structure for $L^p_k(E)$ such that, for each VBN ξ of E, $L^p_k(\xi)$ is an open submanifold of $L^p_k(E)$. Moreover if $f: E \to F$ is a C^∞ fiber bundle morphism then $s \mapsto f \circ s$ defines a C^∞ map $L^p_k(f): L^p_k(E) \to L^p_k(F)$. This establishes L^p_k as a functor from $FB(M)$ to the category of C^∞ Banach manifolds and C^∞ maps.*

DEFINITION. Let E be a C^∞ fiber bundle over M, $pk > n$ and $\sigma \in L^p_k(E)$. We define a subset $L^p_k(E)_{\partial\sigma}$ of $L^p_k(E)$, called the Dirichlet subspace of $L^p_k(E)$ defined by σ. Namely $L^p_k(E)_{\partial\sigma}$ is the closure in $L^p_k(E)$ of the set of $s \in L^p_k(E)$ which agree with σ in some neighborhood U (depending on s) of ∂M.

THEOREM. *If $pk > n$ and E is a C^∞ fiber bundle over M then for each $\sigma \in L^p_k(E)$, $L^p_k(E)_{\partial\sigma}$ is a closed C^∞ submanifold of $L^p_k(E)$.*

REMARK. Of course if $\partial M = \varnothing$ then $L^p_k(E)_{\partial\sigma} = L^p_k(E)$.

3. The calculus of variations. In what follows we suppose that there is given a strictly positive smooth measure μ on M and we let E denote a C^∞ fiber bundle over M. The space of k-jets of local sections of E, regarded as a C^∞ fiber bundle over E, will be denoted by $J^k_0(E)$. It can also be regarded as a C^∞ fiber bundle over E in which case we denote it by $J^k(E)$. As usual $j_k : C^\infty(E) \to C^\infty(J^k(E))$ denotes the k-jet extension map.

If F is a C^∞ real valued function on $J^k(E)$ then for each $s \in C^\infty(E)$ we get a C^∞ real valued function $L(s)$ on M by $L(s)(x) = F(j_k(s)_x)$. Such a function $L : C^\infty(E) \to C^\infty(\boldsymbol{R}_M)$ (where $\boldsymbol{R}_M = M \times \boldsymbol{R}$ is the product line bundle over M) is called a kth order Lagrangian for E. The set of all kth order Lagrangians for E is a vector space $\operatorname{Lgn}_k(E)$.

DEFINITION. If $pk > n$ then we shall call an element $L \in \operatorname{Lgn}_k(E)$ L^p_k-smooth if $L : C^\infty(E) \to C^\infty(\boldsymbol{R}_M)$ extends to a C^∞ map (clearly unique) $L : L^p_k(E) \to L^1(\boldsymbol{R}_M)$. Since the linear map $f \mapsto \int f(x)\, d\mu(x)$ of $L^1(\boldsymbol{R}_M) \to \boldsymbol{R}$ is continuous it follows that such a Lagrangian defines a C^∞ map $J^L = J : L^p_k(E) \to \boldsymbol{R}$ by $J(s) = \int L(s)(x)\, d\mu(x)$ and then by restriction we have a C^∞ function $J : L^p_k(E)_{\partial\sigma} \to \boldsymbol{R}$ on each Dirichlet subspace of $L^p_k(E)$.

Roughly speaking the "Dirichlet Problem" associated to a given L^p_k-smooth Lagrangian L and "boundary conditions" $\sigma \in L^p_k(E)$ is to describe the critical locus of $J : L^p_k(E)_{\partial\sigma} \to \boldsymbol{R}$. In particular one wants to find criteria for the following:

(a) LUSTERNIK-SCHNIRELMAN EXISTENCE THEOREM. This means that on each component of $L^p_k(E)_{\partial\sigma}$ there should be at least as many critical points as the Lusternik-Schnirelman category of that component (i.e. the smallest number of closed sets, each contractible in the component, needed to cover the component).

(b) MORSE EXISTENCE THEOREMS (*in case* $p = 2$). This means the critical points should all be nondegenerate and of finite index and the type numbers should satisfy the Morse relations.

(c) SMOOTHNESS THEOREMS. This means that certain smoothness hypotheses on σ should imply corresponding smoothness conclusions for critical points (e.g. if $\sigma \in L^p_{k+r}(E)$) we would like all critical points to be in $L^p_{k+r}(E)$ so in particular if $\partial M = \varnothing$ or $\sigma \in C^\infty(E)$ then we should have the generalized "Weyl Lemma," that all critical points are automatically C^∞.

Before going on to see what can be said in this respect we first consider the basic question of finding criteria for when a Lagrangian is L^p_k smooth. It turns out that the Lagrangians that occur naturally in geometric and physical applications have a certain polynomial-like character which we shall now explain, and that the degree (or rather the "weight") of this polynomial determines the p for which the Lagrangian is L^p_k-smooth. Thus the nature of the Lagrangian itself picks out for us the manifolds $L^p_k(E)$ on which it is natural to consider the corresponding Dirichlet problem.

Consider first the case when M is the n-disc D^n and E is the product *vector* bundle $E = D^n \times V$. Then $J^k(E) = (D^n \times V) \times \bigoplus^{|\alpha| \le k} V$, the sum being over all n-multi-indices $\alpha = (\alpha_1, \ldots, \alpha_n)$ with $|\alpha| = \alpha_1 + \cdots + \alpha_n \le k$. If $s \in C^\infty(E)$ then s is given by a C^∞ map $x \mapsto (x, \sigma(x))$ of D^n into $D^n \times V$ and its k-jet at x is given by

$$j_k(s)_x = ((x, \sigma(x)), D^\alpha\sigma(x))$$

where $D^\alpha = \partial^{|\alpha|}/\partial x_1^{\alpha_1} \cdots \partial x_n^{\alpha_n}$. Let $((x, p_0), p^\alpha)$ denote the natural "coordinate projections" of $J^k(E)$. A C^∞ function $F(x, p_0, p^\alpha)$ on $J^k(E)$ will be called a "monomial function of weight $\leq \omega$" if it is of the form

$$F(x, p_0, p^\alpha) = L(x, p_0)(p^{\alpha_1}, \ldots, p^{\alpha_j})$$

where L is a C^∞ map of $D^n \times V$ into the vector space $L^j(V; \boldsymbol{R})$ of j-linear maps of V into $\boldsymbol{R}$ and $|\alpha_1| + \cdots + |\alpha_j| \leq \omega$. A finite sum of monomials of weight $\leq \omega$ will be called a *polynomial function of weight* $\leq \omega$.

Let $L \in \mathrm{Lgn}_k(E)$ be a kth order Lagrangian for E, say $L(s)(x) = F(j_k(s)_x)$. Then L will be said to be polynomial of weight $\leq \omega$ if F is a polynomial function of weight $\leq \omega$. The polynomial kth order Lagrangians of weight $\leq \omega$ clearly form a vector subspace of the vector space of all k order Lagrangians for E which we denote by $\mathrm{Lgn}_k^\omega(E)$. Now at first glance the space $\mathrm{Lgn}_k^\omega(E)$ seems to depend on the choice of coordinates in D^n and, even more, on the vector space structure of V, so the following may be a little surprising.

THEOREM ON INVARIANCE OF POLYNOMIAL LAGRANGIANS. *Let $\varphi: D^n \to D^n$ be a C^∞ map and let $f: E \to E$ be a C^∞ fiber bundle morphism* (i.e. *a C^∞ map $(x, v) \to (x, h(x, v))$ of $D^n \times V$ into $D^n \times V$), and define $T: C^\infty(E) \to C^\infty(E)$ as follows; if $s: x \to (x, \sigma(x))$ is in $C^\infty(E)$ then $Ts(x) = (x, h(\varphi(x), s(\varphi(x))))$. Then if $L \in \mathrm{Lgn}_k^\omega(E)$ so also is $L \circ T$.*

Of course the proof is just a simple inductive application of the "chain rule." The above theorem justifies the following definition.

DEFINITION. Let E be a C^∞ fiber bundle over M and $L \in \mathrm{Lgn}_k(E)$. We say $L \in \mathrm{Lgn}_k^\omega(E)$ if given any chart $\varphi: D^n \to M$, any VBN ξ of $E \mid \varphi(D^n)$ and any trivialization $f: D^n \times V \approx \varphi^*\xi$ the map $L^*: C^\infty(D^n, V) \to \boldsymbol{R}$ defined by $L^*(s)(x) = L(f \circ s \circ \varphi^{-1}) \times (\varphi(x))$, is in $\mathrm{Lgn}_k^\omega(D^n \times V)$. Equivalently it suffices that given any $e \in E$ the latter hold for at least one such choice of φ, ξ, and f with $e \in \xi$.

The importance of the classes $\mathrm{Lgn}_k^\omega(E)$ for the calculus of variations, aside from the fact that they include the usual geometric and physically interesting examples, lies in the following:

SMOOTHNESS THEOREMS FOR POLYNOMIAL LAGRANGIANS. *Let $1 \leq p < \infty$, and $pk > n$. Then $L \in \mathrm{Lgn}_k^\omega(E)$ is L_k^p-smooth provided $\omega \leq pk$. In particular, for p integral, all $L \in \mathrm{Lgn}_k^{pk}(E)$ are L_k^p-smooth.*

Now for some examples. Let W be a C^∞ manifold and let E denote the product bundle $M \times W$ over M. We identify $C^\infty(E)$ with the C^∞ maps of M into W. If $s \in C^\infty(E)$ then ds_x is a linear map $T(M)_x \to T(W)_{s(x)}$. Now suppose both M and W are Riemannian. Then the space of linear maps of $T(M)_x$ into $T(W)_{s(x)}$ has a natural quadratic "Hilbert-Schmidt" norm (given by $\|T\|^2 = trT^*T$) so in particular we can form $\|ds_x\|^p$ for $1 \leq p < \infty$, and if p is an *even* integer then it is easily checked that $L: C^\infty(E) \to C^\infty(M, \boldsymbol{R})$ defined by $L(s)(x) = \|ds_x\|^p$ belongs to $\mathrm{Lgn}_1^p(E)$. We therefore have a natural C^∞ real valued function J on $L_1^p(E)$, called the energy function of degree p, defined by $J(s) = \int \|ds_x\|^p \, d\mu(x)$ where for μ we take the natural Riemannian measure on M. If $n = 1$, so that M is (up to isometry) either the interval $[0, l]$ or the circle of length l, then we can take $p = 2$ (so that $L_1^p(E)$ is a Hilbert manifold) and we get for J the classical energy function $J: L_1^2(E) \to \boldsymbol{R}$, given by $J(s) = \int_0^l \|s'(t)\|^2\, dt$. The generalized Dirichlet problem mentioned above reduces in this case to the classical theory of geodesics of W (with given end points or closed depending on whether M is an interval or a circle) and in particular the classical smoothness theorems and Lusternik-Schnirelman and Morse existence theorems for geodesics become special cases of the more general theorems of this sort for the Dirichlet Problems associated to Lagrangians of the "coercive linearly-embedded" type which we consider in the next section.

We now give a very general method for constructing polynomial Lagrangians (which in particular includes the above example). We assume our bundle E is embedded as a closed sub-bundle of a C^∞ vector bundle ξ over M. This in itself is no restriction; for example we can find a proper C^∞ embedding f of E, considered as a C^∞ manifold, into a vector space V and then if $\pi: E \to M$ is the projection, $e \mapsto (\pi(e), f(e))$ is a C^∞ proper fiber bundle isomorphism of E onto a closed C^∞ sub-bundle of the product vector bundle $M \times V$. We note that $L_k^p(E)$ $(pk > n)$ is a C^∞ submanifold of the Banach space $L_k^p(\xi)$. If $L: C^\infty(\xi) \to C^\infty(M, \boldsymbol{R})$ is an element of $\mathrm{Lgn}_k^\omega(\xi)$ then it is easily checked that $L \mid C^\infty(E): C^\infty(E) \to C^\infty(M, \boldsymbol{R})$ is an element of $\mathrm{Lgn}_k^\omega(E)$. From this we see easily the following.

LINEARLY EMBEDDED LAGRANGIAN THEOREM. *Let the C^∞ fiber bundle E over M be a closed sub-bundle of a C^∞ vector bundle ξ. Let $\eta_1, \ldots, \eta_r$ be C^∞ vector bundles over M and let $A_i: C^\infty(\xi) \to C^\infty(\eta_i)$ be C^∞ linear differential operators of order $k_i \le k$ Let T be a C^∞ section of the bundle $L^r(\eta_1, \ldots, \eta_r; \boldsymbol{R}_M)$ of r-linear functionals on $\eta_1 \oplus \cdots \oplus \eta_r$. Then $L: C^\infty(E) \to C^\infty(M, \boldsymbol{R})$ defined by $L(s)(x) = T(x)(A_1 s(x), \ldots, A_r s(x))$ is an element of $\mathrm{Lgn}_k^\omega(E)$ where $\omega = k_1 + \cdots + k_r$. In particular if $\| \ \|_i$ is a C^∞ Riemannian structure for η_i and p is an even positive integer then $L: C^\infty(E) \to C^\infty(M, \boldsymbol{R})$ defined by $Ls(x) = \sum_{i=1}^r \|A_i s(x)\|_i^p$ is in $\mathrm{Lgn}_k^{pk}(E)$.*

Note that if W is a Riemannian manifold and we embed W isometrically in an orthogonal vector space V (so $E = M \times W$ is a closed sub-bundle of $\xi = M \times V$) then $d: C^\infty(\xi) \to C^\infty(T^*(M) \otimes \xi)$ is a linear differential operator of order one so by the above $Ls(x) = \|ds_x\|^p$ is an element of $\mathrm{Lgn}_1^p(E)$, giving our previous example.

In the next section we will give some conditions on the linear operators A_i of the above theorem that lead to Lusternik-Schnirelman, and Morse type existence theorem and also smoothness theorems for the corresponding calculus of variations problem.

4. Coercive, linearly embedded, Dirichlet problems.

In this section E is a C^∞ fiber bundle which is a closed sub-bundle of a C^∞ vector bundle ξ over M, μ is a strictly positive smooth measure on M, k is a positive integer, and p an even integer with $pk > n$. Let $\sigma_0 \in L_k^p(E)$ and let $X = L_k^p(E)_{\partial\sigma_0}$. Let $L \in \mathrm{Lgn}_k^{pk}(E)$ and define a C^∞ function $J: X \to \boldsymbol{R}$ by $J(s) = \int L(s)(x)\, d\mu(x)$. We shall describe below conditions on L which guarantee that this Dirichlet problem satisfies

Lusternik-Schnirelman and Morse existence theorems and also various smoothness theorems.

Let $\| \ \|^p_{L_k}$ be an admissible norm for the Banach space $L^p_k(\xi)$. Since X is a closed C^∞ submanifold of $L^p_k(\xi)$, this choice of norm induces on X the structure of a complete Finsler manifold. In particular if $s \in X$ then $\|dJ_s\| = \text{Sup}\,\{|dJ_s(v)| \mid \|v\|_{L_k{}^p} = 1,\, v \in T(X)_s\}$ is well defined.

DEFINITION. J satisfies condition (C) if for each subset S of X such that J is bounded on S and $\|dJ\|$ is not bounded away from zero on S, there exists a critical point of J adherent to S.

The following theorem represents a combination of results due to the author S. Smale, and J. T. Schwartz.

THEOREM. *If J satisfies condition* (C) *and is bounded below, then for each component X_0 of X there exists $x_0 \in X_0$ such that $J(x_0) = \text{Inf}\,\{J(x) \mid x \in X_0\}$ and moreover J has at least as many critical points on X_0 as* cat (X_0) *the Lusternik-Schnirelman category of X_0. Assuming J is not anywhere locally constant* (*a condition always satisfied by "reasonable" calculus of variations problems*) *there exists $X_0 \in X$ such that $J(x_0) = \text{Inf}\,\{J(x) \mid x \in X\}$. Finally assuming X is Riemannian* (i.e. $p = 2$) *and that the critical points of J are nondegenerate, then the Morse inequalities are satisfied.*

For full details on the above the reader is referred to the author's *Morse theory on Hilbert manifolds* (Topology **2** (1963), 299–340), *Lusternik-Schnirelman theory on Banach manifolds* (Topology **5** (1966), 115–132), S. Smale's *Morse theory and a nonlinear generalization of the Dirichlet problem* (Ann. of Math. **80** (1964), 382–396) and J. T. Schwartz' *Generalizing the Lusternik-Schnirelman theory of critical points* (Comm. Pure Appl. Math. **17** (1964), 307–315).

We shall now define a class of Lagrangians L, which we call p-coercive, and state theorems to the effect that the functions J defined by sueh Lagrangians satisfy condition (C) and are bounded below and hence satisfy the conclusions of the above theorem. These definitions and theorems represent joint work by the author and Mrs. Karen Uhlenbeck.

Let $\eta_1, \ldots, \eta_r$ be C^∞ Riemannian vector bundles over M and let $\eta = \eta_1 \oplus \cdots \oplus \eta_r$. Let $A_i : C^\infty(\xi) \to C^\infty(\eta_i)$ denote kth order C^∞ linear differential operators and let $A = A_1 \oplus \cdots \oplus A_r : C^\infty(\xi) \to C^\infty(\eta)$. We define explicit norms $\| \ \|_{L^p}$ for the Banach spaces $L^p(\eta_i)$ (and $L^p(\eta)$) by $\|s\|^p_{L^p} = \int \|s(x)\|^p_x \, d\mu(x)$, where $\| \ \|_x$ is the Riemannian norm in the appropriate fiber.

DEFINITION. We say $\{A_i\}$ is an *ample set of kth order linear differential operators for ξ* if $\|s\|_{L^p} + \sum_{i=1}^r \|A_i s\|_{L^p}$ is an admissible norm for the space $L^p_k(\xi)$. If $\sum_{i=1}^r \|A_i s\|_{L^p}$ is an admissible norm for $L^p_k(\xi)$ then we shall say that $\{A_i\}$ is a *strongly ample* set of kth order linear differential operators for ξ.

REMARK. It is easily seen that $\{A_i\}$ is (strongly) ample if and only if $\{A\}$ is (strongly) ample.

Now *a priori* it would seem that the definition of ample depends on p. However, this is in fact not the case, and indeed it is a classical result of the theory of linear differential operators that $\{A\}$ is ample if and only if it is "over determined elliptic" i.e. if and only if for each nonzero cotangent vector (v, x) of M the symbol of A at (v, x), $\sigma_k(A)(v, x): \xi_x \to \eta_x$, is injective, and moreover that in this case A is strongly ample if and only if in addition the equation $Au = 0$ has only the zero solution in $L_k^p(\xi)_0 =$ closure in $L_k^p(\xi)$ of C^∞ sections of ξ with support disjoint from ∂M. Thus

THEOREM. *A necessary and sufficient condition that $\{A_i\}$ be ample is that for each nonzero cotangent vector (v, x) of M the intersection of the kernels of the linear maps $\sigma_k(A_i)(v, x): \xi_x \to \eta_{ix}$ be zero. If in addition there is no $u \in L_k^p(\xi)_0$ such that $A_i u = 0$, $i = 1, \ldots, r$ then $\{A_i\}$ is strongly ample.*

EXAMPLE. Let ξ be a product bundle $M \times V$, η the bundle $L(T(M), \xi) = T^*(M) \otimes \xi$ and $d: C^\infty(\xi) \to C^\infty(\eta)$ the usual differential. Then d is a first order linear differential operator and $\sigma_1(d)(v, x)e = v \otimes e$, so clearly $\{d\}$ is ample. Moreover $du = 0$ if and only if u is constant on each component of M. Since elements of $L_k^p(\xi)_0$ vanish on ∂M it follows that $\ker (d \mid L_k^p(\xi)_0) = 0$ (and hence d is strongly ample) if and only if each component of M has a nonempty boundary.

Now let $F: J^k(\xi) \to \boldsymbol{R}$ be a C^∞ real valued function. Then for each $s \in M$, $F \mid J^k(\xi)_x$ is a C^∞ real valued function on the vector space $J^k(\xi)_x$, so if $s \in C^\infty(\xi)$ we can form the second differential of $F \mid J^k(\xi)_x$ at $j_k(s)_x$, a bilinear functional on $J^k(\xi)_x$ which we denote by $\delta^2 F_{j_k}(s)_x$. Then if $u \in C^\infty(\xi)$ we get a C^∞ real valued function $\delta^2 F_{j_k(s)}(j_k(u), j_k(u))$ on M whose value at x is $\delta^2 F_{j_k(s)_x}(j_k(u)_x, j_k(u)_x)$.

DEFINITION. Let $L \in \mathrm{Lgn}_k^{pk}(\xi)$ be defined by $L(s)(x) = F(j_k(s)_x)$ where F is a C^∞ function on $J^k(\xi)$. We say that L is (strictly) p-coercive if there exists a (strongly) ample family $\{A_i\}$ of kth order linear differential operators for ξ such that given s, $u \in C^\infty(\xi)$

$$\delta^2 F_{j_k(s)}(j_k(u), j_k(u)) \geq \sum_{i=1}^{r} \|A_i s\|^{p-2} \|A_i u\|^2.$$

The following is easily verified.

THEOREM. *If $\{A_i\}$ is a (strongly) ample set of* kth *order linear differential operators for ξ then $L \in \mathrm{Lgn}_k^{pk}(\xi)$ defined by*

$$L(u) = \sum_{i=1}^{r} \|A_i u\|^p$$

is (strictly) p-coercive.

COROLLARY. *If $\xi = M \times V$ is a product bundle and $d: C^\infty(\xi) \to C^\infty(T^*(M) \otimes \xi)$ is the usual differential, then $L \in \mathrm{Lgn}_1^p(\xi)$ defined by $L(s) = \|ds\|^p$ is p-coercive and is strictly p-coercive provided each component of M has nonempty boundary.*

There is a generalization of the above theorem in case $p = 2$, which reduces to the preceding theorem if we take $T = A^*A$.

THEOREM. *Let ξ be a Riemannian bundle with inner product $\langle\ ,\ \rangle_x$ on ξ_x and let $T: C^\infty(\xi) \to C^\infty(\xi)$ be a selfadjoint linear differential operator of order $2k$ on ξ which* *is strongly elliptic (i.e.* $\sigma_{2k}(T)(v, x): \xi_x \to \xi_x$ *is positive definite for each nonzero cotangent vector* (v, x) *of* M*). Then the function* $J: L_k^2(E)_{\partial\sigma} \to \boldsymbol{R}$ *defined by* $J(s) = \int \langle Ts(x), s(x)\rangle_x \, d\mu(x)$ *for* $s \in C^\infty(\xi)$ *can be written in the form* $J(s) = \int L(s)(x)\, d\mu(x)$ *where* $L \in \mathrm{Lgn}_k^{2k}(\xi)$ *is 2-coercive, and is strictly 2-coercive if* T *is strictly positive on* $L_k^2(\xi)_0$.

An interesting application of the above theorem is obtained by taking $\xi = M \times V$ and letting T be the kth power of the Laplace Beltrami operator for M (perhaps adding a constant larger than the smallest eigenvalue on $L_k^2(\xi)_0$ so as to make it positive). This gives rise to the kth order energy function J whose extremals are called polyharmonic maps. See Eells and Sampson, *Energie et déformations en géométrie differentielle*, Ann. Inst. Fourier **14** (1964), 61–69.

The following is our basic theorem in the direction of giving sufficient conditions on a linearly embedded Lagrangian in order that the associated Dirichlet problems should satisfy condition (C) and be bounded below, and hence satisfy the conclusions of Lusternik-Schnirelman theory as stated in the first theorem of this section.

THEOREM. *Let* $L \in \mathrm{Lgn}_k^{pk}(\xi)$ *be p-coercive, and if the fiber of the bundle E is not compact assume L is even strictly p-coercive. Then* $J: L_k^p(E)_{\partial\sigma} \to \boldsymbol{R}$ *defined by* $J(s) = \int L(s)(x)\, d\mu(x)$ *is bounded below and satisfies condition* (C).

The details of the proof of this theorem will be found in §19 of *Foundations of global non-linear analysis.*

An interesting problem, which seems not to have been attacked in any generality, is to find conditions in the case $p = 2$ that insure the nondegeneracy of the critical points of a Dirichlet problem of the above type. Generalizing from a well-known theorem of M. Morse in the geodesic case one would be led to conjecture that for "almost all" choices of Dirichlet boundary conditions σ_0 (in some appropriate sense) all the critical points of $J: L_k^2(E)_{\partial\sigma_0} \to \boldsymbol{R}$ are nondegenerate. Presumably this might follow from some transversality argument. If $J(s) = \int \langle Ts(x), s(x)\rangle_x \, d\mu(x)$ as above then S. Smale has given a beautiful generalization of the Morse Index Theorem that allows one to compute the index of a nondegenerate[2] critical point of J. (*On the Morse index theorem*, J. Math. Mech. **14** (1965), 1049–1055 and Corrigenda **16** (1967), 1069–1070).

We now comment briefly on smoothness theorems for critical points in the present context. We first consider the case $p = 2$ and $J: L_k^2(E)_{\partial\sigma} \to \boldsymbol{R}$ defined by $J(s) = \int \langle Ts(x), s(x)\rangle \, d\mu(x)$ where $T: C^\infty(\xi) \to C^\infty(\xi)$ is a selfadjoint strongly elliptic linear differential operator of order $2k$. In this case Mrs. Karen Uhlenbeck has proved in her thesis that if $\sigma \in L_{k+r}^2(E)$ then every critical point of J is in $L_{k+r}^2(E)$, so in particular we have the generalized Weyl Lemma, that if $\partial M = \varnothing$ or if $\sigma \in C^\infty(E)$ then all critical points of J are C^∞. The proof will be found in §19 of *Foundations of global analysis* for the case that T is "quasi scalar" i.e. where $\sigma_{2k}(T)(v, x): \xi_x \to \xi_x$ is always multiplication by a scalar (e.g. where T is the kth power of the Laplace Beltrami operator on M, in which case $\sigma_{2k}(T)(v, x)$ is multiplication by $\|v\|^{2k}$). In this generality the theorem was first proved by Dr. John Saber in his thesis (Brandeis 1965). Saber also proved that condition (C) was satisfied in this case.

In her thesis Karen Uhlenbeck has obtained interesting smoothness theorems for the case $p > 2$. For simplicity we state them in something less than their full generality. We assume $\partial M = \varnothing$ and that $J : L_k^p(E) \to \mathbf{R}$ is of the form $J(s) = \int \|As(x)\|^p \, d\mu(x)$ where A is an elliptic kth order linear differential operator $A : C^\infty(\xi) \to C^\infty(\xi)$ having scalar symbol. Then every critical point of J lies in the Hölder space $C^{k+\alpha}(E)$ where $\alpha = 1/p - 1$. In particular this covers the interesting geometric case where M is Riemannian, $E = M \times W$, $\xi = M \times W$ (V an orthogonal vector space) and A is a power of the Laplace-Beltrami operator of M.

5. **Beyond the linearly embedded case.** The problems we have been considering are of a basically nonlinear character, and it is somewhat less than satisfying to have to treat them by the "linear embedding" technique (i.e. embedding E in a vector bundle ξ and describing admissible Lagrangians on E in terms of linear differential operators on ξ).

The situation is in many ways parallel to that which obtains in the study of Riemannian geometry. One can develop the whole theory of Riemannian manifolds either intrinsically, or alternatively one can consider Riemannian manifolds M isometrically embedded in an orthogonal vector space V (by Nash's embedding theorem this is no loss of generality) and use the linear structure of V to define such concepts as curvature, parallel translation etc. in M. There are even certain technical advantages in this second approach, in that one can avoid developing all the complicated machinery of connections in principal bundles etc. that usually go along with the intrinsic approach. Yet there seems to be virtually unanimous agreement that both for aesthetic reasons and in order to get a deeper understanding of what is going on, the intrinsic approach is preferable and indeed that the technical machinery that one develops in the intrinsic approach is worth studying for its own sake.

In the present approach to global nonlinear analysis and the calculus of variations we appear to be in a stage of development analogous to that of Riemannian geometry just before E. Cartan. We can formulate the foundations of the theory and the questions we would like to attack intrinsically, but we simply do not have the machinery necessary to carry out the details of the theory intrinsically, so to prove theorems we are forced to fall back on embedding our problems in a linear situation.

Needless to say the next and most exciting stage of the theory lies before us; gaining the insights, the techniques, and the machinery necessary to handle global nonlinear problems intrinsically. It is already becoming clear that this will involve a study of the intrinsic Finsler geometry of the manifolds $L_k^p(E)$ that are induced by certain differential-geometric structures on E that come from Lagrangians on $J^k(E)$.

Both W. Klingenberg and H. Eliasson, for example, have handled the geodesic problem intrinsically from this point of view. And in her thesis Karen Uhlenbeck has made considerable progress in handling quite general classes of calculus of variations problems in several independent variables intrinsically. As one would hope, there are definite indications that the machinery involved is independently interesting and suggests new and interesting global questions in nonlinear analysis.

BRANDEIS UNIVERSITY

NONLINEAR EQUATIONS INVOLVING NONCOMPACT OPERATORS[1]

W. V. Petryshyn

Introduction. The purpose of this paper is to develop the *approximation-solvability theory*[2] (i.e., to obtain approximation and existence results) for nonlinear functional equations involving a class of mappings from a normed linear space X to a normed space Y which are *A-proper*[2] with respect to general approximation schemes. The schemes used here are such that they include both the finite difference schemes and the projection or Galerkin-type methods when applied to special situations.

To put our discussion of A-proper mappings in proper perspective and to trace the development of this concept we first outline briefly the main results that have been recently obtained for various classes of nonlinear mappings. It is clear now that the theory of *monotone operators*[2] initiated independently by Kachurowsky [**23**] (see also Vainberg [**47**]) and Zarantonello [**50**] has undergone recently a significant development. The first interesting existence result obtained independently in [**48**] and [**50**] asserts that if T is a monotone and Lipschitzian mapping of a Hilbert space H into H, then $I + T$ is onto. This result was extended to continuous mappings T by Minty [**30**] and then under much weaker conditions by Browder [**3**]. The theory of monotone operators was extended to reflexive Banach spaces independently by Browder [**4**] and Minty [**31**] (see also Vainberg-Kachurowsky [**49**]) and to densely defined operators by Browder [**5**] and the author [**33**]. Applications to various classes of differential equations were given by Browder, Kato, Lions, Leray and Lions, Lions and Strauss, Dubinsky, Aubin, Raviart, Pohodjayev and others.[3] Results for *complex monotone operators* were given by Zarantonello [**51**] and the author [**33**] for complex Hilbert spaces and were strengthened and extended to complex Banach spaces by Browder [**6**]. The nonlinear generalization of the theory of Friedrich's extensions to densely defined monotone and complex monotone operators in Hilbert spaces were obtained by

[1] The preparation of this paper was partially supported by the National Science Foundation under NSF Grant GP-8556.

[2] For the precise definitions of the concepts and the statements of the results mentioned in the Introduction see §1 of this paper.

[3] For the survey and the exposition of the theory of monotone mappings and its application by many authors to various classes of differential equations as well as for a complete list of references see the excellent recent expository papers by Opial [**32**], Dubinsky [**19**] and Kachurowsky [**26**].

the author [**33**] and were extended to Banach spaces by Browder [**7**]. Studies on J-*monotone* and *accretive operators* were initiated and studied by Browder [**9**], [**10**] and further investigated in [**11**], [**14**], [**28**], [**39**], [**42**], [**44**]. We add that the problem of constructing general approximation methods for the solution of equations involving monotone operators has recently been also treated by Aubin [**1**] and Brezis-Sibony [**2**].

The main result of the theory of monotone, complex monotone, J-monotone, accretive and related classes of operators is that, under weak continuity assumptions and some additional conditions on the behavior at infinity, the mapping T is onto. This basic property is first proved for finite-dimensional spaces and then it is carried over to Banach spaces. As was noted in [**32**], the transition from finite-dimensional spaces to the whole space is nonconstructive in nature. It becomes constructive, however, under additional conditions on the underlying spaces and acquires in this case many features of the projection methods for solving linear functional equations.

This constructive aspect of the theory of monotone and complex monotone operators was developed by the author [**33**], [**35**], [**36**], [**37**] for Hilbert spaces and extended by Browder [**7**] and later by Kachurowsky [**25**] to Banach spaces. The study of the constructive approach to the solution of nonlinear equations by projection methods was extended by the author [**35**], [**36**], [**37**] to equations involving a more general class of *projectionally-compact* (P-compact) mappings in general Banach spaces with a projectionally complete system. The main result of the theory of P-compact operators was a new fixed point theorem obtained for bounded operators in [**35**], [**36**] and for unbounded in [**37**]. Since the class of P-compact mappings includes, among others, completely continuous, quasi-compact [**27**] and continuous, weakly continuous and demicontinuous monotone mappings, from our fixed point theorems for P-compact mappings, we deduced, on the one hand, the fixed point theorems of Schauder, Rothe, Tikhonoff, Krasnoselsky, Altman, Kaniel and others and, on the other hand, we rederived in constructive fashion the basic existence theorems for monotone operators. We mention that further studies of P-compact and other related classes of operators were carried by DeFigueiredo [**21**], Lees-Schultz [**29**], Petryshyn-Tucker [**44**], Pahodjayev [**45**], the author [**42**] and others.

Continuing his study of the constructive approach to the solution of nonlinear equations the author noted that, although the class of P-compact mappings is well suited for the study of fixed-point and eigenvalue-problem theories, the concept of a P-compact mapping should be somewhat modified in order to be also well suited for the study of nonhomogeneous equations. In [**38**], [**39**] the author presented a general theory of the projection method for the solution of equations in reflexive Banach spaces involving nonlinear operators satisfying the so-called *condition* (c). This constructive theory unified the recent results on monotone, complex monotone and J-monotone mappings with the earlier results on linear equations obtained by various authors (see [**39**]). In [**10**], [**12**] Browder showed that if instead of condition (c) we assume the existence of solutions, then somewhat weaker conditions insure the convergence of approximants. In [**40**] it was proved that the solvability condition is equivalent to condition (c) at least for maps in reflexive Banach spaces.

It turned out that to apply the constructive techniques used in [**38**], [**40**], [**12**] to equations involving nonlinear mappings from one normed space to another, the most suitable and natural way from the approximation point of view is to modify condition (c) so as to embody the essential feature of a P-compact mapping. The resultant of this fusion was a new class of operators, introduced and studied by the author in [**41**] to which we referred there as *operators, satisfying condition* (H). At the suggestion of Browder, such operators were called *A-proper* in our paper [**43**] dealing with generalized Fredholm Alternative, where it was shown that if X and Y are reflexive Banach spaces and T is a bounded linear A-proper mapping from X to Y with the null space $N(T)$, then $\dim N(T) = \dim N(T^*)$ if and only if the adjoint operator T^* is also A-proper. In passing we note that in [**16**], [**17**] Browder and Petryshyn succeeded in developing a generalized topological degree theory for A-proper mappings which extends considerably the classical Leray-Schauder theory. We add that our degree is, in general, a multi-valued function which, however, for most practical uses of a degree theory serves as well as a single-valued degree.

Thus, as was already stated, the object of this paper is to develop a constructive approximation-solvability theory for equations involving A-proper mappings with respect to general approximation schemes. Briefly, the following are the results obtained in this paper.

In §1 we first define an admissible approximation scheme Γ_n, an A-proper mapping with respect to Γ_n, strong and feeble approximation-solvability of functional equations and different continuity concepts. The basic approximation and existence results for A-proper mappings are contained in Theorems 1.1 and 1.2 which are then applied to obtain new results for odd (Theorem 1.3) and homogeneous (Theorem 1.4) A-proper mappings. We add that Theorem 1.1 generalizes the corresponding results in [**40**], [**41**], [**12**], [**13**] while Theorems 1.3 and 1.4 strengthen and extend the corresponding results in [**45**].

In §2 we present two sufficient conditions for a continuous mapping to be A-proper. One is the condition (c) for K-monotone mappings while the other is a modification of condition (S) introduced by Browder [**13**]. An interesting feature of our discussion is that T is not required to be defined on all of X. These results are then used to establish (Theorem 2.3) the A-properness of continuous K-monotone mappings under "bounded from below" conditions similar to those used by Browder and Brezis-Sibony.

In §3 we apply theorems of §2 to obtain further new results on feeble and strong approximation-solvability of equations involving A-proper and K-monotone mappings under various continuity assumptions.

In §4 we consider some special cases of the theory developed in §§1, 2, and 3. In particular we discuss some special cases of admissible approximation schemes such as abstract finite difference schemes, projection schemes and injection schemes. These are applied to monotone and accretive mappings and the relation of our results to those of other authors is indicated.

1. A-proper mappings and the approximation-solvability. Let X and Y be two (real or complex) normed linear spaces and let $\{X_n\}$ and $\{Y_n\}$ be two sequences of (real or complex) finite-dimensional spaces. Let T be a (possible nonlinear) mapping with domain $D = D(T)$ in X and range $R(T)$ in Y and let $\{P_n\}$, $\{R_n\}$ and $\{Q_n\}$ be three sequences of (possibly nonlinear) mappings with P_n mapping X_n into X, R_n mapping X into X_n and Q_n mapping Y into Y_n.

REMARK 1.0. For the sake of notational simplicity we use the same symbol $\| \ \|$ to denote the different norms $\| \ \|_X$, $\| \ \|_Y$, $\| \ \|_{X_n}$ and $\| \ \|_{Y_n}$ in the respective spaces X, Y, X_n and Y_n. We hope that at each step it will be clear to the reader which norm is meant. We also use the symbols "$\rightarrow$" and "$\rightharpoonup$" to denote *strong* and *weak* convergence, respectively.

DEFINITION 1.1. *A quintuple of sequences* $\Gamma_n = (\{X_n\}, \{Y_n\}, \{P_n\}, \{R_n\}, \{Q_n\})$ *is called an admissible approximation scheme for mappings T from subsets of X to Y if for each n the dimension of X_n equals the dimension of Y_n and if the mappings P_n, R_n and Q_n satisfy the following conditions*:

(C1) *$\{P_n x_n\}$ is bounded in X whenever the sequence $\{x_n \mid x_n \in X_n\}$ is uniformly bounded.*

(C2) *$P_n R_n x \rightarrow x$ for each x in X.*

(C3) *$\|P_n u_n - P_n v_n\| \rightarrow 0$ whenever the sequences $\{u_n \mid u_n \in X_n\}$ and $v_n \mid v_n \in X_n\}$ are such that $\|u_n - v_n\| \rightarrow 0$.*

(C4) *$\|Q_n f\| \leq k(f)$ for each f in Y, where $k(f)$ (>0) is independent of n.*

Let $\{T_n\}$ be the sequence of mappings of $D_n = P_n^{-1}(D) = \{x \in X_n \mid P_n x \in D\}$ into Y_n defined by $T_n = Q_n T P_n / D_n$. As in our recent papers [**39**], [**40**], [**41**], under very general conditions on T or on T_n's, we use here the approximation scheme

$$\begin{array}{ccc} X \supseteq D & \xrightarrow{T} & Y \\ R_n \big\downarrow \big\uparrow P_n & & \big\downarrow Q_n \\ X_n \supseteq D_n & \xrightarrow{T_n} & Y_n \end{array}$$

given by Definition 1.1 to obtain approximation as well as existence results for the solution of the equation

$$(1) \qquad Tx = f \qquad (f \in Y, x \in D)$$

by utilizing the solutions of the approximate equations

$$(2) \qquad T_n x_n = Q_n(f) \qquad (Q_n f \in Y_n, x_n \in D_n).$$

We would like to add that for particular functional spaces X and Y and for given spaces X_n and Y_n the mappings P_n, R_n and Q_n can be chosen in such a way that the corresponding abstract approximation scheme Γ_n is realizable, for example, by *finite difference schemes* or *Galerkin-type methods*. Usually X_n and Y_n will be the spaces of scalar sequences and in this case one may say that $P_n : X_n \rightarrow X$ defines "the process of approximation" (e.g. an interpolation operator) showing how one associates with a given sequence of scalars (i.e. with an element in X_n)

an element of the space X which will be a space of functions or distributions while the mappings $R_n: X \to X_n$ and $Q_n: Y \to Y_n$ may, for example, associate with each function its values at points of a given net or its Fourier coefficients in case R_n and Q_n are determined by elements from given bases. For the process of constructing these special operators see Aubin [**1**], Brezis-Sibony [**2**] and others.

In this section it will always be assumed, unless additional condition is stipulated, that X and Y are normed linear vector spaces and that the mappings P_n, R_n and Q_n are such that at least the conditions of Definition 1.1 are satisfied.

For later use we recall the following concepts: $T: D(\subseteq X) \to Y$ is *completely continuous* if T is continuous and maps bounded sets from D into relatively compact sets in Y; T is *weakly continuous* if $x_n \rightharpoonup x$ with x_n and x in D implies $Tx_n \rightharpoonup Tx$; T is *demicontinuous* if $x_n \to x$ with x_n and x in D implies $Tx_n \rightharpoonup Tx$; T is *bounded* if T maps bounded sets into bounded sets.

The crucial property of T which we shall apply in this paper is that T be *A-proper with respect to* Γ_n.

DEFINITION 1.2. *T is said to be A-proper with respect to the scheme Γ_n given by Definition* 1.1 *if the following condition holds*:

(H) *If* $\{x_n \mid x_n \in D_n\}$ *is any bounded sequence with* $P_n x_n$ *in* D *such that* $\|T_n x_n - Q_n f\| \to 0$ *for some* f *in* Y, *then there exists an infinite subsequence* $\{x_{n_i}\}$ *and an element* x *in* D *such that* $P_{n_i} x_{n_i} \to x$ *in* X (*as* $i \to \infty$) *and* $Tx = f$.

The concept of an A-proper mapping (i.e., a mapping satisfying condition (H)) was first introduced and studied by the author in [**41**] (in the case when $\{X_n\} \subset X$, $\{Y_n\} \subset Y$, P_n is an identity on X_n and, for each x in X and y in Y, $R_n x \to x$ and $Q_n y \to y$). This concept is a modification of the definition of the P-compact mapping studied in [**35**], [**36**], [**37**] for projectional schemes. A similar definition (also a modification of P-compactness) for projectional schemes has been given in [**45**]. Related classes of mappings were later considered by Browder. Definition 1.1 is essentially Definition 1.2 in Browder-Petryshyn [**17**] (see also [**16**]).

We remark in passing that, as it is not hard to see, if T is A-proper, then $T(D_0)$ is closed in Y if D_0 is a bounded closed subset of D and T satisfies certain continuity conditions and $T \pm C$ is A-proper if C is a completely continuous mapping of D into Y and Q_n's are linear.

DEFINITION 1.3. *For a given f in Y, equation* (1) *is said to be strongly* (*resp. feebly*) *approximation-solvable if and only if there exists an integer* $N \geq 1$ *such that equation* (2) *has a solution* x_n *in* D_n *for each* $n \geq N$ *which are such that* $P_n x_n \to x$ *in* X *for some* x *in* D (*resp.* $P_{n_i} x_{n_i} \to x$ *with* x *in* D *for some subsequence* $\{x_{n_i}\}$ *of* $\{x_n\}$) *and* $Tx = f$.

1.1. *Mappings defined on all of* X. Our first new result in this section is the following generalization of Theorems 1 and 2 in [**41**].

THEOREM 1.1. *Suppose X and Y are normed linear spaces, T a mapping of* $D = X$ *into* Y, Γ_n *an approximation scheme for equation* (1) *given by Definition* 1.1 *with the additional property that* $P_n(0) = 0$ *for each* n, $N_0 \geq 1$ *a fixed integer and* $\alpha(r)$ *a continuous function of* $R^+ = \{r \mid r \geq 0\}$ *into* R^+ (*with* $r_i \to 0$ *whenever*

$\alpha(r_i) \to 0$ and $\{r_i\}$ bounded whenever $\alpha(r_i)$ is bounded) such that the following hypotheses hold:

(H1) *T_n is a continuous mapping of X_n into Y_n for each $n \geq N_0$.*
(H2) *$\|T_n R_n x - Q_n T x\| \to 0$ for each x in X.*
(H3) *$\|T_n x - T_n y\| \geq \alpha(\|x - y\|)$ holds for all x and y in X_n and $n \geq N_0$.*

Then under the hypotheses (H1), (H2), *and* (H3) *the following assertions are equivalent:*

(A1) *T is A-proper with respect to Γ_n.*
(A2) *Equation* (1) *is strongly approximation-solvable for each f in Y.*
(A3) *Equation* (1) *possesses a solution for each f in Y.*

PROOF OF THEOREM 1.1. To prove Theorem 1.1 it suffices to show that under our hypotheses, (A1) $\Rightarrow$ (A2) $\Rightarrow$ (A3) $\Rightarrow$ (A1).

(A1) $\Rightarrow$ (A2). We shall first prove that for a given f in Y, equation (1) can have at most one solution. Suppose instead that $x \neq y$ and $Tx = Ty$. Since $R_n x$ and $R_n y$ lie in X_n, (H3) implies that for all $n \geq N_0$

$$\|T_n R_n x - T_n R_n y\| \geq \alpha(\|R_n x - R_n y\|)$$

whence, since $Tx = Ty$, we obtain the relation

$$\alpha(\|R_n x - R_n y\|) \leq \|T_n R_n x - Q_n T x\| + \|Q_n T y - T_n R_n y\|.$$

In virtue of (H2), the right-hand side in the above inequality approaches zero as $n \to \infty$ and, consequently, $\alpha(\|R_n x - R_n y\|) \to 0$ as $n \to \infty$. Hence the properties of $\alpha(r)$ imply that $\|R_n x - R_n y\| \to 0$ as $n \to \infty$. This and the property of $\{P_n\}$ imply that $\|P_n R_n x - P_n R_n y\| \to 0$ as $n \to \infty$ whence, in virtue of (C2) of Definition 1.1, it follows that $x = y$. This contradiction shows that T is one-to-one.

To establish the constructive existence of a solution of equation (1) we first note that since, as one easily sees from (H3), T_n is a one-to-one continuous mapping of X_n into Y_n for each $n \geq N_0$, the Brouwer theorem on invariance of domain implies that the range $R(T_n)$ is an open set in Y_n; furthermore, it follows from (H1) and (H3) that $R(T_n)$ is also a closed set in Y_n for each $n \geq N_0$. Indeed, if $\{y_m\} \subset R(A_n)$ and $y_m \to y$ in Y_n as $m \to \infty$, then there exists a sequence $\{x_m\} \subset X_n$ such that $y_m = T_n x_m$ and by (H3)

$$\alpha(\|x_i - x_j\|) \leq \|T_n x_i - T_n x_j\| = \|y_i - y_j\| \to 0$$

as $i, j \to \infty$. Thus $\{x_m\}$ is a Cauchy sequence in X_n. Since X_n is complete, there exists x in X_n such that $x_m \to x$ and, by continuity of T_n, $T_n x_m \to T_n x = y$, i.e., $R(T_n)$ is closed. Since $R(T_n)$ is a nonempty set in Y_n, which is both open and closed in Y_n, it follows that $R(T_n) = Y_n$ for each $n \geq N_0$.

Hence for each $n \geq N_0$ and each f in Y there exists a unique element x_n in X_n such that $T_n x_n = Q_n(f)$. For the sequence $\{x_n\}$ thus determined, (C4) and (H3) and the additional property of $\{P_n\}$ imply that

$$k(f) \geq \|Q_n f\| = \|T_n x_n\| \geq \|T_n x_n - T_n(0)\| - \|T_n(0)\| \geq \alpha(\|x_n\|) - k(T(0)),$$

i.e., $\alpha(\|x_n\|) \leq k(f) + k(T(0))$ for each n. Hence our properties of $\alpha(r)$ imply that $\{x_n \mid x_n \in X_n\}$ is a bounded sequence and $\|T_n x_n - Q_n f\| = 0 \to 0$ as $n \to \infty$. Since T is A-proper, there exists a subsequence $\{x_{n_j}\}$ of $\{x_n\}$ and element x_0 in X such that $P_{n_j} x_{n_j} \to x_0$ in X and $Tx_0 = f$, i.e., x_0 is a solution of equation (1). Because, as was shown above, x_0 is the unique solution of equation (1) it follows that the entire sequence $\{P_n x_n\}$ converges strongly to x_0 in X, i.e., equation (1) is strongly approximation-solvable.

(A2) $\Rightarrow$ (A3). This implication follows from Definition 1.3 since the approximation-solvability of equation (1) implies, in particular, the solvability of equation (1).

(A3) $\Rightarrow$ (A1). Let $\{x_{n_j} \mid x_{n_j} \in X_{n_j}\}$ be any bounded sequence such that $\|T_{n_j} x_{n_j} - Q_{n_j} f\| \to 0$ as $j \to \infty$ for some f in Y. Since, by (A3), for any given f in Y, equation (1) is solvable, there exists $x_0 \in X$ such that $Tx_0 = f$. Thus, (H3) and (H2) imply that

$$\begin{aligned}\alpha(\|x_{n_j} - R_{n_j} x_0\|) &\leq \|T_{n_j} x_{n_j} - T_{n_j} R_{n_j} x_0\| \\ &\leq \|T_{n_j} x_{n_j} - Q_{n_j} f\| \\ &\quad + \|Q_{n_j} T x_0 - T_{n_j} R_{n_j} x_0\| \to 0 \qquad (j \to \infty)\end{aligned}$$

from which, by the properties of $\alpha(r)$, it follows that $\|x_{n_j} - R_{n_j} x_0\| \to 0$ as $j \to \infty$. This and the properties of $\{P_n\}$ imply that

$$\|P_{n_j} x_{n_j} - P_{n_j} R_{n_j} x_0\| \to 0 \qquad (j \to \infty)$$

from which, in view of (C2), it follows that $P_{n_j} x_{n_j} \to x_0$ as $i \to \infty$ and $Tx_0 = f$, i.e., T is A-proper.

Q.E.D.

REMARK 1.1. We would like to underline the practical usefulness of Theorem 1.1. It allows us to construct a strongly convergent sequence $\{x_n\}$ of approximate solutions of equation (1) when T satisfies hypotheses (H1)–(H3) and we know already that equation (1) has a solution no matter what argument was used to obtain an existence theorem.

REMARK 1.2. The hypothesis (H2) holds, in particular, when it is assumed that T is continuous and Q_n satisfies also condition (C3) of Definition 1.1.

But, when T is assumed to be only demicontinuous or weakly continuous, then in general the hypothesis (H2) is not verifiable without further assumptions and, therefore, Theorem 1.1 is not immediately applicable in this case to the solution of equation (1). Nevertheless, looking over the proof of the implication (A1) $\Rightarrow$ (A2) we see that the following practically useful corollary is valid.

COROLLARY 1.1. *Suppose that all conditions of Theorem* 1 *are satisfied except for the hypothesis* (H2). *Then, for each f in Y, equation* (1) *is feebly approximation-solvable if T is A-proper and strongly approximation-solvable if T is A-proper and one-to-one.*

REMARK 1.3. In the proof of Corollary 1.1 one does not use the condition (C2).

COROLLARY 1.2. *Let* $\{P_n\}$, $\{R_n\}$ *and* $\{Q_n\}$ *be sequences of continuous linear mappings for which conditions of Definition* 1.1 *hold. Let* T *be a bounded linear mapping of* X *into* Y. *Then equation* (1) *is strongly approximation-solvable if* T *is a one-to-one* A*-proper mapping and the norm in* X_n *is such that when* $\{x_n \mid x_n \in X_n\}$ *is a sequence with* $\|x_n\| = 1$ *and* $P_n x_n \to x$ *in* X *then* $x \neq 0$.

PROOF. First note that, in view of our conditions, (H1) and (H2) are trivially satisfied. Thus, by Theorem 1.1 (case (A1) $\Rightarrow$ (A2)) to prove Corollary 1.2 it suffices to show that our present conditions imply the validity of hypothesis (H3) which in this case reduces to the inequality

$$\text{(H3°)} \qquad \|T_n x\| \geq \alpha(\|x\|) \quad \text{for all } x \text{ in } X_n \text{ and } n \geq N_0$$

with $\alpha(r) = cr$ for some constant $c > 0$. Indeed, if (H3°) were not true, there would exist a sequence $\{x_m \mid x_m \in X_m\}$ with $\|x_m\| = 1$ for each m (because of the linearity of T_n) and $\|T_m x_m\| \to 0$. Now, since T is A-proper, there exists a subsequence $\{x_{m_i}\}$ and an element x in X such that $P_{m_i} x_{m_i} \to x$ in X and $Tx = 0$. Our assumption on the norms in X_n imply that $x \neq 0$ in contradiction to the one-to-one property of T. This proves Corollary 1.2.

REMARK 1.4. It was shown in [**43**] that when X and Y are Banach spaces, $\{X_n\} \subset X$, $\{Y_n\} \subset Y$, P_n is an identity injection of Y_n into X and R_n and Q_n are linear projections such that $R_n x \to x$ and $Q_n y \to y$ for x in X and y in Y, then the converse to Corollary 1.2 is also true.

1.2. *Mappings defined only on subsets of* X. In the rest of this paper it is assumed that X and Y are *real* spaces and that $\{P_n\}$ and $\{Q_n\}$ are sequences of *continuous linear* mappings. In this case, condition (C4) of Definition 1.1 reduces to the requirement

$$\text{(C4°)} \qquad \|Q_n f\|_{Y_n} \leq M_0 \|f\| \quad \text{for each } f \text{ in } Y \text{ and some } M_0 > 0$$

while (C1) and (C3) are replaced by a single condition

$$\text{(C5)} \qquad \|P_n x\| \leq M_1 \|x\|_{X_n} \quad \text{for } x \in X_n.$$

In §§1.2 and 1.3 we also assume that $\{X_n\}$ and $\{Y_n\}$ are sequences of oriented finite-dimensional spaces while in §1.2 we assume that $D = D(T)$ is an open bounded set in X and T is a mapping of $\bar{D}$ into Y, where $\bar{D}$ and ∂D denote the closure and the boundary of D, respectively. As before, we assume that the set $D_n = P_n^{-1}(D) = \{x \in X_n \mid P_n x \in D\}$ is bounded and note that $D_n \cap \partial D_n = \varnothing$ and $P_n^{-1}(\bar{D})$ is a closed subset of X_n containing D_n. Hence $\bar{D}_n \subset P_n^{-1}(\bar{D})$ and $\partial D_n \subset P_n^{-1}(\partial D)$. In what follows we set $T_n = Q_n T P_n \mid_{\bar{D}_n}$.

THEOREM 1.2. *Let* T *be an* A*-proper mapping of* $\bar{D}$ *into* Y *with* T_n *continuous on* $\bar{D}_n$ *and such that for all sufficiently large* n *the degree* deg $(T_n, D_n, 0) \neq 0$ *whenever it is defined, where* deg $(T_n, D_n, 0)$ *is the classical Brouwer degree* (*see* [**18**]) *for mappings of oriented finite-dimensional Euclidean spaces of the same dimension. If*

f is a given vector in Y such that

(H4) $$\|Tx\| > \|f\| \quad \text{for all } x \text{ in } \partial D,$$

then equation (1) *is feebly approximation-solvable in D.*

If, in addition, we assume that equation (1) *has at most one solution in D, then equation* (1) *is strongly approximation-solvable.*

PROOF OF THEOREM 1.2. To prove Theorem 1.2 we need the following lemma which we establish by using the same arguments as in the proof of Lemma 1 in [**36**] (see also [**17**], [**44**], [**45**]).

LEMMA 1.1. *Let T be an A-proper mapping of $\bar{D}$ into Y and g a given element in Y such that*

$$Tx - tg \neq 0 \quad \text{for all } t \text{ in } [0, 1] \text{ and all } x \text{ in } \partial D.$$

Then there exists an integer $N_1 \geq 1$ and a constant $c > 0$ such that

$$\|T_n x - tQ_n g\| \geq c \quad \text{for all } n \geq N_1,\ t \in [0, 1] \text{ and } x \in \partial D_n.$$

PROOF OF LEMMA 1.1. Suppose the assertion of Lemma 1.1 were false. Then there would exist a sequence of vectors $\{x_{n_j} \mid x_{n_j} \in \partial D_{n_j}\}$ and a sequence of numbers $\{t_j\} \subset [0, 1]$ such that

$$\|T_{n_j} x_{n_j} - t_j Q_{n_j} g\| \to 0 \qquad (j \to \infty).$$

Since $\{t_j\} \subset [0, 1]$, it has a limit point, say, $t_0 \in [0, 1]$. Hence there exists a subsequence of $\{t_j\}$, again denoted by $\{t_j\}$, such that $t_j \to t_0$ and, since Q_n's are linear and $\{Q_n g\}$ bounded,

$$T_{n_j} x_{n_j} - Q_{n_j}(t_0 g) = T_{n_j} x_{n_j} - t_j Q_{n_j} g + (t_j - t_0) Q_{n_j} g \to 0$$

as $j \to \infty$. Thus, by A-properness of T, there exists a subsequence, again denoted by $\{x_{n_j}\}$, and an element x_0 in $\bar{D}$ such that $P_{n_j} x_{n_j} \to x_0$ and $Tx_0 = t_0 g$ with $x_0 \in \partial D$ because $P_{n_j} x_{n_j} \in \partial D$ and ∂D is closed. This, however, contradicts the hypothesis that $Tx - tg \neq 0$ for each $x \in \partial D$ and each $t \in [0, 1]$. Q.E.D.

THE PROOF OF THEOREM 1.2 (CONTINUED). Since, for each given f in Y for which the inequality (H4) holds,

$$\|Tx - tf\| \geq \|Tx\| - t\|f\| > \|f\| - t\|f\| \geq (1 - t)\|f\|$$

for each x in ∂D and $t \in [0, 1]$ it follows from Lemma 1.1 and our condition on the degree of T_n that $\deg(T_n, D_n, 0) \neq 0$ for each $n \geq N_1$ and that the mappings T_n and $T_n - Q_n f$ are homotopic on ∂D_n. Hence, for each $n \geq N_1$,

$$\deg(T_n - Q_n f, D_n, 0) = \deg(T_n, D_n, 0) \neq 0.$$

Therefore, for each fixed $n \geq N_1$, there exists an element x_n in D_n such that

$$T_n x_n = Q_n f \qquad (n \geq N_1).$$

Since $\{x_n \mid x_n \in X_n\}$ is bounded and $T_n x_n - Q_n f \to 0$ as $n \to \infty$, the A-properness of T implies the existence of a subsequence $\{x_{n_j}\}$ and an element x_0 in $\bar{D}$ such that $P_{n_j} x_{n_j} \to x_0$ in X and $Tx_0 = f$. Clearly $x_0 \in D$ since x_0 in ∂D would violate the condition (H4). This proves the first part of Theorem 1.2.

The second part follows from the first part of Theorem 1.2, which gives the existence of a solution of equation (1), the additional uniqueness assumption and the A-properness of T.

THEOREM 1.3. *Let D be a symmetric bounded open set in X with $0 \in D$ and let T be an odd (i.e., $T(-x) = -Tx$ for all x in $\bar{D}$) A-proper mapping of $\bar{D}$ into Y such that T_n is continuous on $\bar{D}_n$. If f is a given element in Y for which the inequality* (H4) *holds, then the conclusions of Theorem* 1.2 *remain valid.*

PROOF OF THEOREM 1.3. In view of Theorem 1.2, to prove Theorem 1.3, it suffices to show that there exists an integer $N_1 \geq 1$ such that $T_n x \neq 0$ for all x in ∂D_n and deg $(T_n, D_n, 0) \neq 0$ for all $n \geq N_1$. Now (H4) and Lemma 1.1 imply the existence of an integer $N_1 \geq 1$ such that

$$T_n x - Q_n f \neq 0 \quad \text{for all } t \in [0, 1], x \in \partial D_n \text{ and } n \geq N_1,$$

i.e., T_n and $T_n - Q_n f$ are homotopic on ∂D_n. Since T is odd, D a symmetric open bounded set with $0 \in D$, and P_n and Q_n are linear, D_n is also a symmetric bounded open set in X_n with 0 in D_n and T_n is an odd mapping on $\bar{D}_n$ with $T_n x \neq 0$ on ∂D_n for each $n \geq N_1$. Hence by Borsuk Theorem, deg $(T_n, D_n, 0)$ is an odd number and, in particular, deg $(T_n, D_n, 0) \neq 0$ for each $n \geq N_1$. Q.E.D.

Since P_n and Q_n are continuous linear mappings and in finite-dimensional Banach spaces the strong and the weak convergences coincide, an immediate consequence of Theorem 1.3 is the following practically useful corollary.

COROLLARY 1.3. *Let D, T and f satisfy the conditions of Theorem* 1.3 *except that the continuity assumption on T_n in $\bar{D}_n$ is replaced by the requirement that T be either continuous, demicontinuous or weakly continuous on $\bar{D}$. Then the conclusions of Theorem* 1.3 *hold.*

1.3. *Homogeneous operators.* It is known [**46**] that if A is a linear compact operator of a normed linear space X into itself and $\gamma \neq 0$ is not an eigenvalue of A, then $(\gamma I - A)$ is onto. The purpose of this section is to generalize this fact to nonlinear A-proper mappings of X into Y.

DEFINITION 1.4. *Let X and Y be two normed real linear spaces. A mapping T of X into Y is said to be positively homogeneous of order $\alpha > 0$ if $T(tx) = t^\alpha T(x)$ for all $t > 0$ and all x in X.*

DEFINITION 1.5. *Let T and S be positively homogeneous of order $\alpha > 0$ mappings of X into Y with $T(0) = S(0) = 0$. A number $\gamma \neq 0$ is said to be an eigenvalue of S relative to T if and only if there exists a vector $x \neq 0$ in X such that $\gamma Tx - Sx = 0$.*

Our main result of this section is the following theorem for positive homogeneous operators derived under rather weak conditions.

THEOREM 1.4. *Let T be a mapping of X into Y which is odd, A-proper with with respect to Γ_n and positively homogeneous of order $\alpha > 0$. For each fixed n, let T_n be a continuous mapping of X_n into Y_n. Let S be an odd completely continuous mapping of X into Y which is also positively homogeneous of order $\alpha > 0$. Then, if $\gamma \neq 0$ is not an eigenvalue of S relative to T, for each f in Y the equation*

$$\gamma Tx - Sx = f \qquad (f \in Y)$$

is feebly approximation-solvable (in particular, $\gamma T - S$ is onto).

PROOF OF THEOREM 1.4. Since Q_n's are linear, $\gamma \neq 0$ and T is A-proper, the mapping γT is also A-proper. This and the complete continuity of S imply that $F = \gamma T - S$ is also A-proper; furthermore, F is odd and positively homogeneous of order α. Suppose that $\gamma \neq 0$ is not an eigenvalue of S relative to T. Then $Fx \neq 0$ for all $x \neq 0$; in particular, $Fx \neq 0$ for all x with $\|x\| = 1$. Put $B_1 = \{x \in X \mid \|x\| < 1\}$ and $B_{1n} = P_n^{-1}(B_1)$. Then, because P_n is a continuous linear mapping, $\bar{B}_{1n} \subseteq P_n^{-1}(\bar{B}_1)$ and $\partial B_{1n} \subseteq P_n^{-1}(\partial B_1)$. Applying Lemma 1.1 to F with $g = 0$, we obtain

(a) $$\|Q_n F P_n x\| \geq c \qquad (x \in \partial B_{1n}, n \geq N)$$

for some constant $c > 0$ and integer $N \geq 1$. Let $R \geq 1$ be an arbitrary real number, $B_R = \{x \in X \mid \|x\| < R\}$ and $B_{R^n} = P_n^{-1}(B_R)$ and note that $\bar{B}_{R^n} = B_{R^n} \cup \partial B_{R^n} \subseteq P_n^{-1}(\bar{B}_R)$ and $\partial B_{R^n} \subseteq P_n^{-1}(\partial B_R)$. For any x in ∂B_{R^n}, put $y = x/\|P_n x\| = x/R$ and note that $y \in \partial B_{1n}$ since $P_n y = P_n x/\|P_n x\| = P_n x/R \in \partial B_1$. Hence, by (a) and the positive homogeneity of order α of F,

$$c \leq \|Q_n F P_n y\| = \left\| Q_n F P_n \left(\frac{x}{R}\right) \right\| = \frac{1}{R^\alpha} \|Q_n F P_n x\|,$$

i.e.,

(b) $$\|Q_n F P_n x\| \geq cR^\alpha \quad \text{for all } n \geq N \text{ and all } x \text{ in } \partial B_R n.$$

Now, let f be an arbitrary but fixed element in Y and $M(>0)$ a constant such that $\|Q_n f\| \leq M\|f\|$ for all n. Choose $R > 1$ so that $cR^\alpha > M\|f\| \geq \|Q_n f\|$. This and (b) imply that

$$\|Q_n F P_n x\| \geq cR^\alpha > \|Q_n f\| \quad \text{for all } n \geq N \text{ and all } x \text{ in } \partial B_{R^n}.$$

Consequently, for any $t \in [0, 1]$ and x in ∂B_{R^n},

(c) $$\|Q_n F P_n x - tQ_n f\| > (1 - t)\|Q_n f\| \qquad (n \geq N).$$

Since $Q_n F P_n x = F_n x \neq 0$ for x in ∂B_{R^n} and F_n is odd and continuous for each n, Borsuk Theorem implies that deg $(F_n, B_{R^n}, 0) \neq 0$ for $n \geq N$. By virtue of (c), the homotopy argument shows that deg $(F_n - Q_n f, B_{R^n}, 0) \neq 0$ for each $n \geq N$. Hence, for each $n \geq N$, there exists $x_n \in B_{R^n}$ such that $F_n x_n - Q_n f = 0$ and, consequently, the A-properness of F implies the existence of a subsequence $\{x_{n_i}\}$ and an element x in B_R such that $P_{n_i} x_{n_i} \to x_0$ and $Fx_0 = 0$, i.e., $\gamma T x_0 - Sx_0 = 0$. Q.E.D.

COROLLARY 1.4. *Let X, Y, T and S satisfy all the conditions of Theorem* 1.4 *except that the continuity assumption on T_n is replaced by the requirement that T be either continuous, demicontinuous or weakly continuous. Then the conclusion of Theorem* 1.4 *holds.*

REMARK 1.5. When X is a real reflective Banach space, $Y = X^*$, $\{X_n\} \subset X$, $\{R_n\}$ linear projections such that $R_n x \to x$ for each x in X, $P_n = I$ on X_n, $Q_n = R_n^*$ and $\{Y_n\} = \{R(R_n^*)\} \subset X^*$, Corollaries 1.3 and 1.4 were obtained by Pohodjayev [**45**] for the case when T is a continuous A-proper[4] mapping which is also bounded.

2. Sufficient conditions for a mapping to be A-proper.

The results obtained in the preceding section indicate that the introduction of the concept of an A-proper mapping is not only natural but also a very convenient and useful tool in the constructional study of the solvability of various classes of nonlinear equations. In this section we give two sufficient conditions for a mapping $T: D \subseteq X \to Y$ to be A-proper. The first is the *condition* (c) introduced by the author [**38**], [**39**] while the second is the modification of *condition* (s) introduced by Browder [**13**].

DEFINITION 2.1. *T is said to satisfy condition* (c) *if it has the property that whenever Γ_m is an arbitrary subscheme of the scheme Γ_n for equation* (1) *and $\{x_m \mid x_m \in D_m\}$ is a sequence so that $P_m x_m \rightharpoonup x$ for some x in D and*

$$\|T_n x_n - Q_n g\| \to 0$$

for some g in Y, then $Tx = g$.

It turns out that to obtain useful properties of T (e.g., A-properness of T) when its domain D is a proper subset of X, in general we need to impose some further conditions on X. It was shown in [**42**] that when D is a closed ball, then it suffices to assume that X is a real Banach space with Property (H) introduced by Fan and Glickberg [**20**], where X is said to have Property (H) if X is strictly convex (i.e., for any pair x and y in X the relation $\|x + y\| = \|x\| + \|y\|$ implies that $x = \lambda y$ with $\lambda > 0$) and if for any sequence $\{x_n\} \subset X$ the relations $x_n \rightharpoonup x$ in X and $\|x_n\| \to \|x\|$ imply that $x_n \to x$ in X. It was shown in [**20**] that Hilbert spaces, uniformly convex and locally uniformly convex Banach spaces are examples of spaces having Property (H).

THEOREM 2.1. *Let Y be a normed real space and X a real reflexive Banach space with Property* (H). *Let T be a continuous mapping of $D = B = \{x \in X \mid \|x\| \leq r\}$ into Y with $B_n = P_n^{-1}(B)$. Let $\alpha(r)$ be a continuous function of R^+ into R^+ and $N \geq 1$ a fixed integer such that $r_i \to 0$ whenever $\alpha(r_i) \to 0$ as $i \to \infty$ and*

$$\|T_n x - T_n y\| \geq \alpha(\|x - y\|)(x, y \in B_n, n \geq N). \tag{2.1}$$

[4] Modifying the concept of the P-compact operator, Pohodjayev introduced in [**45**] the definition of a *strongly closed* mapping T from a reflexive Banach space X to X^* which under his conditions is equivalent to the definition of the A-proper mapping in terms of projective schemes.

If T satisfies also condition (c) *on B, then T is A-proper.*

PROOF OF THEOREM 2.1.   Let $\{x_n \mid x_n \in B_n\}$ be a sequence so that

$$\| T_n x_n - Q_n g \| \to 0$$

for some g in Y. Since X is reflexive, $\{P_n x_n\} \subset B$ and B is weakly closed without loss of generality we may assume that $P_n x_n \rightharpoonup x_0$ in X for some x_0 in B. Hence, by condition (c), $T x_0 = g$.

Now, if $x_0 \in \partial B$, then, since X has Property (H), Lemma 1 in [42] implies $P_n x_n \to x_0$ and thus the validity of Theorem 2.1. If, on the other hand, $x_0 \in \text{Int } (B)$ then, since $P_n R_n x_0 \to x_0$ in X, there exists an integer $N_0 \geq 1$ such that $P_n R_n x_0 \in \text{Int } (B)$ for all $n \geq N_0$ and by (2.1)

$$\| T_n x_n - T_n R_n x_0 \| \geq \alpha(\| x_n - R_n x_0 \|) \quad \text{for all } n \geq \max \{N, N_0\}.$$

This, the continuity of T and the relations $T x_0 = g$ and $\| T_n x_n - Q_n g \| \to 0$ imply that

$$\alpha(\| x_n - R_n x_0 \|) \leq \| T_n x_n - Q_n g \| + \| Q_n g - Q_n T P_n R_n x_0 \| \to 0.$$

Hence $\| x_n - R_n x_0 \| \to 0$ and, consequently, $\| P_n x_n - P_n R_n x_0 \| \to 0$, i.e., $P_n x_n \to x_0$ in X.   Q.E.D.

REMARK 2.1.   When $D = X$, $\{X_n\} \subset X$ and $\{Y_n\} \subset Y$ with $P_n x \to x$ for x in X and $Q_n y \to y$ for y in Y, Theorem 2.1 and its converse were first proved in [41] without the assumption that X has Property (H) (see also [17]).

By virtue of Theorem 2.1, it is useful to know when a given operator satisfies condition (c). Some results in this direction were obtained in [42] for the case when $Y = X$, Γ_n is a projective scheme and T is a J-monotone mapping of X into X. Here we propose to extend these results to K-monotone mappings of X into Y with general approximation schemes.

DEFINITION 2.2.   *Let T be a mapping of $D \subseteq X$ into Y and K a (in general nonlinear) mapping of X into Y^*, where Y^* denotes the adjoint of Y. T is said to be K-monotone on D if*

$$(Tx - Ty, K(x - y)) \geq 0 \quad \text{for all } x, y \text{ in } D,$$

where (z, F) denotes the value of F in Y^* at z in Y.

Now, our admissible scheme Γ_n and its auxiliary are schematically represented as follows:

$$
\begin{array}{ccc}
X \supset D & \xrightarrow{\;T\;} & Y \\
R_n \downarrow \uparrow P_n & & \downarrow Q_n \\
X_n \supset D_n & \xrightarrow{\;T_n\;} & Y_n
\end{array}
\qquad\qquad
\begin{array}{ccc}
X & \xrightarrow{\;K\;} & Y^* \\
\uparrow P_n & & \downarrow \\
X_n & \xrightarrow{\;K_n\;} & Y_n^*
\end{array}
$$

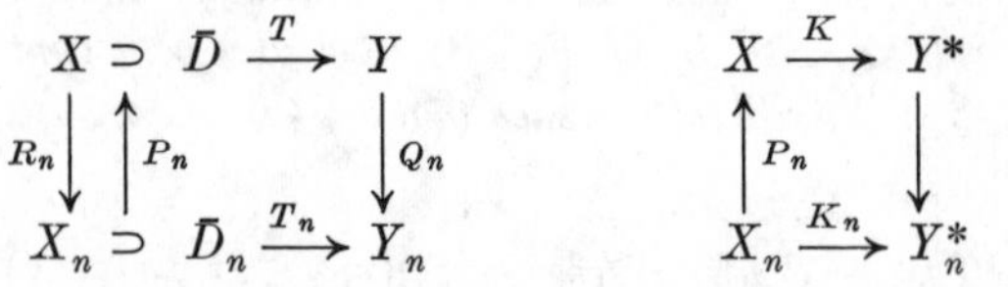

REMARK 2.2.   Definition 2.2 in this generality is essentially due to Kato [28] for nonlinear K and to Kachurowsky [24] and the author [33] for linear K. For the case when $Y = X$ and $K = J$, a duality mapping, it was first introduced by Browder [9] (see also [39] for $K \neq J$). Its special cases will be discussed later.

LEMMA 2.1. *Let X and Y be spaces with X reflexive and having Property* (H). *Let T be a continuous mapping of B into Y and K a mapping of X into Y^* such that the following hypotheses hold*:

(H2a) *T is K-monotone on B.*

(H2b) *K is onto.*

(H2c) *For $t > 0$ and $x \in X$, $K(tx) = k_t K(x)$ with $k_t > 0$.*

(H2d) *$(Q_n g, K_n x) = (g, KP_n x)$ for $x \in X_n$ and $g \in Y$.*

(H2e) *If $\{z_n \mid z_n \in B_n\}$ and $\{y_n \mid y_n \in B_n\}$ are sequences such that $p_n z_n \rightharpoonup z$ in X and $\|T_n y_n - Q_n h\| \to 0$ for some h in Y, then $(T_n y_n, K_n z_n) \to (h, Kz)$.*

Then, under the hypotheses (H2a) — (H2e), *the mapping T satisfies condition* (c).

PROOF. Let $\{x_n \mid x_n \in B_n\}$ be any sequence so that

$$\|T_n x_n - Q_n g\| \to 0 \quad \text{for some } g \text{ in } Y.$$

Since X is reflexive, $\{P_n x_n\} \subset B$ and B is weakly closed, we may assume that $P_n x_n \rightharpoonup x_0$ for some x_0 in B and either $x_0 \in \text{Int (B)}$ or $x_0 \in \partial B$. Suppose first that $x_0 \in \partial B$. Then, since X has Property (H), $P_n x_n \to x_0$ in X. Thus, the continuity of T and the properties of $\{Q_n\}$ imply that $TP_n x_n \to Tx_0$ and $\|Q_n TP_n x_n - Q_n Tx_0\| \to 0$. Hence, since $P_n R_n x \to x$ in X for each x in X, (H2e) implies that $(T_n x_n, K_n P_n x) \to (Tx_0, Kx)$ for each x in X. But, by our hypothesis,

$$\|T_n x_n - Q_n g\| \to 0$$

and therefore, again by (H2e), $(T_n x_n, K_n P_n x) \to (g, Kx)$. Consequently, $(g - Tx_0, Kx) = 0$ for each x in X. Hence, since K is onto, it follows that $Tx_0 = g$.

Suppose now that $x_0 \in \text{Int } (B)$ and let x be any fixed point in Int (B). Since $P_n R_n x \to x$, there exists an integer $N_0 \geq 1$ such that $P_n R_n x \in B$ for all $n \geq N_0$ and, in view of (H2a) and (H2d), for all $n \geq \max \{N, N_0\}$,

$$(TP_n R_n x - TP_n x_n, K(P_n R_n x - P_n x_n)) = (T_n R_n x - T_n x_n, K_n(R_n x - x_n)) \geq 0.$$

Since $\|T_n x_n - Q_n g\| \to 0$ and $P_n R_n x - P_n x_n \rightharpoonup x - x_0$ it follows from (H2e) that

$$(T_n x_n, K_n(R_n x - x_n)) \to (g, K(x - x_0)).$$

Similarly, since $P_n R_n x \to x$ and T is continuous, $\|T_n R_n x - Q_n Tx\| \to 0$ and

$$(T_n R_n x, K_n(R_n x - x_n)) \to (Tx, K(x - x_0)).$$

Consequently, the passage to the limit in the last inequality (as $n \to \infty$) yields the relation

$$(\dagger) \qquad (Tx - g, K(x - x_0)) \geq 0 \quad \text{for each } x \text{ in Int} (B).$$

The inequality (†) implies that $Tx_0 = g$ because assuming to the contrary that $Tx_0 \neq g$ we arrive at the contradiction. Indeed, if $Tx_0 - g \neq 0$, then since K maps X onto Y^*, there exists a vector z in X such that

$$(*) \qquad (Tx_0 - g, Kz) < 0.$$

Since $x_0 \in \operatorname{Int}(B)$, for $t > 0$ and sufficiently small $x_t = x_0 + tz \in \operatorname{Int}(B)$ and, in view of (H2c), the replacement of x in (†) by x_t gives the inequality

$$(\dagger\dagger) \qquad (Tx_t - g, Kz) \geq 0.$$

Adding $-(Tx_0 - g, Kz)$ to both sides of (††) and taking into account (*) we get

$$(**) \qquad (Tx_t - Tx_0, Kz) \geq -(Tx_0 - g, Kz) > 0.$$

Since the right-hand side in (**) is independent of t and T is continuous, the passage to the limit in (**) as $t \to 0^+$ gives the contradiction

$$0 = (Tx_0 - Tx_0, Kz) \geq -(Tx_0 - g, Kz) > 0$$

and thus establishes the validity of Lemma 2.1. Q.E.D.

COROLLARY 2.1. *Let X and Y satisfy the conditions of Theorem* 2.1 *and let T be a continuous mapping of $B \subset X \to Y$ for which* (2.1) *holds. Let K be a mapping of X into Y^* such that the hypotheses* (H2a)-(H2e) *of Lemma* 2.1 *hold. Then T is A-proper.*

PROOF. In view of Lemma 2.1, Corollary 2.1 follows from Theorem 2.1.

Modified condition (s). It was shown in [**36**] that if X is a Hilbert space, $\{X_n\} \subset X$, P_n a linear projection of X onto X_n such that $P_n x \to x$ for each x in X, and A a bounded demicontinuous mapping of X into X such that for each fixed $c > 0$ the mapping $T = A + cI$ satisfies the inequality

$$(\#) \qquad (Tx - Ty, x - y) \geq c\|x - y\|^2 \quad \text{for all } x \text{ and } y \text{ in } X,$$

then A is P-compact which is equivalent (see [**41**]) to the assertion that $T = A + cI$ is A-proper for each $c > 0$ with respect to the scheme $(\{X_n\}, \{P_n\})$. It was noted by Browder [**13**] that the proof of the above fact, instead of the strong monotonicity requirement (#) on T, uses only the condition that the sequence $\{x_n\}$ in X be such that if $x_n \rightharpoonup x$ in X and $(Tx_n - Tx, x_n - x) \to 0$, then $x_n \to x$. The last condition, referred to in [**13**] as *condition* (s), is much weaker than condition (#) on T, and as was shown in [**13**], is easily verifiable for a rather general class of partial differential operators considered as mappings of a suitable Banach space into its dual X^*.

It turns out, however, that when T is not bounded or when T maps $D \subseteq X$ into $Y \neq X^*$ condition (s) cannot be used directly to establish the A-properness of T. Nevertheless, for continuous mappings T, the proof of Lemma 7A in [**39**] suggests the following modification of condition (s) which is quite suitable for our purposes since it depends essentially on the fact that the sequence $\{x_n\} \subset X$ be such that $x_n \in X_n$ for each n, the fact which is embodied in the theory of A-proper mappings.

DEFINITION 2.3. *The mapping $T : D \subseteq X \to Y$ is said to satisfy modified condition* (s) *if for any sequence $\{x_n \mid x_n \in D_n\}$ such that $P_n x_n \rightharpoonup x_0$ in X for some x_0 in D and $(TP_n x_n - TP_n R_n x_0, K(P_n x_n - P_n R_n x_0)) \to 0$ we have $P_n x_n \to x_0$ in X.*

THEOREM 2.2. (a) *Let X and Y be two normed real linear spaces with X a reflexive Banach space. Let T be a continuous mapping of X into Y and K a mapping of X onto Y^* and K_n a mapping of X_n into Y_n^* such that $K(0) = 0$ and the hypotheses* (H2d) *and* (H2e) *of Lemma* 2.1 *hold. Suppose further that T satisfies modified condition* (s). *Then T is an A-proper mapping of X into Y with respect to Γ_n.*

(b) *If in addition to the above conditions we assume that X has Property* (H), *then T defined only on B is also an A-proper mapping of B into Y.*

PROOF OF THEOREM 2.2. (a) Let $\{x_n \mid x_n \in X_n\}$ be any bounded sequence such that

$$(*) \qquad \|T_n x_n - Q_n g\| \to 0 \quad \text{for some } g \text{ in } Y.$$

Since X is reflexive and $\{P_n x_n\}$ is bounded, we may assume that $P_n x_n \rightharpoonup x_0$ for some x_0 in X. Since $P_n R_n x_0 \to x_0$ in X and T is continuous, it follows that $P_n x_n - P_n R_n x_0 \rightharpoonup 0$ and $\|T_n R_n x_0 - Q_n T x_0\| \to 0$; therefore, in view of (H2d) and (H2e) we have

$$\begin{aligned}(TP_n x_n - TP_n R_n x_0, K(P_n x_n - P_n R_n x_0))\\ = (T_n x_n, K_n(x_n - R_n x_0))\\ - (T_n R_n x_0, K_n(x_n - R_n x_0)) \to (g, K(0)) - (Tx_0, K(0)) = 0.\end{aligned}$$

Hence, by modified condition (s), $P_n x_n \to x_0$ in X and, therefore, by continuity of T and the properties of $\{Q_n\}$, $\|T_n x_n - Q_n T x_0\| \to 0$. This, (*) and (H2e) imply that for each x in X

$$(g, Kx) = \lim_n (T_n x_n, K_n R_n x) = (Tx_0, Kx)$$

from which, since K is onto, it follows that $Tx_0 = g$, i.e., T is an A-proper mapping of X into Y.

(b) Let $\{x_n \mid x_n \in B_n\}$ be such that (*) holds. Since X is reflexive and B is weakly closed, there is $x_0 \in B$ such that $P_n x_n \rightharpoonup x_0$ with x_0 belonging either to ∂B or to Int (B). If $x_0 \in \partial B$, then by Lemma 1 in [**42**], $P_n x_n \to x_0$ and the continuity of T and property of Q_n imply that $\|T_n x_n - Q_n T x_0\| \to 0$. Using this, (*), (H2d) and (H2e) we derive, as above, that $(Tx_0, Kx) = (g, Kx)$ for each x in X from which, since K is onto, we derive the equality $Tx_0 = g$.

Suppose now that $x_0 \in \text{Int}(B)$. Then, since $P_n R_n x_0 \to x_0$ in X and T is continuous, it follows that there exists $N \geq 1$ such that $P_n R_n x_0 \in \text{Int}(B)$ for $n \geq N$, $TP_n R_n x_0 \to Tx_0$ and $\|T_n R_n x_0 - Q_n T x_0\| \to 0$. This implies that, as before, we have the relation

$$(TP_n x_n - TP_n R_n x_0, K(P_n x_n - P_n R_n x_0)) \to 0$$

from which, in view of modified condition (s) satisfied by T, we obtain the convergence $P_n x_n \to x_0$ in X. Repeating the argument of the previous paragraph we get the equality $Tx_0 = g$ and thus the validity of Theorem 2.2(b). Q.E.D.

REMARK 2.3. It is easy to see that if we assume that K is weakly continuous and $\|K_n z_n\| \leq M$ for some constant $M > 0$ whenever $\{z_n \mid z_n \in D_n\}$ is bounded, then the hypothesis (H2e) holds. Hence Lemma 2.1 and Theorem 2.2 remain valid if instead of (H2e) we assume that K is a weakly continuous mapping of X into Y^* and K_n is a mapping of X_n into Y_n^* such that $\{K_n z_n\}$ is bounded whenever $\{z_n \mid z_n \in D_n\}$ is bounded.

Our next result gives sufficient conditions for continuous K-monotone mappings to be A-proper.

THEOREM 2.3 (a). *Let X and Y be as in Theorem 2.2(a), T a continuous mapping of X into Y and K a weakly continuous mapping of X onto Y^* such that*

$$(2.2)\qquad (Tx - Ty, K(x - y)) \geq c(\|x - y\|) \qquad (x, y \in X),$$

where $c(r)$ is a continuous function of R^+ into R^+ such that $c(0) = 0$, $c(r) > 0$ for $r > 0$ and $r_i \to 0$ whenever $c(r_i) \to 0$ as $i \to \infty$. Suppose that K_n is a mapping of X_n into Y_n^* such that (H2d) *holds and*

$$(2.3)\qquad \|K_n x\|_{Y_n^*} \leq \|KP_n x\|_{Y^*} = \|P_n x\|_X = \|x\|_{X^n} \quad \textit{for each } x \textit{ in } X_n.$$

Then T is A-proper with respect to Γ_n.

(b) *Let X and Y be as in Theorem* 2.2(b), *T a continuous mapping of X into Y and K a weakly continuous mapping of X onto Y^* such that*

$$(2.4)\quad (Tx - Ty, K(x - y)) \geq (\phi(\|x\|) - \phi(\|y\|))(\|Kx\| - \|Ky\|) \qquad (x, y \in X),$$

where $\phi(r)$ is a continuous strictly increasing function of R^+ into the reals R such that $\phi(r) \to +\infty$ as $r \to \infty$. Suppose that K_n is a mapping of X_n into Y_n^* such that (H2d) *and* (2.3) *hold. Then T is A-proper with respect to Γ_n.*

PROOF OF THEOREM 2.3. The part (a) follows essentially from Theorem 2.2(a) since our present conditions on T, K and K_n imply that $K(0) = 0$ and that (H2e) and the modified condition (s) hold. Indeed, the equality $K(0) = 0$ follows from (2.3) while the modified condition (s) follows from (2.3) provided we establish the validity of (H2e). Now, let $\{z_n \mid z_n \in X_n\}$ and $\{y_n \mid y_n \in X_n\}$ be bounded sequences such that $P_n z_n \rightharpoonup z$ in X and $\|T_n y_n - Q_n h\| \to 0$ for some h in Y. Since, by virtue of (H2d),

$$(T_n y_n, K_n z_n) = (T_n y_n - Q_n h, K_n z_n) + (h, KP_n z_n)$$

and, by our conditions on $\{z_n\}$ and $\{y_n\}$ and the weak continuity of K,

$$(h, KP_n z_n) \to (h, Kz)$$

and

$$|(T_n y_n - Q_n h, K_n z_n)| \leq \|T_n y_n - Q_n h\| \, \|K_n z_n\| = \|T_n y_n - Q_n h\| \, \|P_n z_n\| \to 0$$

it follows that $(T_n y_n, K_n z_n) \to (h, Kz)$, i.e., (H2e) holds and thus Theorem 2.2(a) implies the validity of Theorem 2.3(a).

To prove (b) it suffices to show that under our conditions on X, T, K and K_n

the inequality (2.4) implies the validity of modified condition (s). Now, let $\{x_n \mid x_n \in X_n\}$ be a sequence so that $P_n x_n \rightharpoonup x_0$ in X and

$$(TP_n x_n - TP_n R_n x_0, K(P_n x_n - P_n R_n x_0)) \to 0.$$

In view of (2.4) the last relation and the equality $\|KP_n x_n\| = \|P_n x_n\|$ imply that

$$(\dagger) \qquad (\phi(\|P_n x_n\|) - \phi(\|P_n R_n x_0\|)(\|P_n x_n\| - \|P_n R_n x_0\|) \to 0.$$

Put $a_n = \|P_n x_n\|$ and $b_n = \|P_n R_n x_0\|$ for each n and observe that $b_n \to b_0 = \|x_0\|$ since $P_n R_n x_0 \to x_0$ in X. Since $\{a_n\}$ is bounded and $\phi(r)$ is continuous, it follows from (†) that

$$(\phi(a_n) - \phi(b_0))(a_n - b_0) \to 0$$

because (†) and the relations $b_n \to b_0$ and $\phi(b_n) \to \phi(b_0)$ imply that

$$\begin{aligned}(\phi(a_n) - \phi(b_0))(a_n - b_0) &\\ = (\phi(a_n) - \phi(b_n))(a_n - b_n)&\\ + (\phi(a_n) - \phi(b_n))(b_n - b_0) + (\phi(b_n) - \phi(b_0))(a_n - b_0) &\to 0.\end{aligned}$$

Hence Lemma 2.1 in [**2**] implies that $a_n \to b_0$, i.e., $\|P_n x_n\| \to \|x_0\|$. Since we also have that $P_n x_n \rightharpoonup x_0$ in X and X has Property (H), it follows that $P_n x_n \to x_0$, i.e., T satisfies modified condition (s).

3. **Further results on approximation-solvability and A-proper mappings; K-monotone mapping.** We start this section with the following known result (see, for example, [**36**], [**44**]) to be used below.

LEMMA 3.1. *Let V be a finite-dimensional real Banach space and D a bounded open convex subset of V with $0 \in D$. Let A be a continuous mapping of $\bar{D}$ into V such that if $Ax = \alpha x$ for some x in ∂D, then $\alpha \le 1$. Then A has a fixed point in $\bar{D}$.*

THEOREM 3.1. *Let X and Y be real normed linear spaces and D a bounded open convex subset of X with $0 \in D$. Let T be an A-proper mapping of $\bar{D}$ into Y with $T_n = Q_n T P_n \mid_{\bar{D}_n}$ continuous from $\bar{D}_n$ to Y_n. Let K be a mapping of X into Y^* and K_n a mapping of X_n into Y_n^* such that* (H2d) *holds. Let M_n be a linear isomorphism of X_n onto Y_n such that*

(H3a) $\qquad (M_n x, K_n x) \ge 0$ *for each x in ∂D_n.*

Then, if f is a given vector in Y such that

(H3b) $\qquad (Tx, Kx) \ge (f, Kx)$ *for all x in ∂D,*

equation (1) *is feebly approximation-solvable. If, in addition, we assume that for f satisfying* (H3b) *equation* (1) *has at most one solution in $\bar{D}$, then it is strongly approximation-solvable.*

PROOF. For each n and each x in $\bar{D}_n$ consider the mapping A_n defined by $A_n x = T_n x - Q_n f$. By virtue of (H2d), the inequality (H3b) and the fact that

$P_nx \in \partial D$ for x in ∂D_n, it follows that for all x in D_n we have

$$(A_nx, K_nx) = (Q_nTP_nx, K_nx) - (Q_nf, K_nx) = (TP_nx, KP_nx) - (f, KP_nx) \geq 0.$$

Now, since M_n is a linear isomorphism of X_n onto Y_n, $D'_n \equiv M_nD_n$ is a bounded open convex set in Y_n with $0 \in D'_n$, $D'_n \cap \partial D'_n = \varnothing$, and M_n maps ∂D_n homeomorphically onto $\partial D'_n$. Let G_n be the mapping of $\bar{D}'_n$ $(=M_n\bar{D}_n \subset Y_n)$ into Y_n defined by $G_n = I_n - A_nL_n$, where $L_n = M_n^{-1}$ and I_n is the identity mapping in Y_n. The above discussion implies that, by virtue of Lemma 3.1, to establish the existence of solutions $x_n \in \bar{D}_n$ of the equation $T_nx = Q_nf$, it suffices to show that if the equation $G_ny = \alpha y$ holds for some y in $\partial D'_n$, then $\alpha \leq 1$. Now suppose that $G_ny_0 = \alpha y_0$ for some y_0 in $\partial D'_n$. First, let x_0 be the unique point in ∂D_n such that $y_0 = M_nx_0$. Then

$$\begin{aligned}\alpha(M_nx_0, K_nx_0) &= (\alpha y_0, K_nL_ny_0) = (G_ny_0, K_nL_ny_0)\\ &= (y_0, K_nL_ny_0) - (A_nL_ny_0, K_nL_ny_0)\\ &= (M_nx_0, K_nx_0) - (A_nx_0, K_nx_0)\end{aligned}$$

from which, in view of (H3a), it follows that $\alpha \leq 1$. Hence, by Lemma 3.1, for each n there exists y_n in $\bar{D}'_n$ such that $G_ny_n = y_n$ or $T_nx_n = Q_nf$, where $x_n = L_ny_n \in \bar{D}_n$. Thus, by A-properness of T, there exists a subsequence $\{x_{n_i}\}$ and a vector x_0 in $\bar{D}$ such that $P_{n_i}x_{n_i} \to x_0$ and $Tx_0 = f$, i.e., equation (1) is feebly approximation-solvable.

If, in addition, we assume that equation (1) has at most one solution in $\bar{D}$, then x_0 constructed above is the unique solution of equation (1) and, therefore, the A-properness of T implies that the entire sequence $\{x_n\}$ converges to x_0, i.e., equation (1) is strongly approximation-solvable.

COROLLARY 3.1. *Let D, T, K, K_n, M_n and f satisfy the conditions of Theorem* 3.1 *except that the continuity assumption on $T_n: \bar{D}_n \to Y_n$ is replaced by the requirement that T be either continuous, demicontinuous or weakly continuous. Then the conclusions of Theorem* 3.1 *hold.*

An immediate consequence of Theorem 2.2 and Theorem 3.1 is the following useful result.

PROPOSITION 3.1. *Let X and Y be real normed spaces with X reflexive and having Property* (H). *Let T be a continuous mapping of B $(\subset X)$ into Y, K a mapping of X onto Y^* with $K(0) = 0$, K_n a mapping of X_n into Y_n^* satisfying* (H2d) *and* (H2e) *and M_n a linear isomorphism of X_n onto Y_n such that* (H3a) *holds. Suppose further that T satisfies modified condition* (s). *Then for each fixed f in Y such that* (H3b) *holds for each x in ∂B, the conclusions of Theorem* 3.1 *hold.*

Another consequence of Theorems 1.1 and 2.3 is the following proposition.

PROPOSITION 3.2. *Suppose X, Y, T, K and K_n satisfy the conditions of Theorem* 2.3(a). *Suppose further that $c(r)/r \to \infty$ as $r \to \infty$. Then equation* (1) *is strongly approximation-solvable for each f in Y.*

PROOF. By Theorem 2.3(a), T is an A-proper mapping of X into Y. Furthermore, it follows from (2.2), (2.3) and (H2d) that for all x and y in X_n

$$\begin{aligned} c(\|x-y\|) &= c(\|P_nx - P_ny\|) \\ &\leq (TP_nx - TP_ny, K(P_nx - P_ny)) \\ &= (T_nx - T_ny, K_n(x-y)) \leq \|T_nx - T_ny\| \, \|x-y\|. \end{aligned}$$

Hence setting $\alpha(r) = c(r)/r$ for $r > 0$ we see that $\alpha(r)$ is a continuous function of R^+ into R^+ such that $r_j \to 0$ whenever $\alpha(r_j) \to 0$ (as $j \to \infty$), $\{r_j\}$ is bounded whenever $\{\alpha(r_j)\}$ is bounded and

$$\|T_nx - T_ny\| \geq \alpha(\|x-y\|) \quad \text{for all } x \text{ and } y \text{ in } X_n.$$

Moreover, it is also obvious that under our conditions the hypotheses (H1) and (H2) of Theorem 1.1 hold. Hence, Proposition 3.2 follows from Theorem 1.1.

PROPOSITION 3.3. *Let X, Y, T, K and K_n satisfy the conditions of Theorem 2.3(b). Let M_n be a linear isomorphism of X_n onto Y_n such that $(M_nx, K_nx) \geq 0$ for all $x \in \partial B_{rn}$ and each $r > 0$. Then equation (1) is feebly approximation-solvable.*

PROOF. Proposition 3.3 follows from Theorems 2.3(b) and 3.1. To see this note first that, by Theorem 2.3(b), T is A-proper and that for any fixed f in Y we can choose r so large that $\phi(r) > \phi(0) + \|T(0) - f\|$. Now it is easy to see from (2.3) and (2.4) that for each x in ∂B_r

$$\begin{aligned} (Tx - f, Kx) &= (Tx - T(0), Kx) + (T(0) - f, Kx) \\ &\geq (\phi(r) - \phi(0) - \|T(0) - f\|)r > 0. \end{aligned}$$

Thus f satisfies (H3b) on ∂B_r and, consequently, our assertion follows from Theorem 3.1.

DEFINITION 3.1. *Let $W (\subseteq X)$ be a convex subset of X. A mapping F of $W \times W$ into R is said to satisfy condition (J) if, for each fixed u in W, either $F(\cdot, u)$ satisfies Jensen's equality $F(\frac{1}{2}(v + w), u) = \frac{1}{2}(F(v, u) + F(w, u))$ for all v and w in W or $F(u, \cdot)$ satisfies Jensen's inequality $F(u, \frac{1}{2}(v + w)) \leq \frac{1}{2}(F(u, v) + F(u, w))$ for all v and w in W.*

Before stating a proposition concerning the strong approximation-solvability of equation (1) involving an operator satisfying the conditions of Proposition 3.3 we first prove a lemma which, we believe, is interesting in its own right.

DEFINITION 3.2. *A mapping T of D ($\subseteq X$) into Y is said to be hemicontinuous if $u \in D$, $v \in X$ and $u + t_nv \in D$ with $t_n > 0$ and $t_n \to 0$ as $n \to \infty$, together imply that $T(u + t_nv) \rightharpoonup Tu$ in Y.*

LEMMA 3.2. *Let X be a strictly convex Banach space and let T be a hemicontinuous mapping of $B = \{x \in X \mid \|x\| \leq r\}$ into Y with the property that for a given f in Y the null set N_f of $T_f = T - f$ in B is not empty. Let K be a mapping of X onto Y^* such that T is K-monotone on B,* (H2c) *holds, and for each fixed u in B,* *the functional* $F(u, v) = (T_f u, Kv)$ *defined on* B *is continuous and satisfies Jensen's inequality in its second variable. Then* N_f *is a convex set.*

PROOF. Let x_1 and x_2 be any two points in N_f such that $x_1 \neq x_2$ and for s such that $0 < s < 1$ let $x_s = sx_2 + (1 - s)x_1$. Since K-monotonicity of T implies the same for T_f, it follows that for each y in Int (B)

$$(3.1) \qquad (T_f y, K(x_1 - y)) \leq 0, \qquad (T_f y, K(x_2 - y)) \leq 0.$$

Now, since for each fixed u in B the functional $F_u(v) = F(u, v)$ is continuous and satisfies Jensen's inequality on B, the same arguments as those used in [**22**] for convex functions show that $F_u(v)$ is convex on B, i.e., for any s with $0 < s < 1$

$$F_u(sw + (1 - s)v) \leq sF_u(w) + (1 - s)F_u(v) \qquad (v, w \in B).$$

Hence putting $u = y$, $v = x_1 - y$ and $w = x_2 - y$ we have

$$\begin{aligned}(T_f y, K(sx_2 + (1 - s)x_1 - y)) &= (T_f y, K(s(x_2 - y) + (1 - s)(x_1 - y))) \\ &= F(u, sw + (1 - s)v) \\ &\leq sF(u, w) + (1 - s)F(u, v) \\ &= s(T_f y, K(x_2 - y)) + (1 - s)(T_f y, K(x_1 - y)),\end{aligned}$$

from which, by virtue of (3.1) and the relations $s > 0$ and $1 - s > 0$ we get the needed inequality

$$(3.2) \qquad (T_f y, K(x_s - y)) \leq 0 \qquad (y \in \text{Int}\,(B)).$$

The inequality (3.2) implies that $T_f(x_s) = Tx_s - f = 0$ because assuming to the contrary that $T_f x_s \neq 0$ we arrive at the contradiction. Indeed, if $T_f x_s \neq 0$ then because K maps X onto Y^*, there is z in X such that $(T_f x_s, Kz) > 0$. Since X is strictly convex and s is such that $0 < s < 1$, it follows that $x_s \in \text{Int}\,(B)$. Hence for $t_n > 0$ and sufficiently small, $x_{t_n} = x_s - t_n z \in \text{Int}\,(B)$. If now in (3.2) we replace y by x_{t_n} and use the fact that $K(t_n z) = k_{t_n} K(z)$ with $k_{t_n} > 0$ we get

$$(3.3) \qquad (T_f(x_s - t_n z), Kz) \leq 0.$$

Adding $-(T_f x_s, Kz)$ to both sides of (3.3) we get

$$(3.4) \qquad (T_f(x_s - t_n z) - T_f x, Kz) \leq -(T_f x_s, Kz) < 0.$$

Since the right-hand side is independent of t_n and T is demicontinuous in B, the passage to the limit in (3.4) as $t_n \to 0$ gives the contradiction

$$0 = (T_f x_s - T_f x_s, Kz) \leq -(T_f x_s, Kz) < 0$$

and thus establishes the validity of Lemma 3.2.

REMARK 3.1. In case X is a Hilbert space, $Y = X$ and T is a demicontinuous (and hence hemicontinuous) mapping of B ($\subseteq X$) into X, Lemma 3.2 (for $Y^* = X$ and $K = I$) reduces to Lemma 1 in [**34**] which asserted for the first time that the null set of a monotone mapping acting in a Hilbert space is a convex set. In case X is a strictly convex and reflexive Banach space, Lemma 3.2 includes Browder's Theorem 2 [8] asserting essentially the convexity of the null set of a demicontinuous mapping T of X with X^*; indeed, since in this case $Y^* = (X^*)^* = X$, it suffices to take $K = I$ in our Lemma 3.2.

PROPOSITION 3.4. *Suppose the conditions of Proposition* 3.3 *hold. Suppose further that the functional* $F(u, v) = (T_f u, Kv)$ *of* $X \times X$ *into* R *satisfies condition* (J). *Then equation* (1) *is strongly approximation-solvable for each* f *in* Y.

PROOF. In view of Proposition 3.3, to prove Proposition 3.4 it suffices to prove that for each f in Y equation (1) has a unique solution since in that case the entire sequence $\{x_n \mid x_n \in X_n\}$, constructed by Proposition 3.3, will converge strongly to that unique solution. To do this observe first that, by (2.4), if $Tx_1 = Tx_2 = f$, then $\|x_1\| = \|x_2\|$, i.e., all solutions of equation (1) (for a given f in Y) lie on some sphere, say, B_r.

Now let x_1 and x_2 be any two solutions of equation (1). Suppose first that, for each fixed v in X, $F(\cdot, v)$ satisfies Jensen's equality. Then for each fixed v in X we have

$$\begin{aligned}(T_f(\tfrac{1}{2}(x_1 + x_2)), Kv) &= F(\tfrac{1}{2}(x_1 + x_2), v) = \tfrac{1}{2}(F(x_1, v) + F(x_2, v)) \\ &= \tfrac{1}{2}(Tx_1 - f, Kv) + \tfrac{1}{2}(Tx_2 - f, Kv) = 0.\end{aligned}$$

Since K maps X onto Y^* it follows that $\bar{x} = \frac{1}{2}(x_1 + x_2)$ is a solution of equation (1) and, therefore, $\|\bar{x}\| = r$, i.e., $\|x_1 + x_2\| = \|x_1\| + \|x_2\|$. Since X is strictly convex, $x_1 = \gamma x_2$ with $\gamma > 0$ from which we deduce that $x_1 = x_2$ because $\|x_1\| = \|x_2\|$, i.e., equation (1) has a unique solution.

Suppose now that, for each fixed u in X, $F(u, \cdot)$ satisfies Jensen's inequality and let x_1 and x_2 be any two solutions of equation (1). Repeating the arguments of Lemma 3.2 for $s = \frac{1}{2}$ we find that $\bar{x} = \frac{1}{2}(x_1 + x_2)$ is also a solution of equation (1). Hence, as in the proof of the first part, we derive the uniqueness of a solution given by Proposition 3.3.

4. Special cases and examples of approximation schemes Γ_n. In this section we consider some special cases of the theory developed in the preceding sections by specifying the spaces X, Y, X_n and Y_n and the operators T, P_n, R_n, Q_n, K, K_n and M_n. The purpose is not to exhaust the various known and new special cases but to illustrate the generality of our theory and to show how one chooses the auxiliary operators K, K_n and M_n for the theory to be applicable to various classes of functional equations. Some special approximation schemes were also discussed in [**17**]. For the actual construction of particular schemes for various functional (e.g. Sobolev) spaces see Aubin [**1**].

4.1. *Abstract finite difference schemes for maps from X to X^*.* Let X be a real Banach space, $Y = X^*$ and T be a mapping with domain $D(T)$ in X and range $R(T)$ in X^*.

DEFINITION 4.1. *An abstract finite difference scheme* $(A^*\Gamma_n)$ *for mappings from* D *to* X^* *consists of a sequence* $\{X_n\}$ *of finite-dimensional spaces, two sequences* $\{R_n\}$ and $\{P_n\}$ *(with* P_n *linear and continuous for each* n*) of mappings with* $R_n:X \to X_n$ *and* $P_n:X_n \to X$ *and* $P_nR_nx \to x$ *for each* x *in* X, *and sequences* $\{Y_n\}$ *and* $\{Q_n\}$ *with* $Y_n = X_n^*$ *and* $Q_n = P_n^*$, *where* $P_n^*:X^* \to X_n^*$ *denotes the dual of* P_n.

It is not hard to see that if, for example, the norms in X_n and X_n^* are defined by

$$(4.1)\quad \|x\|_{X_n} = \|P_nx\|_X \quad (x \in X_n), \qquad \|x^*\|_{X_n^*} = \sup_{x\in X_n} \frac{|(x, x^*)|}{\|x\|_{X_n}} \quad (x^* \in X_n^*),$$

then condition (C5) is trivially satisfied while (C4°) also hold since for each f in X^*, by virtue of (4.1),

$$\|Q_nf\| = \|P_n^*f\| = \sup_{x\in X_n} \frac{|(x, P_n^*f)|}{\|x\|} = \sup_{x\in X_n} \frac{|(P_nx, f)|}{\|x\|} \leq \|f\|_{X^*}.$$

Hence the abstract finite difference scheme $(A^*\Gamma_n)$ is a special case of the admissible approximation scheme Γ_n given by Definition 1.1. Consequently, all the approximation and existence results obtained in §1 are valid for finite difference schemes $(A^*\Gamma_n)$ provided, of course, that the mapping $T:D(\subseteq X) \to X^*$ satisfies the corresponding conditions.

To deduce the constructive results for monotone mappings $T:D \to X^*$ from those on K-monotone mappings we take (as usual) X to be reflexive and observe that in this case $K:X \to X\ (=Y^*)$ and the K-monotonicity of T means that T is monotone in the usual sense if we let $K = I$, identity on X. The obvious choice for $K_n:X_n \to X_n\ (=Y_n^*)$ is to take $K_n = I_n$, identity on X_n for each n. Clearly, K and K_n thus chosen satisfy all the conditions used in §§2 and 3. To construct a linear isomorphism $M_n:X_n \to X_n^*\ (=Y_n)$ such that $(M_nx, K_nx) = (M_nx, x) \geq 0$ for each x in ∂D_n we note first that if dim $X_n = m_n$ and $\{\phi_1, \ldots, \phi_{m_n}\}$ is a basis for X_n, then there exists a linearly independent set $\{f_1, f_2, \ldots, f_{m_n}\}$ in X_n^* such that $(f_i, \phi_j) = f_i(\phi_j) = \delta_{ij}$ $(1 \leq i, j \leq m_n)$. If we define M_n by

$$(4.2)\qquad M_n(x) = \sum_{i=1}^{m_n} f_i(x)f_i \qquad (x \in X_n),$$

then it is easy to show that M_n, thus defined, is a linear one-to-one mapping of X_n onto X_n^* and $(M_nx, x) = \sum_{i=1}^{m_n} f_i^2(x) \geq 0$ for each x in X_n.

Thus, in view of the above discussion, we deduce, for example, from §3 the following new results concerning the applicability of abstract finite difference schemes $(A^*\Gamma_n)$ to the solution of equation (1) with T mapping its domain in X into its range in X^*.

COROLLARY 4.1. *Let* D *be a bounded open convex set in* X *with* $0 \in D$ and T *an A-proper mapping with respect to* $(A^*\Gamma_n)$ *of* $\bar{D}$ *into* X^* *with* T_n *continuous on* $\bar{D}_n$. *If* f *is a given vector in* X^* *such that*

$$(4.3)\qquad (Tx, x) \geq (f, x) \qquad (x \in \partial D),$$

then equation (1) *is feebly approximation-solvable. If we additionally assume that equation* (1) *has at most one solution in* $\bar{D}$, *then equation* (1) *is strongly approximation-solvable.*

COROLLARY 4.2. *Suppose* X *has also Property* (H) *and* T *is a continuous mapping of* $\bar{D} = B(\subset X)$ *into* X^* *which satisfies modified condition* (s). *Then for each* f *in* X^* *for which* (4.3) *holds the conclusions of Corollary* 4.1 *hold.*

COROLLARY 4.3. *Let* T *be a continuous mapping of* X *into* X^* *with*

$$(Tx - Ty, x - y) \geq c(\|x - y\|) \qquad (x, y \in X), \tag{4.4}$$

where $c(r)$ *has the properties of Proposition* 3.2. *Then equation* (1) *is strongly approximation-solvable for each* f *in* X^*.

COROLLARY 4.4. *Suppose* X *has also Property* (H) *and* T *is a continuous mapping of* X *into* X^* *such that*

$$(Tx - Ty, x - y) \geq (\phi(\|x\|) - \phi(\|y\|))(\|x\| - \|y\|) \qquad (x, y \in X). \tag{4.5}$$

Then equation (1) *is strongly approximation-solvable for each* $f \in X^*$.

REMARK 4.1. In case X is a uniformly convex Banach space and T is a bounded demicontinuous mapping from X to X^* for which (4.5) holds, Corollary 4.4 was proved in [**2**]. We add that boundedness of T was essential in the type of proof used in [**2**].

4.2. *Projection schemes for mappings from* X *to* X^*. Let X be a reflexive Banach space with a Schauder basis $\{\phi_1, \phi_2, \ldots\} \subset X$, $Y = X^*$ and let $X_n = \operatorname{span}\{\phi_1, \ldots, \phi_n\}$ for each n.

DEFINITION 4.2. *A projection scheme* $(P^*\Gamma_n)$ *for mappings from subsets of* X *to* X^* *consists of a sequence* $\{X_n\}$ *of subspaces of* X *and a sequence* $\{R_n\}$ *of linear projections of* X *into* X *with* $R(R_n) = X_n$; *we let* P_n *be an (identity) injection of* X_n *into* X *and set* $Q_n = R_n^*$ *with* $Y_n = R(Q_n) = R(R_n^*) \subset X^*$.

It is clear that the scheme $(P^*\Gamma_n)$ satisfies all the conditions used in the preceding sections. We set $D_n = X_n \cap D$, $T_n = R_n^* T P_n|_{\bar{D}_n}$ and for $K: X \to X(=Y^*)$ we take $K = I$. Since $K_n: X_n \to Y_n^*$ and $Y_n^* \neq X_n$ we cannot set $K_n = I_n$ as in §4.1 but construct K_n as well as $M_n: X_n \to Y_n$ as follows. Let $\{f_i\}$ be a sequence in X^* which satisfies the biorthogonality relation $(f_i, \phi_j) = \delta_{ij}$ $(i, j = 1, 2, \ldots)$. Then, for each n, $Y_n = \operatorname{span}\{f_1, \ldots, f_n\}$ and M_n defined by $M_n(x) = \sum_{i=1}^n f_i(x) f_i$ $(x \in X_n)$ is linear, one-to-one and onto. Now, for each n, there exists a unique dual basis $\{\phi_1^*, \ldots, \phi_n^*\}$ in Y_n^* such that $(\phi_i^*, \phi_j) = \delta_{ij}$ $(1 \leq i, j \leq n)$ and $\phi_i^* = f_i|_{X_n}$ (see [**43**]). Let K_n be the linear mapping of $X_n = Y_n' \equiv R(Q_n^*)$ into Y_n^* such that for each $x \in X_n$ the element $x^* = K_n x$ in Y_n^* is given by

$$(x^*, f) = (x, f) = (K_n x, f) \qquad \text{for each } f \text{ in } Y_n.$$

It follows (see [**43**]) that K_n thus defined is a one-to-one mapping of X_n onto Y_n^* with $\|K_n x\|_{Y_n^*} \leq \|x\|$ for x in X_n. Furthermore, K_n satisfies the hypotheses

(H2d), (H2e) and for each x in X_n

$$(K_n x, M_n x) = (x, M_n x) = \sum_{i=1}^{n} f_i^2(x) \geq 0.$$

Hence every result in §§1, 2 and 3 can be reformulated in terms of the projectional schemes $(P * \Gamma_n)$ for equation (1) involving operators from D $(\subseteq X)$ into X^*.

4.3. *Injective schemes for mappings from X to X^*.*

DEFINITION 4.3. *Let X be a reflective Banach space. An injective scheme $(I * \Gamma_n)$ for mappings from X to X^* consists of a sequence $\{X_n\}$ of finite-dimensional subspaces of X and a sequence $\{R_n\}$ of linear projections of X into X with $R(R_n) = X_n$ and $R_n x \to x$ for each x in X; we let P_n be a linear injection of X_n into X and set $Q_n = P_n^*: X^* \to X_n^* = R(Q_n) = Y_n$.*

It is easy to see that the scheme $(I * \Gamma_n)$ satisfies all the conditions used above. Consequently, all the results of §1 hold in case of injective schemes. Since $Y^* = X$ and $Y_n^* = X_n$, the results for monotone operators from X to X^* in case of injective schemes are deduced from §§2 and 3 by letting $K = I$, $K_n = I_n$ and defining M_n as in §4.1.

4.4. *Abstract finite difference schemes for maps from X to X.* When X is a Banach space or a Hilbert space and $Y = X$, then our theory reduces to mappings acting in the same space. Two special classes of such mappings attracted recently a considerable attention. One is the class of mapping T of X into X introduced in [**38**], [**39**] such that

$$\|T_n x - T_n y\| \geq c(\|x - y\|) \quad \text{for all } x \text{ and } y \text{ in } X_n, \tag{4.6}$$

where T_n maps X_n into X_n and $c(r)$ is a function from R^+ to R^+ satisfying certain conditions, which proved to be very suitable for approximation purposes, and the other is the class of J-monotone and accretive mappings introduced in [**9**], [**10**], where T is accretive or J-monotone if

$$(Tx - Ty, J(x - y)) \geq 0 \quad \text{for all } x \text{ and } y \text{ in } X \tag{4.7}$$

with J a duality mapping of X into X^* defined by the relations

$$Jx = \{w \mid w \in X^*; (x, w) = \|x\| \, \|w\|; \|w\| = \|x\|\}.$$

It is known that when X is reflexive and X^* strictly convex then J is a single-valued mapping of X onto X^* and $J(tx) = tJx$ for $t \geq 0$ and $x \in X$. Clearly, since $Y^* = X^*$, when $Y = X$, J-monotone mappings form a subclass of K-monotone mappings with $K = J$.

DEFINITION 4.4. *An abstract finite difference scheme $(A\Gamma_n)$ for mappings from X to X consists of a sequence $\{X_n\}$ of finite-dimensional spaces and two sequences $\{P_n\}$ and $\{R_n\}$ of continuous linear mappings such that P_n injects X_n into X, R_n maps X onto X_n, $P_n R_n x \to x$ for each x in X, $Y_n = X_n$ and $Q_n = R_n$.*

It is easy to see that if, for example, the norm in X_n is defined by $\|x\|_{X_n} = \|P_n x\|_X$ for each x in X_n, then (C5) is trivially satisfied while (C4°) follows from the uniform boundedness principle and the requirement that $P_n R_n x \to x$ for each x in

X. Hence the scheme $(A\Gamma_n)$ is a special case of the scheme (Γ_n). Consequently, all the approximation and existence results of §1 are valid when finite difference methods are used. However, it appears that in order to apply the results involving abstract finite difference schemes $(A\Gamma_n)$ to J-monotone operators we need to impose an additional condition on P_n and R_n.

CONDITION (B). Let X be a Banach space with X^* strictly convex and let the linear mappings P_n and R_n of Definition 4.4 be such that $S_n w = w$ and $\|S_n^* Jw\| \le \|Jw\|$ for each w in $\tilde{X}_n = P_n X_n (\subset X)$, where $S_n = P_n R_n$ for each n.

Using Condition (B) and the techniques of [**15**] one shows that the equality $S_n w = w$ for $w \in \tilde{X}_n$ is equivalent to the equality $R_n P_n x = x$ for $x \in X_n$ and that

$$S_n^* J P_n x = J P_n x \quad \text{for all } x \text{ in } X_n. \tag{4.8}$$

Now, if Condition (B) holds and if $K = J$, then a suitable choice for $K_n : X_n \to X_n^*$ is $K_n = P_n^* J P_n$. It is easy to see that for this choice of K_n and for each x in X_n and g in X

$$\|K_n x\|_{X_n^*} \le \|x\|_{X_n} \quad \text{and} \quad (R_n g, K_n x) = (g, J P_n x), \tag{4.9}$$

i.e., (H2d) holds. Furthermore, if J is also assumed to be weakly continuous, then (H2e) is likewise true. Since $Y_n = X_n$, the simplest choice for $M_n : X_n \to X_n$, for which (H3a) is satisfied, is $M_n = I_n$ for each n.

Thus, a number of results obtained in preceding sections are valid when finite difference schemes $(A\Gamma_n)$ are applied to accretive or J-monotone operators provided Condition (B) holds. As an example we get the following practically useful result.

PROPOSITION 4.1. *Let X be a Banach space with X^* strictly convex, $Y = X$ and T a Lipschitzian mapping of X into X such that*

$$(Tx - Ty, J(x - y)) \ge c_0 \|x - y\|^2 \qquad (x, y \in X,\ c_0 > 0). \tag{4.10}$$

Let $(A\Gamma_n)$ be the abstract finite difference scheme given by Definition 4.4 *for which Condition* (B) *holds. Then equation* (1) *is strongly approximation-solvable for each f in X.*

PROOF. Let $T_n = R_n T P_n$ and $K_n = P_n^* J P_n$ for each n. First, it follows from (4.8) and (4.10) that for each x and y in X_n

$$c_0 \|P_n x - P_n y\|^2 \le (TP_n x - TP_n y, K(P_n x - P_n y)) = (T_n x - T_n y, K_n(x - y))$$

from which, in view of (4.9) we deduce the inequality

$$c_0 \|x - y\|_{X_n} \le \|T_n x - T_n y\|_{X_n} \quad \text{for all } x \text{ and } y \text{ in } X_n,$$

i.e., condition (H3) of Theorem 1.1 holds. Because T is Lipschitzian (and hence continuous) it follows that (H1) and (H2) also hold. Moreover, by Theorem 4 in [**11**], T is onto, i.e., assertion (A3) of Theorem 1.1 holds. Hence, by Theorem 1.1 (see Remark 1.1) equation (1) is strongly approximation-solvable for each f in X, i.e., Proposition 4.1 is true.

Note that in Proposition 4.1 the mapping $J : X \to X^*$ is not assumed to be weakly continuous or even continuous.

4.5. *Projective schemes for mappings from X to X.*

DEFINITION 4.5. *A projective scheme* $(P\Gamma_n)$ *for mappings from* X *to* X *consists of a sequence* $\{X_n\}$ *of finite-dimensional subspaces of* X *and a sequence* $\{R_n\}$ *of linear projections of* X *into* X *with* $R(R_n) = X_n$, $R_n x \to x$ *for each* x *in* X *and* $\|R_n\| = 1$. *We take* P_n *to be an (identity) injection of* X_n *into* X *and set* $Y_n = X_n$ *and* $Q_n = R_n$.

In this case $T_n = R_n T P_n : X_n \to X_n$, $K_n = P_n^* J P_n : X_n \to X_n^*$ and $M_n = I_n$ on X_n. It is obvious that $(P\Gamma_n)$ is an admissible scheme, the mappings R_n and P_n satisfy trivially Condition (B) and K_n satisfies (4.9). Hence all results are valid for projectional schemes $(P\Gamma_n)$.

REFERENCES

1. J. P. Aubin, *Approximation des espaces de distributions et des opérateurs différentiels*, Bull. Soc. Math. France, Mémoire **12** (1967), 139 pp.

2. H. Brezis and M. Sibony, *Méthodes d'Approximation et d'Itération pour les Opérateurs Monotones*, Arch. Rational Mech. Anal. **28** (1968), 59–82.

3. F. E. Browder, *Variational boundary value problems for quasilinear elliptic equations*. II, III. Proc. Nat. Acad. Sci. U.S.A. **50** (1963), 594–598, 794–798.

4. ———, *Nonlinear elliptic boundary value problems*, Bull. Amer. Math. Soc. **69** (1963), 862–874

5. ———, *Strongly nonlinear parabolic boundary value problems*, Amer. J. Math. **86** (1964), 339–357.

6. ———, *Remarks on nonlinear fractional equations*. II, III, Illinois J. Math. **9** (1965), 608–616, 617–662.

7. ———, *Further remarks on nonlinear functional equations*, Illinois J. Math. **10** (1966), 275–286.

8. ———, *Mapping theorems for noncompact nonlinear operators in Banach spaces*, Proc. Nat. Acad. Sci. U.S.A. **54** (1965), 337–342.

9. ———, *Fixed point theorems for nonlinear semicontractive mappings in Banach spaces*, Arch. Rational Mech. Anal. **21** (1966), 259–270.

10. ———, *Nonlinear accretive operators in Banach spaces*, Bull. Amer. Math. Soc. **73** (1967), 470–476.

11. ———, *Nonlinear mappings of nonexpansive and accretive type in Banach spaces*, Bull. Amer. Math. Soc. **73** (1967), 875–887.

12. ———, *Approximation-solvability of nonlinear functional equations in normed linear spaces*, Arch. Rational Mech. Anal. **26** (1967), 33–42.

13. ———, *Nonlinear eigenvalue problems and Galerkin approximations*, Bull. Amer. Math. Soc. **74** (1968), 651–656.

14. ———, *Nonlinear elliptic eigenvalue problems in Banach spaces* (to appear).

15. F. E. Browder and D. G. De Figueiredo, *J-monotone nonlinear operators in Banach spaces*, Nederl. Akad. Wetensch. Proc. Ser. A **69** = Indag. Math. **28** (1966), 412–420.

16. F. E. Browder and W. V. Petryshyn, *The topological degree and Galerkin approximations for noncompact operators in Banach spaces*, Bull. Amer. Math. Soc. **74** (1968), 641–646.

17. ———, *Approximation methods and the generalized topological degree for nonlinear mappings in Banach spaces*, J. Functional Analysis (to appear).

18. J. Cronin, *Fixed points and topological degree in nonlinear analysis*, Math Surveys, no. 11, Amer. Math. Soc., Providence, R.I., 1964.

19. Yu. A. Dubinsky, *Quasilinear elliptic and parabolic equations of an arbitrary order*, Uspehi Mat. Nauk **23** No. 1 (1968), 45–90. (Russian)

20. K. Fan and I. Glickberg, *Some geometric properties of the spheres in a normed linear space*, Duke Math. J. **25** (1958), 533–568.

21. D. G. De Figueiredo, *Fixed point theorems for nonlinear operators and Galerkin approximations*, J. Differential Equation **3** (1967), 271–281.

22. J. L. W. V. Jensen, *Sur les fonctions convexes et les inèqualities entre les valeurs moyennes*, Acta. Math. **30** (1906), 175–193.

23. R. I. Kačurovskiĭ, *On monotone operators and convex functionals*, Uspehi Mat. Nauk **15** No. 4 (1960), 213–215. (Russian)

24. ———, *Monotonic nonlinear operators in Banach spaces,* Dokl. Akad. Nauk S.S.S.R. **163** (1965), 559–562 = Soviet Math. Dokl. **6** (1965), 953–956.

25. ———, *Approximation methods in the solution of nonlinear operator equations,* Izv. Vysš. Učebn. Zaved Matematika **12** (1967), 27–37.

26. ———, *Nonlinear monotone operators in Banach spaces,* Uspehi Mat. Nauk **23** No. 2 (1968), 121–168.

27. S. Kaniel, *Quasicompact nonlinear operators in Banach space and applications,* Arch. Rational Mech. Anal. **20** (1967), 259–278.

28. T. Kato, *Demicontinuity, hemicontinuity and monotonicity.* II. Bull. Amer. Math. Soc. **73** (1967), 470–476.

29. M. Lees and M. H. Schultz, *A Leray-Schauder principle for A-compact mappings and the numerical solution of nonlinear two-point boundary value problems,* Wiley, New York, 1967, pp. 167–179.

30. G. J. Minty, *Monotone (nonlinear) operators in Hilbert space,* Duke Math. J. **29** (1962), 341–346.

31. ———, *On a "monotonicity" method for the solution of nonlinear equations in Banach spaces,* Proc. Nat. Acad. Sci. U.S.A. **50** (1963), 1038–1041.

32. M. Opial, *Nonexpansive and monotone mappings in Banach spaces,* Lecture Notes, Division of Appl. Math., Brown University, Providence, R.I., 1967.

33. W. V. Petryshyn, *On the extension and the solution of nonlinear operator equations,* Illinois J. Math. **10** (1966), 255–274.

34. ———, *Construction of fixed points of demicompact mappings in Hilbert spaces,* J. Math. Anal. Appl. **14** (1966), 274–284.

35. ———, *On a fixed point theorem for nonlinear P-compact operators in Banach space,* Bull. Amer. Math. Soc. **72** (1966), 329–334.

36. ———, *On nonlinear P-compact operators in Banach space with applications to constructive fixed-point theorems,* J. Math. Anal. Appl. **15** (1966), 228–242.

37. ———, *Further remarks on nonlinear P-compact operators in Banach space,* Proc. Nat. Acad. Sci. U.S.A. **55** (1966), 684–687 (for its expanded version see J. Math. Anal. Appl. **16** (1966), 243–253).

38. ———, *On the projectional solution of nonlinear operator equations.* Inform. Bull. No. 5, Intern. Congress of Mathematicians, Moscow, 1966.

39. ———, *Projection methods in nonlinear numerical functional analysis,* J. Math. Mech. **17** (1967), 353–372.

40. ———, *Remarks on the approximation-solvability of nonlinear functional equations,* Arch. Rational Mech. Anal. **26** (1967), 43–49.

41. ———, *On the approximation-solvability of nonlinear equations,* Math. Ann. **177** (1968), 156–164.

42. ———, *Fixed point theorems involving P-compact, semicontractive and accretive operators not defined on all of a Banach space,* J. Math. Anal. Appl. **23** (1968), 336–354.

43. ———, *On the projectional solvability and the Fredholm alternative for equations involving linear A-proper operators,* Arch. Rational Mech. Anal. **30** (1968), 270–284.

44. W. V. Petryshyn and T. S. Tucker, *On the functional equations involving nonlinear generalized P-compact operators,* Trans. Amer. Math. Soc. **135** (1969), 343–374.

45. S. I. Pohožaev, *The solvability of nonlinear equations with odd operators,* Funkcional Anal. i Priložen **1** (1967) = Functional Anal. Appl. **1** (1967), 66–73.

46. A. E. Taylor, *Introduction to functional analysis,* Wiley, New York, 1958.

47. M. M. Vainberg, *New theorems for nonlinear operators and equations,* Uch. Zap. Mosk. Reg. Pet. Inst., **1959**, 131–145.

48. ———, *On some new principle in the theory of nonlinear equations,* Uspehi Mat. Nauk U.S.S.R. **15** (1960), 243–244.

49. M. M. Vainberg and R. I. Kačurovskiĭ, *On the variational theory of nonlinear operators and equations,* Dokl. Akad. Nauk S.S.S.R. **129** (1959), 1199–1202. (Russian)

50. E. H. Zarantonello, *Solving functional equations by contractive averaging,* Tech. Rep. No. 160, U.S. Army Res. Center, Madison, Wisconsin, 1960.

51. ———, *The closure of the numerical range contains the spectrum,* Bull. Amer. Math. Soc. **70** (1964), 781–787.

RUTGERS UNIVERSITY

THE BÉNARD PROBLEM[1]

P. H. Rabinowitz

1. **Introduction.** This talk concerns nonuniqueness and bifurcation phenomena for a problem in fluid dynamics. In an earlier talk in this symposium Dr. Berger discussed similar questions arising in the calculus of variations. One of the interesting features of the case we consider is that it is not variational and, therefore, the powerful techniques of the calculus of variations are not available.

The Bénard problem is a problem in thermal convection. Mathematically it involves the system of five nonlinear partial differential equations:

$$(1)\qquad \begin{aligned} \Delta \boldsymbol{u} - \nabla p + R\theta \boldsymbol{e} &= (\boldsymbol{u} \cdot \nabla)\boldsymbol{u} \\ \Delta\theta + w &= (\boldsymbol{u} \cdot \nabla)\theta \\ \nabla \cdot \boldsymbol{u} &= 0. \end{aligned}$$

Here $\boldsymbol{u} = (u, v, w)$ is the fluid velocity, p is the pressure, θ the temperature, $\boldsymbol{e} = (0, 0, 1)$, and R is a dimensionless parameter called the Rayleigh number. (Another dimensionless parameter, the Prandtl number, has been equated to one for convenience.)

To make what follows more meaningful, let me briefly describe the physical situation from which these equations originate. Consider an idealized experiment: An infinite horizontal layer of an incompressible fluid lies between two rigid perfectly conducting walls. A constant temperature gradient is maintained between the walls, the lower being warmer. If the temperature gradient is small, the fluid remains at rest and heat is transported through the fluid solely by conduction. However, exceeding a certain critical temperature gradient causes the onset of steady, i.e., time independent, motions in the fluid called convection currents. Heat is transported through the fluid by convection, as well as conduction.

In actual experiments the fluid arranges itself in a regular cellular pattern with fluid motions and heat transport occurring only within the cells. The shape of the cells seems to be determined largely by the shape of the fluid container. The first experiments of this nature were carried out by Bénard (1900) who found hexagonal cells. His fluids, however, had a free upper surface and for this case surface tension effects are important, which is not the case in the situations considered here.

[1] This work was partially supported by the U.S. Air Force under contract AF 49(638)1345.

We want to show these physical phenomena are implicit in the equations of motion of the fluid. As they have been written in (1), the conduction solution corresponds to

$$u_0 = 0, \qquad \theta_0 = p_0 = 0. \tag{2}$$

The Rayleigh number R is a dimensionless version of the temperature gradient. Note that the conduction solution (2) exists for all values of R. Thus, the problem becomes one of nonuniqueness. We will show that in the neighborhood of certain critical Rayleigh numbers, nontrivial classical solutions of (1) corresponding to rectangular cells bifurcate from the conduction solution. The reason for seeking rectangular cells is that this is the simplest problem to pose mathematically. The solutions will be constructed using an iteration scheme to be described below.

In §2 a more precise mathematical formulation of the problem is given. The linearized equations are studied in §3. Lastly, the existence construction is given in §4. For more details and an extensive bibliography, see [**1**].

Much work has been done on the Bénard problem experimentally, numerically, and using formal mathematical methods. However, few rigorous mathematical results have been obtained. My own interest in bifurcation problems in fluid dynamics arose from reading two outstanding papers of Velte [**2**], [**3**], who obtained such branching of solutions using topological degree of mapping arguments. My attempts to do this constructively led to this work. The method used here can be employed to solve Velte's problems. On the other hand, a degree of mapping argument can be given to obtain an existence theorem for the Bénard problem. However, the constructive method tells you much more about the solutions. The work of Iudovich (Prikl. Mat. Meh. in recent years) should also be mentioned here. He has treated bifurcation problems in fluid dynamics first by degree of mapping arguments and, more recently, by the method of Lyapunov-Schmidt. This latter method is related to our construction.

2. **Formulation of the problem.** As domain for the equations

$$\begin{aligned} \Delta \boldsymbol{u} - \nabla p + R\theta \boldsymbol{e} &= (\boldsymbol{u} \cdot \nabla)\boldsymbol{u} \\ \Delta\theta + w &= (\boldsymbol{u} \cdot \nabla)\theta \\ \nabla \boldsymbol{u} &= 0 \end{aligned} \tag{1}$$

we take

$$-\infty \le x, y \le \infty, \quad |z| < 1. \tag{3}$$

The boundary conditions in the z direction are

$$\boldsymbol{u} = 0, \theta = 0, \quad z = \pm 1. \tag{4}$$

Note that $\nabla \cdot \boldsymbol{u} = 0$ on $z = \pm 1$ implies $w_z = 0$ for $z = \pm 1$. In the x and y

directions periodicity of the dependent variables with periods $2\pi/a$, $2\pi/b$ is required where $a, b \geq 0$, i.e.,

$$
\begin{aligned}
&\boldsymbol{u}(x + 2\pi/a, y, z) = \boldsymbol{u}(x, y, z) = \boldsymbol{u}(x, y + 2\pi/b, z) \\
(5)\qquad &\theta(x + 2\pi/a, y, z) = \theta(x, y, z) = \theta(x; y + 2\pi/b, z) \\
&p(x + 2\pi/a, y, z) = p(x, y, z) = p(x, y + 2\pi/b, z).
\end{aligned}
$$

One but not both of a and b is allowed to equal zero and in this case two dimensional solutions called rolls are obtained. Such solutions have actually been observed in experiments and there is some indication that these are the preferred solutions from the point of view of stability.

The periodicity assumption (5) does not bring in strongly enough the cellular nature of the solutions. As Dr. Lions mentioned in his talk, one of the major difficulties in nonlinear problems is to choose the right space in which to work. To do that here we recall the experimental information showing the fluid motions take place within the cells and the cell walls act as thermal insulators. Mathematically this suggests seeking solutions with

$$
\begin{aligned}
&-u(-x, y, z) = u(x, y, z) = u(x, -y, z) \\
(6)\qquad &+v(-x, y, z) = v(x, y, z) = -v(x, -y, z) \\
&q(-x, y, z) = q(x, y, z) = q(x, -y, z)
\end{aligned}
$$

where $q = w, \theta$ or p.

3. **The linearized equations.** To determine solutions of (1) near the conduction solution (2), we have to study (1) linearized about (2). We want three results for this linear problem:

(α) It's an eigenvalue problem.

(β) The eigenvalues are simple.

(γ) A priori estimates for the inhomogeneous equation.

The linearized equations are:

$$
\begin{aligned}
&\Delta \boldsymbol{U} - \nabla P + R\Theta \boldsymbol{e} = 0 \\
(7)\qquad &\Delta\Theta + W = 0 \\
&\nabla \cdot \boldsymbol{U} = 0
\end{aligned}
$$

where $\boldsymbol{U} = (U, V, W)$, Θ and P satisfy the boundary and periodicity conditions (4), (5), and (6).

(α) R appears in (7) as an eigenvalue, i.e., only for certain values of R can we find nontrivial solutions of (7) satisfying (4)–(6). In contrast to (1), the system (7) with conditions (4)–(6) can be formulated as a variational problem, namely

$$\text{extremize } \frac{\int_{\mathscr{C}} |\nabla \boldsymbol{U}|^2 \, d\boldsymbol{x}}{\int_{\mathscr{C}} |\nabla \Theta|^2 \, d\boldsymbol{x}}$$

subject to the constraints

$$\nabla \cdot \boldsymbol{U} = 0, \quad \Delta\Theta + W = 0$$

and the conditions (4)–(6). Here $\mathscr{C}$ denotes a cell: $|x| \le \pi/a$, $|y| \le \pi/b$, $|z| \le 1$, $|\nabla \boldsymbol{U}|^2 = |\nabla U|^2 + |\nabla V|^2 + |\nabla W|^2$, and $d\boldsymbol{x} = dx\, dy\, dz$.

While this formulation of (7) gives us the existence of a sequence of positive eigenvalues with corresponding smooth eigenfunctions, it does not suffice for our later purpose. The iteration scheme employed to solve (1) requires a simple eigenvalue of (7). Thus we must make a deeper study of the eigenvalues of (7).

(β) Conditions (5) and (6) allow us to expand $\boldsymbol{U}$, Θ, P into Fourier series of the form:

$$U = \sum_{j,k=0}^{\infty} A_{jk}(z) \sin jax \cos kby$$

$$V = \sum_{j,k=0}^{\infty} B_{jk}(z) \cos jax \sin kby$$

$$W = \sum_{j,k=0}^{\infty} C_{jk}(z) \cos jax \cos kby \tag{8}$$

$$\Theta = \sum_{j,k=0}^{\infty} D_{jk}(z) \cos jax \cos kby$$

$$P = \sum_{j,k=0}^{\infty} E_{jk}(z) \cos jax \cos kby.$$

Substituting (8) into (7) leads to an infinite system of ordinary differential equations. After some simplification this can be put into the form:

$$\begin{aligned} L_{jk}^2 C_{jk} &= R\omega_{jk}^2 D_{jk}, \\ L_{jk} D_{jk} &= -C_{jk} \end{aligned} \qquad j, k = 0, 1, \ldots \tag{9}$$

$$C_{jk} = C'_{jk} = D_{jk} = 0, \quad \text{at} \quad z = \pm 1 \tag{10}$$

where $\omega_{jk}^2 = j^2a^2 + k^2b^2$, $L_{jk} = d^2/dz^2 - \omega_{jk}^2$ and $C' = dC/dz$.

Using the Green's functions for L_{jk} and L_{jk}^2 under the boundary conditions (10), the system (9) can be converted to a system of integral equations:

$$C_{jk}(z) = R \int_{-1}^{1} K_{j,k}(z, \zeta) C_{jk}(\zeta)\, d\zeta, \quad j, k = 0, 1, \ldots . \tag{11}$$

The kernels K_{jk} turn out to be totally positive. Karlin has shown [**6**]—under conditions which are satisfied here—integral equations with totally positive kernels possess a sequence of positive simple eigenvalues λ_n and the nth eigenfunction has $n - 1$ interior zeros. Thus (11) for fixed j, k possesses simple eigenvalues $R_n(j, k)$.

Unfortunately this still does not suffice for us. Although for fixed j, k, a given eigenvalue $R_n(j, k)$ is a simple eigenvalue of (11) or (9)–(10), it need not be a simple eigenvalue of (7). It is possible that $R_n(j, k) = R_m(r, s)$, $n \neq m$. To get around this difficulty the remark of Lions is again appropriate. The problem (1) is posed in another space by changing conditions (5)–(6). In this new space $R_n(j, k)$ is a simple eigenvalue of (7) and solutions of (1) are a fortiori solutions of (1), (4)–(6). For details we refer the reader to [**1**].

Let us then assume we have a simple eigenvalue R_c of (7), (4)–(6) with corresponding eigenfunctions $\boldsymbol{U}_c$, Θ_c, P_c normalized by

$$\int_{\mathscr{C}} |\nabla \Theta_c|^2 \, d\boldsymbol{x} = 1. \tag{12}$$

(γ) In order to solve (1), a priori estimates are needed for the inhomogeneous version of (7):

$$\begin{aligned} \Delta \boldsymbol{U} - \nabla P + R_c \Theta \boldsymbol{e} &= \boldsymbol{F} \\ \Delta \Theta + W &= \Phi \\ \nabla \cdot \boldsymbol{U} &= 0 \end{aligned} \tag{13}$$

with the boundary and periodicity conditions (4)–(6). The functions $\boldsymbol{F} = (F, G, H)$ and Φ are also asumed to satisfy (5)–(6). As is well known a necessary condition for (13), (4)–(6) to possess a solution is that its right-hand side satisfy an orthogonality condition, namely it must be orthogonal to the null space of the homogeneous equations adjoint to (13), (4)–(6). It is not difficult to see that this null space is spanned by $(\boldsymbol{U}_c, R_c\Theta_c, P_c)$. Thus the orthogonality condition becomes:

$$\int_{\mathscr{C}} (\boldsymbol{U}_c \cdot \boldsymbol{F} + R_c \Theta_c \Phi) \, d\boldsymbol{x} = 0. \tag{14}$$

The condition (14) is also sufficient for the solvability of (13). To be more precise, consider the set of C^∞ functions on a cell $\mathscr{C}$ which are $2\pi/a$, $2\pi/b$ periodic in x, y. Let H_m denote the completion of this set with respect to the norm

$$|\psi|_m^2 = \sum_{|\sigma|=m} \int_{\mathscr{C}} |D^\sigma \psi|^2 \, d\boldsymbol{x} + \int_{\mathscr{C}} \psi^2 \, d\boldsymbol{x}$$

where the familiar multi-index notation is being employed. In a straightforward but somewhat cumbersome fashion, we can show

THEOREM 1. *If* $(\boldsymbol{F}, \Phi) \in H_m$, *satisfy* (5)–(6), *and* (14), *then there exists* $(\boldsymbol{U}, \Theta, P)$ *satisfying* (13), *and* (4)–(6) *where* $\boldsymbol{U}$, $\Theta \in H_{m+2}$ *and* $P \in H_{m+1}$. *The solution* $(\boldsymbol{U}, \Theta, P)$ *is made unique by requiring* $\int_{\mathscr{C}} W\Theta_c \, d\boldsymbol{x} = 0$ *and satisfies*

$$|\boldsymbol{U}|_{m+2} + |\Theta|_{m+2} + |P|_{m+1} \leq k(|\boldsymbol{F}|_m + |\Phi|_m) \tag{15}$$

where k is a constant.

As earlier by $|\boldsymbol{U}|$ we mean the sum of the norms of its components.

4. **Construction of the solution of (1).** In this section we will prove the existence of a solution of (1), (4)–(6) near the conduction solution (2) provided that R

is close to R_c. We introduce an artifical amplitude ϵ and try for a solution of (1) of the form

$$
\begin{aligned}
\boldsymbol{u}(\boldsymbol{x}, \epsilon) &= \epsilon \boldsymbol{U}_c(\boldsymbol{x}) + \epsilon^2\, \delta \boldsymbol{u}(\boldsymbol{x}, \epsilon) \\
\theta(\boldsymbol{x}, \epsilon) &= \epsilon \Theta_c(\boldsymbol{x}) + \epsilon^2\, \delta\theta(\boldsymbol{x}, \epsilon) \\
p(\boldsymbol{x}, \epsilon) &= \epsilon P_c(\boldsymbol{x}) + \epsilon^2\, \delta P(\boldsymbol{x}, \epsilon) \\
R(\epsilon) &= R_c + \epsilon^2\, \delta R(\epsilon)
\end{aligned}
\tag{16}
$$

where $\boldsymbol{x} = (x, y, z)$. Heuristically this is reasonable since in some sense we are close to the linear theory and therefore linear effects should be dominant.

Substituting (16) in (1) gives

$$
\begin{aligned}
\Delta\, \delta \boldsymbol{u} - \nabla\, \delta p + R_c\, \delta\theta \boldsymbol{e} &= (\boldsymbol{U}_c + \epsilon\, \delta \boldsymbol{u}) \cdot \nabla(\boldsymbol{U}_c + \epsilon\, \delta \boldsymbol{u}) - \epsilon\, \delta R(\Theta_c + \epsilon\, \delta\theta)\boldsymbol{e} \\
\Delta\, \delta\theta + \delta w &= (\boldsymbol{U}_c + \epsilon\, \delta \boldsymbol{U}) \cdot \nabla(\Theta_c + \epsilon\, \delta\theta) \\
\nabla \cdot \delta \boldsymbol{u} &= 0.
\end{aligned}
\tag{17}
$$

$(\delta \boldsymbol{u}, \delta\theta, dp)$ also satisfy (4)–(6).

Now an iteration scheme is set up to solve (17), (4)–(6). Let $\delta \boldsymbol{u}_0 = 0$, $\delta\theta_0 = 0 = \delta p_0$ and

$$
\begin{aligned}
\Delta\, \delta \boldsymbol{u}_{n+1} - \nabla\, \delta p_{n+1} + R_c\, \delta\theta_{n+1} \boldsymbol{e} &= (\boldsymbol{U}_c + \epsilon\, \delta \boldsymbol{u}_n) \cdot \nabla(\boldsymbol{U}_c + \epsilon\, \delta \boldsymbol{u}_n) \\
&\quad - \epsilon\, \delta R_n(\theta_c + \epsilon\, \delta\theta_n)\boldsymbol{e} \\
\Delta\, \delta\theta_{n+1} + \delta w_{n+1} &= (\boldsymbol{U}_c + \epsilon\, \delta \boldsymbol{u}_n) \cdot \nabla(\Theta_c + \epsilon\, \delta\theta_n) \\
\nabla \cdot \delta \boldsymbol{u}_{n+1} &= 0
\end{aligned}
\tag{18}
$$

together with the conditions (4)–(6). The system (18) is of the form (13) with right-hand side satisfying (5)–(6). We cannot solve (18) unless its right-hand side satisfies (14). But the right-hand side of (18) contains a free parameter δR_n. Let us therefore choose δR_n so that (14) is satisfied. This can be done if the coefficient of δR_n obtained on imposing (13) is nonzero. The coefficient in question is

$$
-\epsilon \int_{\mathscr{C}} (\Theta_c + \epsilon\, \delta\theta_n) W_c\, d\boldsymbol{x}. \tag{19}
$$

From (2), and (12), it follows

$$
\int_{\mathscr{C}} \Theta_c W_c\, d\boldsymbol{x} = \int_{\mathscr{C}} |\nabla \Theta_c|^2\, d\boldsymbol{x} = 1.
$$

In the course of the iterations we keep pointwise bounds on $\delta\theta_n$ independent of n for $|\epsilon| \leq \epsilon_0$. Therefore if ϵ is sufficiently small, (19) is nonzero and the iteration scheme is justified.

Using the estimates (15) together with some general interpolation inequalities it is not too difficult to show for $|\epsilon|$ sufficiently small the iterates converge in H_{m+1} to a solution of (18). Thus (16) gives us a nontrivial solution of (1). For m large enough, the Sobolev inequality implies this is a classical solution of (1). A standard bootstrap argument implies the solution is in C^∞.

Actually our procedure gives two nontrivial solutions of (1) because the normalization (12) only determines (U_c, Θ_c, P_c) up to a factor of ± 1. It can be shown that the corresponding $\pm$ solutions have the same $\delta R(\epsilon)$ and $\delta R(\epsilon) = \delta R(-\epsilon)$.

Thus as our main result, we have

THEOREM 2. *For any a, $b > 0$ and any eigenvalue R_c of* (13), *there exists an $\epsilon_0 = \epsilon_0(a, b, R_c)$ such that for $|\epsilon| \leq \epsilon_0$,* (1) *possesses a solution of the form* (16) *satisfying conditions* (4)–(6).

REMARK 1. Theorem 2 indicates a large amount of nonuniqueness for solutions of (1). Which solutions one should see if an experiment is run sufficiently long is probably a stability question. The two dimensional roll seems to be the preferred shape [**5**].

REMARK 2. If $R_c = R_0$ is the smallest eigenvalue of (13), $\delta R > 0$ since for $R \leq R_0$, Joseph has shown [**4**], (1) possesses only the trivial solution. This of course agrees with the experimental evidence. It would be interesting to know more about δR for $R_c \neq R_0$.

REMARK 3. In his talk on bifurcation in the variational case, Dr. Berger showed that one could continue solutions as a function of the eigenvalue parameter. There is some indication that this can be established for (1) also since it is possible to find a priori bounds depending only on R for all solutions of (1).

REMARK 4. The method of this paper can be applied to the Bénard problem in a more general situation, e.g., a bounded cylindrical domain, however it requires a knowledge of the multiplicity of the eigenvalues for the corresponding linearized problem. This seems to be a very difficult question.

BIBLIOGRAPHY

1. P. H. Rabinowitz, *Existence and nonuniqueness of rectangular solutions of the Bénard problem*, Arch. Rational Mech. Anal. **29**, (1968), 32–57.

2. W. Velte, *Stabilitätinverhalten und Verzweigung stationärer Lösungen der Navier-Stokesschen Gleichungen*, Arch. Rational Mech. Anal. **16** (1964), 97–125.

3. ———, *Stabilität und Verzweigung stationärer Lösungen der Navier-Stokesschen Gleichungen beim Taylor Problem*, Arch. Rational Mech. Anal. **22** (1966), 1–14.

4. D. D. Joseph, *On the stability of the Boussinesq equations*, Arch. Rational Mech. Anal. **20** (1965), 59–71.

5. R. Krisnamurthi, *Finite amplitude thermal convection with changing mean temperature: The stability of hexagonal flows and the possibility of finite amplitude instability*, Dissertation, UCLA, 1967.

6. S. Karlin, *The existence of eigenvalues for integral operators*, Trans. Amer. Math. Soc. **113** (1964), 1–17.

STANFORD UNIVERSITY

MONOTONE OPERATORS ASSOCIATED WITH SADDLE-FUNCTIONS AND MINIMAX PROBLEMS

R. T. Rockafellar[1]

1. **Introduction.** Let X be a locally convex Hausdorff topological vector space over the real number system R, and let X^* be the dual of X, with $\langle x, x^* \rangle$ written in place of $x^*(x)$ for $x \in X$ and $x^* \in X^*$. A multivalued mapping $T: X \to X^*$ is called a (nonlinear) *monotone operator* if

$$\langle x_1 - x_2, x_1^* - x_2^* \rangle \geq 0 \tag{1.1}$$

when $x_1^* \in T(x_1)$ and $x_2^* \in T(x_2)$. It is called a *maximal* monotone operator if, in addition, its graph

$$\{(x, x^*) \mid x^* \in T(x)\} \subset X \times X^*, \tag{1.2}$$

is not contained properly in the graph of any other monotone operator $T': X \to X^*$.

One of the main classes of examples of monotone operators from X to X^* consists of the subdifferential mappings ∂f of the proper convex functions f on X. For $T = \partial f$, the solutions x to the relation

$$0 \in T(x), \tag{1.3}$$

which plays a fundamental role in monotone operator theory, are the points where f attains its global minimum on X. It is known that ∂f is a maximal monotone operator when f is finite and continuous throughout X (Minty [**5**]), or when X is a Banach space and f is (proper and) lower semicontinuous throughout X (Rockafellar [**12**], [**15**]).

The purpose of this paper is to present a new class of examples, the monotone operators associated with saddle-functions on X (i.e. functions which are partly convex and partly concave in a sense explained below). For such a monotone operator T, the solutions to (1.3) are the saddle-points in a certain minimax problem. It will be proved that T is maximal when the saddle-function from which it arises satisfies continuity conditions comparable to those in the case of ∂f.

The monotone operators associated with saddle-functions are of theoretical interest because they are closely related to extremum problems, even though they are not actually generalized gradient operators. The maximality theorems to be proved below for such operators open up a new area of applications of the theory

[1] This research was supported in part by the Air Force Office of Scientific Research under grant AF-AFOSR-1202-67.

of evolution equations involving monotone operators. These applications have a significance in mathematical economics.

2. **Saddle-functions.** A *convex function* on X is an everywhere-defined extended-real-valued function f (i.e. a function whose values are real numbers or $\pm\infty$) whose epigraph

$$\{(x, \mu) \mid x \in X, \mu \in R, \mu \geq f(x)\}, \tag{2.1}$$

is a convex set in the space $X \oplus R$. Such a function is said to be *proper* if it is not indentically $+\infty$ and if it nowhere has the value $-\infty$. A *subgradient* of a convex function f at a point $x \in X$ is an $x^* \in X^*$ such that

$$f(x') \geq f(x) + \langle x' - x, x^* \rangle, \quad \forall x' \in X. \tag{2.2}$$

The (possibly empty) set of all such subgradients at x is denoted by $\partial f(x)$, and the multivalued mapping $\partial f : x \to \partial f(x)$ from X to X^* is called the *subdifferential* of f.

It is known that, when f is everywhere Gâteaux differentiable, ∂f reduces to the (single-valued) gradient mapping ∇f from X to X^*. More generally, if f is finite and continuous at x, $\partial f(x)$ is a nonempty weak* compact convex subset of X and

$$f'(x; x') = \max \{\langle x', x^* \rangle \mid x^* \in \partial f(x)\}, \quad \forall x' \in X, \tag{2.3}$$

where

$$\begin{aligned} f'(x; x') &= \lim_{\lambda \downarrow 0} [f(x + \lambda x') - f(x)]/\lambda \\ &= \inf_{\lambda > 0} [f(x + \lambda x') - f(x)]/\lambda. \end{aligned} \tag{2.4}$$

For proofs and further details, see Moreau [**7**], [**8**], [**9**].

An extended-real-valued function g on X is said to be *concave* if $-g$ is convex. It is said to be a *proper* concave function if $-g$ is a proper convex function, i.e. if g is not identically $-\infty$ and g nowhere has the value $+\infty$.

We assume henceforth that $X = Y \oplus Z$, where Y and Z are locally convex Hausdorff topological vector spaces with duals Y^* and Z^*. We identify X^* with $Y^* \oplus Z^*$ and write

$$\langle x, x^* \rangle = \langle y, y^* \rangle + \langle z, z^* \rangle$$

for $x = (y, z) \in X$ and $x^* = (y^*, z^*) \in X^*$.

By a *saddle-function* on X (with respect to the given decomposition $X = Y \oplus Z$), we shall mean an everywhere-defined extended-real-valued function K such that $K(y, z)$ is a concave function of $y \in Y$ for each $z \in Z$ and a convex function of $z \in Z$ for each $y \in Y$. A saddle-function K will be called *proper* if there exists at least one point $x = (y, z)$ such that $K(y', z) < +\infty$ for every $y' \in Y$ and $K(y, z') > -\infty$ for every $z' \in Z$. The set of all such points will be called the *effective domain* of K and denoted by dom K. Obviously K is finite (i.e. real-valued) on dom K, and if K is finite everywhere one has dom $K = X$.

As an example, let L be any finite saddle-function on X, let C and D be non-empty convex sets in Y and Z, respectively, and let

$$K(y, z) = L(y, z) \quad \text{if } y \in C \text{ and } z \in D, \qquad = +\infty \quad \text{if } y \in C \text{ and } z \notin D, \qquad = -\infty \quad \text{if } y \notin C. \tag{2.5}$$

Then K is a proper saddle-function on X with

$$\operatorname{dom} K = C \oplus D. \tag{2.6}$$

In particular, L here could be any function of the form

$$L(y, z) = g(y) + h(z) + b(y, z), \tag{2.7}$$

where g is a finite concave function on Y, h is a finite convex function on Z and b is a bilinear function on $Y \times Z$.

The elementary but fundamental fact which motivates this paper is stated in the following theorem. This fact has previously been observed by Dantzig-Cottle [**4**] in the special case where X is finite-dimensional and K is a (finite) quadratic function (so that T is a linear operator).

THEOREM 1. *Let K be a proper saddle-function on $X = Y \oplus Z$, and for each $x = (y, z)$ in X let $T(x) = T(y, z)$ be the set of all $x^* = (y^*, z^*)$ in $X^* = Y^* \oplus Z^*$ such that y^* is a subgradient of the convex function $-K(\cdot, z)$ at y and z^* is a subgradient of the convex function $K(y, \cdot)$ at z. The multivalued mapping $T: X \to X^*$ is then a monotone operator with*

$$\{x \mid T(x) \neq \varnothing\} \subset \operatorname{dom} K. \tag{2.8}$$

PROOF. Let $(y_1^*, z_1^*) \in T(y_1, z_1)$ and $(y_2^*, z_2^*) \in T(y_2, z_2)$. By definition,

$$-K(y, z_1) \geq -K(y_1, z_1) + \langle y - y_1, y_1^* \rangle, \quad \forall y \in Y, \tag{2.9}$$

$$K(y_1, z) \geq K(y_1, z_1) + \langle z - z_1, z_1^* \rangle, \quad \forall z \in Z, \tag{2.10}$$

$$-K(y, z_2) \geq -K(y_2, z_2) + \langle y - y_2, y_2^* \rangle, \quad \forall y \in Y, \tag{2.11}$$

$$K(y_2, z) \geq K(y_2, z_2) + \langle z - z_2, z_2^* \rangle, \quad \forall z \in Z. \tag{2.12}$$

Since in particular (y, z) could be a point of dom K, we have $-K(y_1, z_1) < +\infty$ by (2.9) and $K(y_1, z_1) < +\infty$ by (2.10). Thus $K(y_1, z_1)$ is finite, and by (2.9) and (2.10) we have $(y_1, z_1) \in \operatorname{dom} K$, establishing (2.8). By the same argument, $K(y_2, z_2)$ is finite. Taking $y = y_2$ in (2.9), $z = z_2$ in (2.10), $y = y_1$ in (2.11) and $z = z_1$ in (2.12), we get, by adding the four inequalities,

$$0 \geq \langle y_2 - y_1, y_1^* \rangle + \langle z_2 - z_1, z_1^* \rangle + \langle y_1 - y_2, y_2^* \rangle + \langle z_1 - z_2, z_2^* \rangle.$$

In other words

$$0 \leq \langle y_1 - y_2, y_1^* - y_2^* \rangle + \langle z_1 - z_2, z_1^* - z_2^* \rangle,$$

and this means that T is a monotone operator.

The mapping T in Theorem 1 will be called the *monotone operator associated with* K. (It should be noted that T depends, not only on K as a function on X, but on the given direct sum decomposition $X = Y \oplus Z$, which must be specified before the concept of "saddle-function" has meaning. There may be some other decomposition $X = Y' \oplus Z'$ with respect to which the same K is a saddle-function but has a different monotone operator $T': X \to X^*$ associated with it. The decomposition $X = Y \oplus Z$ is fixed throughout the present discussion.)

The monotone operator T associated with a saddle-function K is closely related to the *subdifferential mapping* ∂K which we have introduced elsewhere **[10]** in connection with minimax theory: namely, ∂K is given in terms of T by

$$\partial K(y, z) = \{(-y^*, z^*) \mid (y^*, z^*) \in T(y, z)\}. \tag{2.13}$$

For this reason, properties of T have a bearing on certain extremum problems, as we shall now explain.

A point $(y, z) \in X$ is called a *saddle-point* of a saddle-function K if

$$K(y', z) \leq K(y, z) \leq K(y, z'), \quad \forall y' \in Y, \forall z' \in Z, \tag{2.14}$$

i.e. if the concave function $K(\cdot, z)$ attains its maximum at y and the convex function $K(y, \cdot)$ attains its minimum at z. It is well-known that (2.14) implies

$$K(y, z) = \sup_{y' \in Y} \inf_{z' \in Z} K(y', z') = \inf_{z' \in Z} \sup_{y' \in Y} K(y', z'). \tag{2.15}$$

In the case where K is of the form (2.5), it is not hard to see that (y, z) is a saddle-point of K if and only if (y, z) is a saddle-point of L with respect to $C \times D$, i.e.

$$L(y', z) \leq L(y, z) \leq L(y, z'), \quad \forall y' \in C, \forall z' \in D, \tag{2.16}$$

in which event

$$L(y, z) = \sup_{y' \in C} \inf_{z' \in D} L(y', z') = \inf_{z' \in D} \sup_{y' \in C} L(y', z'); \tag{2.17}$$

see **[10]**.

Suppose now that T is the monotone operator associated with a proper saddle-function K. According to the definition of T, the relation $(y^*, z^*) \in T(y, z)$ can be expressed equivalently as

$$\begin{aligned} \langle y', y^* \rangle - \langle z, z^* \rangle + K(y', z) &\leq \langle y, y^* \rangle - \langle z, z^* \rangle + K(y, z) \\ &\leq \langle y, y^* \rangle - \langle z', z^* \rangle + K(y, z'), \quad \forall y' \in Y, \forall z' \in Z. \end{aligned} \tag{2.18}$$

But this means that (y, z) is a saddle-point of the proper saddle-function $\langle \cdot, y^* \rangle - \langle \cdot, z^* \rangle + K$. In particular, the solutions (y, z) to

$$(0, 0) \in T(y, z), \tag{2.19}$$

are just the saddle-points of K, if any. It follows that general results about the domain and range of T can be interpreted as results about the existence of saddle-points, i.e. minimax theorems.

Although we shall not pursue the point here, we should like to mention that, in view of Theorem 1, the theory of monotone operators has a further bearing on minimax theory when X is a Hilbert space, namely through the study of the general "evolution equation"

$$-\dot{x}(t) \in T(x(t)) \quad \text{for almost all } t, \tag{2.20}$$

where $t \to x(t)$ is (in a suitable sense) an absolutely continuous function from $[0, +\infty)$ to X with derivative $t \to \dot{x}(t)$. It can be shown that, in certain cases where X is finite-dimensional and T is the monotone operator associated with a saddle-function K of the form (2.5) with C and D polyhedral and L differentiable, (2.20) reduces to the Arrow-Hurwicz differential equation [**1**, p. 118]. This equation and its generalizations are of interest in mathematical economics and game theory, because they describe evolution towards a state of "competitive equilibrium."

An elegant theory has already been developed concerning the existence and uniqueness of solutions to the "evolution equation" (2.20) and the convergence of such solutions to points satisfying (1.3)—see the papers of Browder and Kato in this volume. This theory requires only that T be a maximal monotone operator. The maximality theorems established below will therefore make it possible to apply this theory to a new area, the study of saddle-points via generalizations of the Arrow-Hurwicz differential equation.

3. **Maximality theorems.** We shall now prove our main results, which give conditions under which the monotone operators in Theorem 1 are maximal.

THEOREM 2. *Let K be a finite (i.e. everywhere real-valued) saddle-function on $X = Y \oplus Z$ such that $K(y, z)$ is everywhere separately continuous in y and z. The monotone operator T associated with K is then maximal. Moreover, for each $(y, z) \in X$, $T(y, z)$ is a nonempty weak* compact convex subset of X^*.*

PROOF. By definition, $(y^*, z^*) \in T(y, z)$ if and only if $y^* \in C(y, z)$ and $z^* \in D(y, z)$, where $C(y, z)$ is the set of all subgradients of $-K(\cdot, z)$ at y and $D(y, z)$ is the set of all subgradients of $K(y, \cdot)$ at z. Since the convex functions $-K(\cdot, z)$ and $K(y, \cdot)$ are finite and continuous by hypothesis, $C(y, z)$ and $D(y, z)$ are nonempty weak* compact convex subsets of Y^* and Z^* respectively, as indicated at the beginning of §2, and hence $T(y, z)$ is a nonempty weak* compact convex subset of X^*. Now fix any $(y_1, z_1) \in X$ and any $(y_1^*, z_1^*) \in X^*$ such that $(y_1^*, z_1^*) \notin T(y_1, z_1)$. We shall show that there exist a $(y_2, z_2) \in X$ and a $(y_1^*, z_1^*) \in T(y_2, z_2)$ such that

$$\langle y_2 - y_1, y_2^* - y_1^* \rangle + \langle z_2 - z_1, z_2^* - z_1^* \rangle < 0, \tag{3.1}$$

and this will establish the maximality of T.

Let k be the real-valued function on $X \times X$ defined by

$$\begin{aligned} k(y, z; y', z') &= \max \{\langle y', y^* \rangle + \langle z', z^* \rangle \mid (y^*, z^*) \in T(y, z)\} \\ &= \max \{\langle y', y^* \rangle \mid y^* \in C(y, z)\} + \max \{\langle z', z^* \rangle \mid z^* \in D(y, z)\}. \end{aligned} \tag{3.2}$$

We note that, in view of formulas (2.3) and (2.4) (applied to the convex functions $-K(\cdot, z)$ and $K(y, \cdot)$),

$$(3.3)\qquad \begin{aligned} k(y, z; y', z') &= \lim_{\lambda \downarrow 0} [-K(y + \lambda y', z) + K(y, z)]/\lambda \\ &\quad + \lim_{\lambda \downarrow 0} [K(y, z + \lambda z') - K(y, z)]/\lambda \\ &= \lim_{\lambda \downarrow 0} [K(y, z + \lambda z') - K(y + \lambda y', z)]/\lambda \\ &= \inf_{\lambda > 0} [K(y, z + \lambda z') - K(y + \lambda y', z)]/\lambda. \end{aligned}$$

Since $T(y_1, z_1)$ is a nonempty weak* closed convex set not containing (y_1^*, z_1^*), we can strictly separate (y_1^*, z_1^*) from $T(y_1, z_1)$ by some weak* closed hyperplane in X^*. Thus by (3.2) there exists some $(y', z') \in X$ such that

$$(3.4)\qquad k(y_1, z_1; y', z') < \langle y', y_1^* \rangle + \langle z', z_1^* \rangle.$$

Consider the function

$$(3.5)\qquad p(\theta) = k(y_1 + \theta y', z_1 + \theta z'; y', z'), \quad \theta \in R.$$

According to (3.3),

$$(3.6)\quad p(\theta) = \inf_{\lambda > 0} [K(y_1 + \theta y', z_1 + \theta z' + \lambda z') - K(y_1 + \theta y' + \lambda y', z_1 + \theta z')].$$

Now, for any λ, the function

$$M_\lambda(\theta, \mu) = K(y_1 + \theta y', z_1 + \mu z' + \lambda z'),$$

is concave in $\theta \in R$ for each $\mu \in R$ and convex in $\mu \in R$ for each $\theta \in R$. Thus M_λ is a finite saddle-function on $R \oplus R$. But a finite saddle-function on a finite-dimensional space is everywhere jointly continuous; this is proved in [**14**, §35]. Therefore $M_\lambda(\theta, \theta)$ is a continuous function of θ. Similarly,

$$K(y_1 + \theta y' + \lambda y', z_1 + \theta z'),$$

is a continuous function of θ. Formula (3.6) thus expresses p as the pointwise infimum of a collection of continuous functions, and it follows that p is upper semi-continuous. Hence by (3.4), since $p(0) = k(y_1, z_1; y', z')$, we must have

$$(3.7)\qquad k(y_1 + \theta y', z_1 + \theta z'; y', z') < \langle y', y_1^* \rangle + \langle z', z_1^* \rangle$$

for all sufficiently small real numbers θ. Fix any $\theta > 0$ for which (3.7) holds, and let

$$(3.8)\qquad y_2 = y_1 + \theta y', \qquad z_2 = z_1 + \theta z'.$$

Take any $(y_2^*, z_2^*) \in T(y_2, z_2)$. The definition of k implies that

$$\langle y', y_2^* \rangle + \langle z', z_2^* \rangle \le k(y_2, z_2; y', z').$$

Combining this with (3.7), we get

$$\langle y', y_2^* - y_1^* \rangle + \langle z', z_2^* - z_1^* \rangle < 0,$$

which is equivalent to (3.1) in view of (3.8). This completes the proof of Theorem 2.

COROLLARY 1. *Let K be a finite saddle-function on $X = Y \oplus Z$, and suppose that X is finite-dimensional. The monotone operator T associated with K is then maximal.*

PROOF. This is immediate from the well-known fact that a finite convex or concave function on a finite-dimensional space is necessarily continuous.

COROLLARY 2. *Let K be a finite saddle-function on $X = Y \oplus Z$ which is everywhere Gâteaux differentiable, and suppose that the spaces Y and Z are barrelled. Express the Gâteaux gradient of K by*

$$\nabla K(y, z) = (\nabla_1 K(y, z), \nabla_2 K(y, z)),$$

where $\nabla_1 K(y, z) \in Y^*$ and $\nabla_2 K(y, z) \in Z^*$. The single-valued mapping

$$(y, z) \to (-\nabla_1 K(y, z), \nabla_2 K(y, z)) \tag{3.9}$$

is then a maximal monotone operator from X to X^*.

PROOF. The monotone operator T associated with K reduces to (3.9), in view of the Gâteaux differentiability of K. For each y, the convex function $K(y, \cdot)$, being Gâteaux differentiable, is the pointwise supremum of a certain collection of continuous affine functions, and hence is lower semicontinuous. But a finite lower semicontinuous convex function on a barrelled space is necessarily continuous (see [**11**]). Thus $K(y, z)$ is continuous in z for each y. By a similar argument $K(y, z)$ is continuous in y for each z, and it follows from the theorem that T is maximal.

To get maximality results in the case of saddle-functions which are not everywhere finite, such as those of the form (2.5), more complicated continuity conditions must be imposed. These are most easily described in terms of the so-called *closure operation* for convex functions.

A convex function on X is said to be *closed* if it is proper and lower semicontinuous, or else if it is one of the constant functions $+\infty$ or $-\infty$. Given any convex function f on X, there exists a unique greatest closed convex function majorized by f (the pointwise supremum of the collection of all closed convex functions majorized by f). This function is called the *closure* of f and denoted by $\mathrm{cl}\, f$.

Given any saddle-function K on $X = Y \oplus Z$, we denote by $\mathrm{cl}_2\, K$ the function on X such that, for each $y \in Y$, $(\mathrm{cl}_2\, K)(y, \cdot)$ is the closure of the convex function $K(y, \cdot)$ on Z. Similarly, we denote by $\mathrm{cl}_1\, K$ the function on X such that, for each $z \in Z$, $-(\mathrm{cl}_1\, K)(\cdot, z)$ is the closure of the convex function $-K(\cdot, z)$ on Y. Two saddle-functions K and K' are called *equivalent* if $\mathrm{cl}_1\, K = \mathrm{cl}_1\, K'$ and $\mathrm{cl}_2\, K = \mathrm{cl}_2\, K'$. A saddle-function K is said to be *closed* if $\mathrm{cl}_1\, K$ and $\mathrm{cl}_2\, K$ are saddle-functions equivalent to K.

These notions of equivalence and closure of saddle-functions have a natural significance in minimax theory, as we have shown in [**10**], [**13**], [**14**]. For present purposes, we shall only mention a few pertinent facts. The proofs are all given in [**14**] in a finite-dimensional context, but the arguments do not actually rely on finite-dimensionality, so that they carry over immediately to arbitrary locally convex Hausdorff topological vector spaces.

The facts are as follows. Given any saddle-function K, the closures $\mathrm{cl}_1 K$ and $\mathrm{cl}_2 K$ are again saddle-functions. Furthermore, $\mathrm{cl}_1 (\mathrm{cl}_2 K)$ and $\mathrm{cl}_2 (\mathrm{cl}_1 K)$ are closed saddle-functions (not necessarily equivalent). If K' is a saddle-function equivalent to K, then $\partial K' = \partial K$ (cf. (2.13)). *Thus the monotone operator T associated with a proper saddle-function K really depends only on the equivalence class containing K.* The most important fact is that the formula

$$F(y, z^*) = \sup \{\langle z, z^*\rangle - K(y, z) \mid z \in Z\}, \tag{3.10}$$

defines a one-to-one correspondence between the equivalence classes of closed proper saddle-functions K on $X = Y \oplus Z$ and the lower semicontinuous proper convex functions F on the space $Y \oplus Z^*$, where the topology on $Y \oplus Z^*$ is taken to be the product of the given topology on Y and the Mackey topology on Z^*. Moreover, under this correspondence one has

$$(y^*, z^*) \in \partial K(y, z) \Leftrightarrow (-y^*, z) \in \partial F(y, z^*), \tag{3.11}$$

where ∂F is the subdifferential of F. (Here the space of all continuous linear functionals on $Y \oplus Z^*$ in the cited topology is identified in the natural way with $Y^* \oplus Z$.) It follows that, if K is a closed proper saddle-function and T is the monotone operator associated with K, one has

$$(y^*, z^*) \in T(y, z) \Leftrightarrow (y^*, z) \in \partial F(y, z^*) \tag{3.12}$$

for the F defined by (3.10). *Thus T can be obtained by partial inversion of the subdifferential mapping of a certain lower semicontinuous proper convex function F on $Y \oplus Z^*$.* If this subdifferential ∂F is maximal, then T itself must be maximal. In particular, we get the following result.

THEOREM 3. *Let K be a closed proper saddle-function on $X = Y \oplus Z$, and suppose that Y and Z are Banach spaces, at least one of which is reflexive. The monotone operator T associated with K is then maximal.*

PROOF. Suppose that Z is reflexive, say. Then $Y \oplus Z^*$ is a Banach space whose dual may be identified with $Y^* \oplus Z$. Since the F defined by (3.10) is a lower semicontinuous proper convex function on a Banach space, its subdifferential ∂F is a maximal monotone operator (Rockafellar [**15**]). Hence, by relation (3.12), T is maximal. The case where Y, rather than Z, is reflexive, can be established similarly by replacing K by $-K$ and reversing the roles of the arguments y and z.

COROLLARY 1. *Let K be a proper saddle-function on $X = Y \oplus Z$ such that $K(y, z)$ is upper semicontinuous in y for each z and lower semicontinuous in z for* *each y. Suppose that Y and Z are Banach spaces, at least one of which is reflexive. The monotone operator T associated with K is then maximal.*

PROOF. The semicontinuity conditions on K imply that K is closed, as is not difficult to verify using the fact that a lower semicontinuous convex function which is not proper (or an upper semicontinuous concave function which is not proper) can have no values other than $+\infty$ and $-\infty$. (Note: not every closed proper saddle-function satisfies these semicontinuity conditions, even in the case where Y and Z are one-dimensional; see [**14**, §34] for counterexamples.)

COROLLARY 2. *Let K be a saddle-function on $X = Y \oplus Z$ of the form (2.5), where C and D are nonempty closed convex sets in Y and Z, respectively, and L is a finite saddle-function such that $L(y, z)$ is upper semicontinuous in y for each z and lower semicontinuous in z for each y. Suppose that Y and Z are Banach spaces, at least one of which is reflexive. The monotone operator T associated with K is then maximal.*

PROOF. Here K satisfies the hypothesis of Corollary 1.

4. **A counterexample.** In view of the many connections between monotone operators and convexity, it might be conjectured that, for every maximal monotone operator $T: X \to X^*$ which is not in fact the subdifferential of some convex function on X, there exists a direct sum decomposition $X = Y \oplus Z$ and a function K on X which is a saddle-function with respect to this decomposition, such that T is the monotone operator associated with K. We shall show that this is not true even when X is two-dimensional.

The counterexample we shall furnish is based on the fact that, in the finite-dimensional case at least, the set of points (y, z) where $\partial K(y, z) \neq \varnothing$ is dense in dom K, and dom K is the direct sum of a convex set in Y and a convex set in Z (see Rockafellar [**10**], [**14**]). This implies that the closure of the set

$$D(T) = \{(y, z) \mid T(y, z) \neq \varnothing\} \tag{4.1}$$

is the direct sum of a closed convex set in Y and a closed convex set in Z. (Incidentally, we do not know whether $D(T)$ is dense in dom K when X is not finite-dimensional, although the situation in the case of purely convex functions [**2**] would suggest that this might always be true when K is closed and Y and Z are Banach spaces.)

Let $X = R^2$, and let A be the linear operator from X to $X^* = R^2$ defined by

$$x = (\xi_1, \xi_2) \to x^* = (-\xi_2, \xi_1).$$

Let B be the closed unit disk in X, and let S be the subdifferential of the *indicator* of B, i.e. the lower semicontinuous proper convex function f such that $f(x) = 0$ for $x \in B$ and $f(x) = +\infty$ for $x \notin B$. (Thus $S(x)$ consists of the zero vector alone when x is an interior point of B, $S(x)$ consists of all the nonnegative multiples of x when x is a boundary point of B and $S(x) = \varnothing$ when $x \notin B$.) Of course A is a continuous single-valued monotone operator, whereas S is a maximal monotone operator whose effective domain is B [**15**]. It follows (see [**3**]) that the mapping $T:X \to X^*$ defined by $T(x) = S(x) + A(x)$ is a maximal monotone operator with $D(T) = B$. Since $D(T)$ cannot be expressed as the direct sum of two line segments, T cannot arise from any saddle-function K, as explained above. On the other hand, T is not the subdifferential of any convex function f on X by [**12**, Theorem 1], because T reduces to A on the interior of B and consequently is not cyclically monotone.

References

1. K. J. Arrow, L. Hurwicz and H. Uzawa, *Studies in linear and non-linear programming*, Stanford University Press, Stanford, Calif., 1958, 117–145.

2. A. Brøndsted and R. T. Rockafellar, *On the subdifferentiability of convex functions*, Proc. Amer. Math. Soc. **16** (1965), 605–611.

3. F. E. Browder, *Nonlinear maximal monotone operators in Banach spaces*, Math. Ann. **175** (1968), 89–113.

4. G. B. Dantzig and R. W. Cottle, "Positive (semi-) definite programming", in *Nonlinear programming*, edited by J. Abadie, Amsterdam, 1967, pp. 53–73.

5. G. J. Minty, *On the monotonicity of the gradient of a convex function*, Pacific J. Math. **14** (1967), 243–247.

6. J. J. Moreau, *Théorèmes 'inf-sup'*, C. R. Acad. Sci. Paris **258** (1964), 2720–2722.

7. ———, *Sur la polaire d'une fonctionelle semi-continue superieurment*, C. R. Acad. Sci. Paris **258** (1964), 1128–1130.

8. ———, *Sous-differentiabilité*, Proc. Colloquium on Convexity (Copenhagen, 1965), Københavns Univ. Mat. Inst., Copenhagen, 1967, 185–201.

9. ———, *Fonctionelles convexes*, mimeographed lecture notes, College de France, 1967.

10. R. T. Rockafellar, *Minimax theorems and conjugate saddle-functions*, Math. Scand. **14** (1964), 151–173.

11. ———, *Level sets and continuity of conjugate convex functions*, Trans. Amer. Math. Soc. **123** (1966), 46–63.

12. ———, *Characterization of the subdifferentials of convex functions*, Pacific J. Math. **17** (1966), 497–510. A gap in the maximal proof in this paper is corrected in [**15**].

13. ——— *A general correspondence between dual minimax problems and convex programs*, Pacific J. Math. **25** (1968), 597–611.

14. R. T. Rockafellar, *Convex analysis*, Princeton Univ. Press, Princeton, N.J., 1969.

15. ———, *On the maximal monotonicity of subdifferential mappings*, Michigan J. Math. (to appear).

University of Washington

SOME REMARKS ON VECTOR FIELDS IN HILBERT SPACE

Erich H. Rothe

1. **Introduction.** Let G be an open bounded connected set in a real Banach space E. Let $\bar{G}$ denote the closure of G, and $\dot{G}$ its boundary. The zero element of E will be denoted by θ. Let f be a map $\bar{G} \to E$ of the form

$$f(x) = x + F(x) \tag{1.1}$$

with completely continuous F. After the mapping degree $A(\theta, G, f)$ was introduced by Leray and Schauder [5], it had been easy to define the related notions of the order $u(\theta, \dot{G}, f)$ of θ with respect to $f(\dot{G})$ and the characteristic $\chi(\dot{G}, f)$ of the "vector field" f defined on $\dot{G}$.

In §2 definitions and facts concerning these notions are collected. At this point we mention the following three statements:

I. If G is a ball B and if f is a vector field of form (1.1) defined on $\dot{B}$ none of whose vectors vanishes or has the direction of the interior normal then

$$\chi(\dot{B}, f) = 1 \tag{1.2}$$

[**9**, Satz 7b].

II. Let $B_0, B_1, \ldots, B_q$ be open balls in E. We assume that for $i = 1, \ldots, q$, $\bar{B}_i \subset B_0$ and that the $\bar{B}_i$ are disjoint. Let $G = B_0 - \bigcup_{i=1}^q \bar{B}_i$, and let the map (1.1) have no zeros on $\dot{G}$, and at most a finite number of zeros, say $x_1, \ldots, x_r$, in G. Then

$$\chi(\dot{B}_0, f) - \sum_{i=1}^{q} \chi(\dot{B}_i, f) = \sum_{\rho=1}^{r} j_\rho \tag{1.3}$$

where j_ρ denotes the Leray-Schauder index at the zero x_ρ [**9**, Satz 5].

III. If in particular f has on none of the $\dot{B}_k$ $(k = 0, 1, \ldots, q)$ the direction of the normal interior to B_k then

$$1 - q = \sum_{\rho=1}^{r} j_\rho \tag{1.4}$$

as is seen from (1.2) and (1.3).

In §3 of the present paper analogues to the above statements are presented: the balls are replaced by star domains (Theorems 3.1 and 3.2). Normals are then not necessarily defined but the notion of an "exteriorly directed vector" can be introduced (Definition 3.3).

In §4 new assumptions are made: (1) E is a Hilbert space, and (2) the B_k above are replaced by "elements", i.e. sets $\psi(B_k)$ where ψ is of the form (1.1) and one to one. If, in addition ψ has certain differentiability properties then the $\psi(B_k)$ turn out to be smooth manifolds in the sense of [**12**]. Consequently unique interior and exterior normals exist, and closer analogues to the statements I–III above are proved provided that the $\psi(B_k)$ are star domains (Theorems 4.1 and 4.2).

If $E = E^n$, a Hilbert space of finite dimension n, then formula (1.4) is closely related to a special case of the Hopf-Lefschetz theorem on vector fields (see the papers by these authors quoted in [**1**, p. 421 bottom]): if each vector of the vector field f on $\dot{G}$ is on B_k exteriorly directed with respect to G (i.e. interiorly directed on $\dot{B}_k$ with respect to B_k for $k = 1, \ldots, q$) then (1.4) has to be replaced by

$$1 - (-1)^n q = \sum_{\rho=1}^{r} j_\rho \tag{1.5}$$

as follows from (1.3) since now $\chi(\dot{B}_k, f) = (-1)^n$. But the left member of (1.5) is the Euler characteristic of G. Thus (1.5) is the Hopf-Lefschetz formula in the special case considered.

Now M. Morse [**6**] showed how to modify the Hopf-Lefschetz formula for the case that at a finite number of points on $\dot{G}$ the vector f has the direction of the interior normal by considering the indices of the singularities of the "tangential" field at those points.

§5 of the present paper deals with the corresponding modification of formulas (1.2) and (1.4) in the Hilbert space case under certain nondegeneracy assumptions concerning the singularities of the tangential field (Theorems 5.1, 5.2). A treatment of this problem has already been given in [**12**] in the case that f is a gradient field.

2. **Prerequisites concerning the notions of order and characteristic.** We use the notations of the introduction. For definition and basic properties of the Leray-Schauder degree $A(\theta, G, f)$ we refer the reader to [**5**], [**4**], [**2**], or [**7**] and [**8**]. (The last two references together form a complete, self-contained exposition of the theory including the finite-dimensional case).

To define the order $u(\theta, G, f)$ we need the following two lemmas.

LEMMA 2.1. *Let F be a completely continuous map $C \to E$ where C is a closed subset of E. Then there exists a completely continuous extension of F to E.*

This is a special case of an extension theorem by Dugundji (see [**4**, p. 113] and [**3**, p. 29]).

LEMMA 2.2. *Let G, $\dot{G}$ be defined as in the first paragraph of the introduction. Let $f: \dot{G} \to E$ be of form* (1.1) *and satisfy*

$$f(x) \neq \theta \quad \text{for } x \in \dot{G}. \tag{2.1}$$

Let F_0 and F_1 be two completely continuous extensions of F to $\bar{G}$. Let $f_i(x) = x + F_i(x)$, $(i = 0, 1)$. Then

$$A(\theta, G, f_0) = A(\theta, G, f_1). \tag{2.2}$$

The proof follows immediately from the homotopy property of the degree by considering the map $(1-t)f_0 + tf_1$, $0 \leqq t \leqq 1$.

DEFINITION 2.1. Let f be a map $\dot{G} \to E$ of the form (1.1) satisfying (2.1). Then the order $u(\theta, \dot{G}, f)$ of θ with respect to $f(\dot{G})$ is defined by

$$u(\theta, \dot{G}, f) = A(\theta, G, f_0), \tag{2.3}$$

where $f_0(x) = x + F_0(x)$ with F_0 being a completely continuous extension of F to $\bar{G}$.

(The definition of u in [**9**] and in [**10**] was restricted to G being a ball or a convex set. The definition for G open connected was given by Krasnosel'skiĭ, see [**4**, p. 121] and the literature mentioned there. The definition given above is somewhat different.)

Often f is thought of as a field of vectors $f(x)$ attached to the point x, (see [**1**, p. 478] for the finite dimensional case, and [**9**, §3] for balls in a Banach space). We then define

DEFINITION 2.2. The characteristic $\chi(\dot{G}, f)$ of the nonvanishing vector field f of form (1.1) defined on $\dot{G}$ is given by

$$\chi(\dot{G}, f) = u(\theta, \dot{G}, f). \tag{2.4}$$

Since many essential properties of vector fields are not changed under multiplication by a nonvanishing scalar factor it is convenient to slightly generalize the preceding definitions as follows (see also [**10**, p. 375]):

DEFINITION 2.3. Let D be a closed bounded set in E. Let $\lambda = \lambda(x)$ be a real valued function defined in D and satisfying for suitable constants M and m the inequality

$$0 < m \leqq \lambda(x) \leqq M, \quad x \in D. \tag{2.5}$$

Let $F(x, \mu)$ be defined and completely continuous in $D \times [m, M]$. Let

$$f(x) = x\lambda(x) + F(x, \lambda(x)). \tag{2.6}$$

We assume that (2.1) holds with $F(x)$ replaced by $F(x, \lambda(x))$ and $\dot{G}$ by $\dot{D}$. We then set

$$A(\theta, G, f) = A(\theta, G, \lambda^{-1}f) \quad \text{if } D = \bar{G},\ \theta \notin f(\dot{G}) \tag{2.7}$$

and, more general by, for any point $a \notin f(\dot{G})$

$$A(a, G, f) = A(a, G, a + \lambda^{-1}(f - a)). \tag{2.8}$$

Moreover

$$\left.\begin{array}{ll} u(\theta, \dot{G}, f) = u(\theta, \dot{G}, \lambda^{-1}f) & \text{if} \quad (2.9) \\ \chi(\dot{G}, f) = \chi(\dot{G}, \lambda^{-1}f) & \text{if} \quad (2.10) \end{array}\right\} D = \dot{G}.$$

[Note that the assumptions made on λ and $F(x, \mu)$ imply the complete continuity of $\lambda^{-1}F(x, \lambda(x))$.]

LEMMA 2.3. *Let*

$$f_t(x) = x\lambda(x, t) + F(x, \lambda(x, t), t) \tag{2.11}$$

where $\lambda(x, t)$ *is continuous in* $D \times [0, 1]$ *and*

$$0 < m \leqq \lambda(x, t) \leqq M, \tag{2.12}$$

and where $F(x, \mu, t)$ *is assumed to be completely continuous in* $D \times [m, M] \times [0, 1]$. *Let* a_t *be a point of* E *depending continuously on* t *for which* $a_t \notin f_t(\dot{D})$. *Then for* $D = \bar{G}$, $A(a_t, G, f_t)$ *is defined and independent of* t. *If* $D = \dot{G}$ *and* $\theta \notin f_t(\dot{G})$, *then* $u(\theta, \dot{G}, f_t)$ *and* $\chi(\dot{G}, f_t)$ *are defined and independent of* t.

We omit the proof since on account of the above definitions it is easy to reduce the assertions to the case where $\lambda(x, t) = 1$ in which case the lemma is well known (see [**8**, Theorem 7]).

LEMMA 2.4. *Let* G_0 *be an open bounded set in* E. *Let* $G_1, \ldots, G_q$ *be open sets whose closures are disjoint and contained in* G_0. *Let* f *be a nonvanishing vector field of form* (2.6) *defined on* $\bigcup_{k=1}^{q} \dot{G}_k$, *and let* f_0 *be an extension of* f *to the closure* $\bar{G}$ *of* $G = G_0 - \bigcup_{i=1}^{q} \bar{G}_i$. *Then*

$$u(\theta, \dot{G}_0, f) = \sum_{i=1}^{q} u(\theta, \dot{G}_i, f) + A\left(\theta, G_0 - \bigcup_{i=1}^{q} \bar{G}_i, f_0\right), \tag{2.13}$$

$$\chi(\dot{G}_0, f) = \sum_{i=1}^{q} \chi(\dot{G}_i, f) + A\left(\theta, G_0 - \bigcup_{i=1}^{q} \bar{G}_i, f_0\right). \tag{2.14}$$

PROOF. By (2.9) and (2.10) it is sufficient to consider fields of the form (1.1). But for any extension f_0 (of the same form) of f to $\bar{G}$ the "addition theorem" [**8**, Theorem 6] states that

$$A(\theta, G_0, f_0) = \sum_{i=1}^{q} A(\theta, G_i, f_0) + A\left(\theta, G_0 - \bigcup_{i=1}^{q} \bar{G}_i, f_0\right). \tag{2.15}$$

This proves our assertions by Definitions (2.1) and (2.2).

LEMMA 2.5. *Let* G *be a bounded open set in* E. *Let* $f: \bar{G} \to E$ *be of the form* (2.6) *and satisfy* $\theta \notin f(\dot{G})$. *We assume that any solution of*

$$f(x) = \theta \tag{2.16}$$

is isolated. Then (2.16) *has at most a finite number of solutions.*

We omit the proof which is quite similar to the proof of Theorem 3.2 in [**11**]. We recall the definition of the Leray-Schauder index:

DEFINITION 2.4. With the notations of the preceding lemma let x^0 be an isolated solution of (2.16). Then the Leray-Schauder index $j(x^0) = j(x^0, f)$ is given by $j(x^0, f) = A(\theta, G^0, f)$, where G^0 is an open subset of G containing x^0 but no other solution of (2.16).

LEMMA 2.6. *Using the notations of Lemma 2.4 we suppose, in addition to the assumptions of Lemma 2.5, that f is defined on $\bar{G}$ and has at most isolated zeros. Denote these by $x_1, \ldots, x_r$* (cf. *Lemma* 2.5). *Then*

$$\chi(\dot{G}_0, f) = \sum_{i=1}^{q} \chi(G_i, f) + \sum_{\rho=1}^{r} j_\rho, \qquad j_\rho = j(x_\rho, f). \tag{2.17}$$

The proof follows immediately from (2.14), Definition 2.4, and the addition theorem for the Leray-Schauder degree.

LEMMA 2.7. *Let G be open and bounded and let the map* (1.1) *map $\bar{G}$ one to one onto $f(\bar{G})$. Then* (i) *$f(G)$ is open,* (ii)

$$A(a, G, f) = \pm 1 \quad \textit{for } a \in f(G), \tag{2.18}$$

(iii)
$$f(\dot{G}) = (f(G))^{\cdot}.$$

For the proof of (i) and (ii) we refer to [8, Theorem 10]; (iii) follows easily from (i).

By the use of Lemma 2.7 the following lemma is easily proved to be a special case of the "product theorem" for the Leray-Schauder degree [8, Theorem 9].

LEMMA 2.8. *Let G be an open connected and bounded set in E. Let*

$$\psi(x) = x + \Psi(x), \quad \Psi \textit{ completely continuous,} \tag{2.19}$$

be a one to one map $\bar{G} \to E$. Let H be an open set containing $\psi(\bar{G})$, and let f be a map $\bar{H} \to E$ of form (1.1). *Moreover let a and b be points in E for which $a \notin f\psi(\dot{G})$, $a \notin f(\dot{H})$, $b \in \psi(G)$. Then*

$$A(a, G, f\psi) = A(a, \psi(G), f) A(b, G, \psi) = \pm A(a, \psi(G), f). \tag{2.20}$$

LEMMA 2.9. *The preceding lemma is still true if f is of form* (2.6).

PROOF. We set

$$f_1(y) = a + \lambda^{-1}(f(y) - a)$$
$$(f\psi(x))_1 = a + \lambda^{-1}(f\psi(x) - a).$$

Then by direct verification $f_1\psi(x) = (f\psi(x))_1$. Using this, definition (2.8) and noting that (2.20) holds if f is replaced by f_1 we see that

$$\begin{aligned} A(a, G, f\psi) &= A(a, G, (f\psi)_1) = A(a, G, f_1\psi) \\ &= \pm A(a, \psi(G), f_1) = \pm A(a, \psi(G), f). \end{aligned}$$

3.1. Vector fields on the boundary of a star domain.

DEFINITION 3.1. The vector field $f(x)$ on the boundary $\dot{G}$ of an open bounded set G is called exteriorly directed (with respect to G) if there exists a $\lambda_0(x) > 0$ such that the points $x + \lambda f(x)$ are exterior to G for $0 < \lambda \leq \lambda_0(x)$. The field is called strictly exterior if $\lambda_0(x)$ can be chosen independent of x.

THEOREM 3.1. *Let G be an open bounded star domain in the Banach space E, and let f be an exteriorly directed field of form* (2.6) *defined on $\dot{G}$. Then*

$$\chi(\dot{G}, f) = 1. \tag{3.1}$$

PROOF. It is seen from Definitions 3.1 and 2.3 that it is sufficient to prove the lemma for the case where f is of the form (1.1). Now by assumption there exists a point $x_0 \in G$ such that G is star shaped with respect to x_0. If then f_0 denotes the identity map of $\bar{G}$ onto itself we have

$$A(x_0, G, f_0) = 1. \tag{3.2}$$

We now extend f in the usual manner (cf. Lemma 2.1) to $\bar{G}$ denoting the extension again by f. Then with λ as in Definition 3.1 we claim

$$A(x_0, G, f_0 + \lambda f) = 1. \tag{3.3}$$

To prove this we consider $f_t = x + t\lambda f(x)$. By Definition 3.1 the point $f_t(x)$ is exterior to G for $x \in \dot{G}$ and $0 < t \leqq 1$, and a boundary point for $t = 0$. Consequently the interior point x_0 of G is not $\in f_t(\dot{G})$ for $0 \leqq t \leqq 1$. This proves (3.3) on account of (3.2) and Lemma 2.3.

We now set

$$\tilde{f}_t(x) = (1 - t)x + \lambda f(x) = x(1 - t + \lambda) + \lambda F(x),$$
$$x_t = (1 - t)x_0, \qquad \tilde{\tilde{f}}_t(x) = x_t + (1 - t + \lambda)^{-1}(\tilde{f}_t(x) - x_t).$$

(Note that $1 - t + \lambda > 0$ for $0 \leqq t \leqq 1$ since $\lambda > 0$). Now by (2.8)

$$A(x_t, G, \tilde{f}_t) = A(x_t, G, \tilde{\tilde{f}}_t). \tag{3.4}$$

We claim that this number is independent of t for $0 \leqq t \leqq 1$. It is, by Lemma 2.3, sufficient to prove that

$$_t(x) \neq x_t \quad \text{for } x \in \dot{G}, \qquad 0 \leqq t \leqq 1. \tag{3.5}$$

Now if for some $x \in \dot{G}$ and $t \in [0, 1]$ equality would hold in (3.5) then for such x and t

$$x + \lambda f(x) = tx + (1 - t)x_0 \tag{3.6}$$

as an elementary calculation shows. But the right member of (3.6) is a linear convex combination of the boundary point x and the point x_0 with respect to which G is star shaped. Therefore the point (3.6) is not an exterior point which contradicts the fact that the left member of (3.6) is exterior to G by Definition 3.1.

Thus $A(x_t, G, \tilde{f}_t)$ is independent of t, and application of Lemma 2.3 in conjunction with (3.3) shows that $A(\theta, G, \lambda f) = 1$. This proves (3.1) on account of (2.7), (2.3), and (2.4).

THEOREM 3.2. *In addition to the hypotheses of Lemma* 2.6 *we assume that each G_k $(k = 0, 1, \ldots, q)$ is a star domain and that on each $\dot{G}_k$ the field f is exteriorly* *directed. Then*

(3.7) $$1 - q = \sum_{\rho=i}^{r} j_\rho.$$

PROOF. The theorem follows immediately from (2.17) and application of Theorem 3.1 to each G_k.

4. **Vector fields on smooth hypermanifolds in Hilbert space.** From now on E will always denote a Hilbert space, and G an open bounded connected set in E whose boundary G is a smooth hypermanifold in the sense of [**12**, Definition 3.2].

We recall briefly the definition and some of the properties of such a manifold: a subset S of E is called a hypermanifold if to each point $x_0 \in S$ there exists a neighborhood $N(x_0)$ of x_0, a linear subspace U of E which is maximal (i.e. which is such that E/U has dimension 1), and a neighborhood U_0 (relative to U) of θ such that there is a one to one correspondence

(4.1) $$x = x(t), \qquad x_0 = x(\theta),$$

between the points t of U_0 and the points x of $S(x_0) = S \cap N(x_0)$. S is called smooth if U_0 and U can be chosen in such a way that for each $t_0 \in U_0$ the map (4.1) has a nonsingular differential $d(t_0; t)$ for which $dx(t_0; t) - t$ is completely continuous. For the rather obvious definitions of an "admissible" parameter space and of S being of class C'' we refer to [**12**, Definition 3.2].

The tangent space $T(x_0)$ at the point x_0 of the smooth hypermanifold S is defined as the image of U under the linear map $dx(\theta, t)$. $T(x_0)$ itself is then an admissible parameter space. Moreover the $t \in T(x_0)$ corresponding by (4.1) to the point $x \in N(x_0) \cap S$ may be chosen as the projection of $x - x_0$ on $T(x_0)$. With this choice

(4.2) $$dx(\theta, t) = t$$

[**12**, Theorems 3.1 and 3.2].

A normal $n = n(x_0)$ at the point x_0 of the smooth hypermanifold S is a unit vector orthogonal to $T(x_0)$. If $S = \dot{G}$ then n is an exterior normal if the vector n is exteriorly directed in the sense of Definition 3.3 of the present paper; an interior normal is defined correspondingly. There exists at every point x_0 of $\dot{G}$ a unique exterior and a unique interior normal [**12**, Theorem 4.1].

We now introduce the following

ASSUMPTION (A). $\dot{G}$ is a smooth hypermanifold, and the field of exterior normals is strictly exterior (Definition 3.1) and of the form (2.6), i.e.

(4.3) $$n(x) = \mu(x)x + N(x), \quad x \in \dot{G},$$

where $N(x)$ is completely continuous and where $\mu(x)$ is a continuous real valued function satisfying

(4.4) $$\mu(x) \geqq m > 0,$$

for some suitable constant m.

We will first draw some easy consequences of this assumption (Theorems 4.1–4.3) using the results of §3. Then we will establish sufficient conditions for the validity of assumption (A) (Theorem 4.4). (A trivial example of a G satisfying (A) is a ball with center θ; here $n(x) = x$).

THEOREM 4.1. *Let G be an open and bounded star domain in E which satisfies assumption* (A). *Then*

$$\chi(n, \dot{G}) = 1. \tag{4.5}$$

PROOF. The theorem is an immediate consequence of Theorem 3.1.

THEOREM 4.2. *Let G be as in Theorem* 4.1. *Let f be a nonvanishing vector field defined on $\dot{G}$ of the form* (2.6). *We assume that f has in no point the direction of the interior normal. Then $\chi(\dot{G}, f) = 1$.*

PROOF. Let $f_t = (1 - t)f + tn$. Then

$$f_t(x) \neq \theta \quad \text{for } x \in \dot{G} \quad \text{and} \quad 0 \leq t \leq 1. \tag{4.6}$$

This is obvious for $t = 0$ and $t = 1$. Suppose now $f_t(x_0) = \theta$ for some t in $(0, 1)$ and some $x_0 \in \dot{G}$. Then $f(x_0) = (1 - t)^{-1}tn(x_0)$, and $f(x_0)$ would have the direction of the interior normal. But from (4.6) we conclude that $\chi(\dot{G}, f) = \chi(\dot{G}, n)$. Thus our assertion follows from (4.5).

THEOREM 4.3. *Let G be as in Theorem* 3.2. *In addition we require that assumption* (A) *be satisfied. Of f we require that in no point of G_k $(k = 0, 1, \ldots, q)$ it has the direction of the interior normal (with respect to G_k). Then* (3.7) *holds.*

We now turn to conditions which are sufficient for the validity of assumption (A).

DEFINITION 4.1. A set $G \subset E$ is called an element if there exists a ball B and a 1-1 map $\psi: \bar{B} \rightarrow E$ of form (2.19) such that $\psi(B) = G$.

THEOREM 4.4. *Let G be an element. Concerning the map ψ in Definition* 4.1 *of an element we make the following additional assumptions: at every point u of B the (Fréchet-) differential $d\psi(u; \eta) = \eta + d\Psi(u; \eta)$ exists and is nonsingular and continuous. Then $\dot{G}$ is a smooth manifold.*

PROOF. Without restriction of generality we assume that the radius of B equals 1. Let $u_0 \in \dot{B}$, let $T(u_0)$ be the tangent space to $\dot{B}$ at u_0, let N_0 be a spherical neighborhood with center u_0 and a radius $< \sqrt{2}$, and for $u \in N_0 \cap \dot{B}$ let $t = t(u)$ be the orthogonal projection of $u - u_0$ on $T(u_0)$. Then the following formulas are easily verified:

$$t(u) = u - u_0(u, u_0), \qquad u(t) = t + u_0\sqrt{1 - (t, t)} \tag{4.7}$$

$$du(t; h) = h - u_0(t, h)(1 - (t, t))^{-1}. \tag{4.8}$$

A parametric representation of the points x on $\dot{G}$ in a neighborhood of $\psi(u_0)$ is then given by

$$x = \xi(t) = \psi(u(t)) = t + U(t) \tag{4.9}$$

where, by (4.8), $U(t) = u_0\sqrt{1 - (t, t)} + \Psi(u(t))$ is completely continuous. (4.9) has a unique inverse. Indeed if $\psi^{-1}(x) = \varphi(x) = x + \Phi(x)$ (with Φ completely continuous) then by (4.7)

$$t = x + \Phi_1(x) = \varphi_1(x) \quad \text{where } \Phi_1(x) = \Phi(x) - u_0(\varphi(x), u_0). \tag{4.10}$$

The existence of $d\xi$ follows from (4.9) and the chain rule (see e.g. [**12**, p. 372]) according to which

$$d\xi(t; h) = d\psi(u(t); du(t; h)). \tag{4.11}$$

The continuity of $d\xi$ follows easily from the assumed continuity of $d\psi$ and (4.11). (For the definition of the continuity of a differential see e.g. [**12**, p. 361].) Moreover $d\xi$ has a bounded inverse. Indeed from the assumed properties of ψ it follows that $\varphi = \psi^{-1}$ has a differential. By (4.10) the same is true for $\varphi_1 = \xi^{-1}$. But the differential of φ_1 is the inverse of the differential of ξ. Finally we see from (4.11), from the form of $d\psi$ given in the statement of theorem 4.4, and from (4.8) that

$$d\xi(t; h) = h - u_0(t, h)(1 - (t, t))^{-1} + d\Psi(u(t); du(t; h)).$$

In particular taking in account (4.2)

$$d\xi(\theta; h) = h + d\Psi(u_0; h). \tag{4.12}$$

Now the complete continuity of Ψ implies the complete continuity of $d\Psi(u_0; h)$ in h (see e.g. [**4**, p. 135]). Therefore (4.12) exhibits the complete continuity of $d\xi(\theta; h) - h$.

LEMMA 4.1. *If, in addition to the assumptions of the preceding theorem, ψ is of class C'' then $\dot{G}$ is of class C''.*

PROOF. Since obviously $u(t)$ has a second differential (see (4.7), (4.8)) the assertion follows from (4.9) and the chain rule for second differentials (see e.g. [**12**, p. 362]).

Theorem 4.4 gives sufficient conditions for the first part of Assumption (A). To obtain sufficient conditions for the second part we introduce

ASSUMPTION (B_1). There exists a constant m such for all $u_0 \in \dot{B}$

$$(n(x_0), d\psi(u_0, u_0)) \geqq m > 0, \quad x_0 = \psi(u_0). \tag{4.13}$$

Motivation of assumption (B_1). Let us assume that ψ is defined, of form (2.19), and 1-1 in a ball $B' \supset B$ concentric to B. If the positive number t is small enough then $u_0 + tu_0$ will be in $B' - B$, and consequently, by Lemma 2.7, the point $\psi(u_0 + tu_0)$ will be in the exterior of G. We claim that this latter fact is implied by assumption B_1 if G is convex: since by definition

$$d\psi(u_0; u_0) = \lim_{t\to 0} \frac{\psi(u_0 + t_0) - \psi(u_0)}{t},$$

(4.13) implies that $(n(x_0), \psi(u_0 + tu_0) - \psi(u_0)) > 0$ for small enough positive t, i.e. that $n(x_0)$ and $\psi(u_0 + tu_0) - \psi(u_0)$ lie in the same of the two half spaces into which E is divided by the tangent space to $\dot{G}$ at $x_0(u_0)$. This obviously proves our claim since $n(x_0)$ is the *exterior* normal.

ASSUMPTION (B_2). $d\Psi(u; h)$ is weakly continuous as function of u, uniformly in $\|h\| \leqq 1$.

LEMMA 4.2. *Let $l(u, h)$ be a map $E_1 \times E$ into E where E_1 is a subset of E. Let l be linear and completely continuous in h, and weakly continuous as a function of u, uniformly for $\|h\| \leqq 1$. For fixed u let $l^*(u, h)$ denote the adjoint of the $l(u, h)$ as linear operator in h. Then $l^*(u, h)$ is completely continuous in (u, h).*

PROOF. Let u_n, h_n be bounded sequences in E_1 and E resp. We have first to show the existence of subsequences u''_n, h''_n such that $l^*(u''_n, h''_n)$ converges. Now by the weak compactness of a closed ball in Hilbert space there exist points u_0 and h_0 in E and subsequences u'_n, h'_n of u_n, h_n converging weakly to u_0, h_0 resp. We will prove that for some subsequence u''_n, h''_n of the sequence u'_n, h'_n

$$\lim_{n\to\infty} \|l^*(u''_n, h''_n) - l^*(u_0, h_0)\| = 0. \tag{4.14}$$

Now with

$$\eta'_n = \frac{l^*(u'_n, h'_n) - l^*(u_0, h_0)}{\|l^*(u'_n, h'_n) - l^*(u_0, h_0)\|} \tag{4.15}$$

we see that

$$\begin{aligned} \|l^*(u'_n, h'_n) - l^*(u_0, h_0)\| &= (\eta'_n, l^*(u'_n, h'_n) - l^*(u_0, h_0)) \\ &= (l(u'_n, \eta'_n), h'_n) - (l(u_0, \eta'_n), h_0) \\ &= (l(u'_n, \eta'_n), - l(u_0, \eta'_n), h'_n) + (l(u_0, \eta'_n), h'_n - h_0). \end{aligned} \tag{4.16}$$

The first term of the right member converges to zero because of the weak continuity of $l(u, \eta)$ in u, uniformly for bounded η, and because of the boundedness of h'_n. As to the second term at of the right member we note that by the complete continuity of $l(u, \eta)$ in η there exists a subsequence η''_n of η'_n and an $l_0 \in E$ such that

$$l(u_0, \eta''_n) \to l_0. \tag{4.17}$$

We denote by u''_n and h''_n the corresponding subsequences of u'_n and h'_n. Then

$$l(u_0, \eta''_n), h''_n - h_0) = (l(u_0, \eta''_n) - l_0, h''_n - h_0) + (l_0, h''_n - h_0).$$

The first term on the right converges to 0 by (4.17), and the second one by the weak convergence of h''_n to h_0. Thus (4.14) holds. It remains to prove the continuity of $l^*(u, h)$ in both variables, i.e. to prove: if u_n, h_n are sequences converging (strongly) to points u_0, h_0 resp., then (4.14) holds with $u''_n = u_n$, $h''_n = h_n$. Let $\eta_n = \eta'_n$ be given by (4.15) with $u'_n = u_n$, $h'_n = h_n$. Then (4.16) holds (with the identifications indicated) and implies our assertion.

THEOREM 4.5. *The following conditions together are sufficient for the validity of Assumption* (A)*: the assumptions on G of Theorem* 4.1, *Assumptions* (B_1) *and* (B_2), *and the assumption that ψ is of class C''.*

PROOF. The tangent space at the point $x_0 = \psi(u_0)$ of $\dot{G}$ is the image of the tangent space $T(u_0)$ to $\dot{B}$ at u_0 under the map $d\xi(\theta, h)$ (cf. (4.9) and (4.11)). Therefore

$$(n(x_0), d\xi(\theta, h)) = 0. \tag{4.18}$$

Let us denote the adjoint to $d\Psi(u_0, h)$ (as linear operator in h) by $d^*(u_0, h)$. We then see from (4.18) and (4.12) that $(n(x_0) + d^*(u_0, n(x_0)), h) = 0$ for all $h \in T(u_0)$. Since $T(u_0)$ is a hyperplane orthogonal to u_0, this implies that

$$n(x_0) + d^*(u_0 n, (x_0)) = \mu(x_0)u_0 \tag{4.19}$$

for some scalar $\mu(x_0)$. $\mu(x_0)$ is continuous. This follows by multiplying (4.19) scalar by the unit vector u_0 upon noting that the assumption $\psi \in C''$ implies by Lemma 4.1 that $\dot{G}$ is of class C'', and therefore the continuity of $n(x_0)$ since by Theorem 4.4 $\dot{G}$ is a smooth manifold [**12**, Theorem 4.2], and by noting further that $u_0 = \varphi(x_0)$ is continuous, and by applying Lemma 4.2.

The scalar multiplication indicated proves also (4.4). Indeed we obtain by this operation:

$$\begin{aligned}\mu(x_0) &= (n(x_0), u_0) + (d^*(u_0, n(x_0)), u_0) \\ &= (n(x_0), u_0) + (n(x_0), d\Psi(u_0, u_0)) = (n(x_0), d\psi(u_0, u_0)) > m\end{aligned}$$

by (4.13).

Finally, taking into account that $u_0 = \varphi(x_0) = x_0 + \Phi(x_0)$, we see from (4.19) that (4.3) holds with

$$N(x_0) = \mu(x_0)\Phi(x_0) - d^*(\varphi(x_0), n(x_0)).$$

Lemma 4.2 assures the complete continuity of $N(x_0)$. Because of Theorem 4.4 this finishes the proof of Theorem 4.5.

5. **General boundary conditions.** As pointed out in the introduction §5 deals with the modification of Theorems 4.2 and 4.3 necessary if boundary vectors in direction of the interior normal are admitted. However, we restrict in this section ourselves to G being a ball in case of Theorem 4.2, and to $G_0, G_1, \ldots, G_q$ being balls in case of Theorem 4.3. Assumption (A) of §4 is then trivially satisfied.

Before stating the main results of this section (Theorems 5.1 and 5.2) we have to define the tangent field f_t attached to the nonvanishing field f on the boundary $\dot{B}$ of the unit ball B, and to define the index of a singularity of f_t.

DEFINITION 5.1. Let $u \in \dot{B}$, and let $T(u)$ denote the tangent space to $\dot{B}$ at u. For any $z \in E$ we denote by $[z]_t$ and by $[z]_u$ the orthogonal projection of z on $T(u)$

and on the linear space spanned by u_0 resp. If u_0 is a fixed point $\in \dot{B}$ we set $T_0 = T(u_0)$ and denote the corresponding orthogonal projections by z_{t_0} and z_{u_0}. If $f = f(u)$ is a nonvanishing vector field on $\dot{B}$ we call $f_t = [f(u)]_t$ the tangential field of f. Let $f(u)$ be of form (1.1) and let $u(t)$ be given by (4.7). We set $\gamma(t) = f(u(t)) = u(t) + F(u(t)) = u(t) + \Gamma(t)$ such that $\gamma(\theta) = f(u_0)$. We then set

(5.1) $$df(u_0; t) = d\gamma(\theta; t), \qquad dF(u_0; t) = d\Gamma(\theta; t)$$

if the right members exist. (Note that by (4.8)

(5.2) $$du(\theta; t) = t,$$

and that therefore both differentials (5.1) exist if one of them exists.)

ASSUMPTION (C). In every point $u_0 \in \dot{B}$ in which f has the direction of the interior normal to $\dot{B}$ the differentials (5.1) exist, and $[dF(u_0; t)]_{t_0} = [dF(\theta; t)]_{t_0}$ has no real eigenvalues $\leqq 0$.

LEMMA 5.1. *Let f be as in Definition 5.1. Let $f(u_0)$ have the direction of the interior normal with respect to B at u_0, and let Assumption (C) be satisfied. Let p_0 be a positive number and let*

(5.3) $$f^p(u) = (1 + p)u + F(u), \quad 0 \leqq p \leqq p_0.$$

Then there exists a positive constant μ such that

(5.4) $$[f^p(u)]_{t_0} \geqq \|t\| (2\mu)^{-1}, \quad t = [u - u_0]_{t_0}$$

for

(5.5) $$0 \leqq p \leqq p_0.$$

PROOF. Let $\gamma^p(t) = (1 + p)u(t) + \Gamma(t)$. Then by (5.3), (5.1), and (5.2)

(5.6) $$[df^p(u_0, t)]_{t_0} = (1 + p)t + [d\Gamma(\theta; t)]_{t_0} \quad \text{for all } p \text{ in (5.5).}$$

By assumption (C), $\lambda t + [d\Gamma(\theta; t)]_{t_0}$ has a bounded inverse R_λ for $\lambda \geqq 0$. Since $\|R_\lambda\|$ is continuous in λ (see e.g. [**13**, p. 259]), it follows from (5.6) that there exists a $\mu > 0$ such that for all p in (5.5)

(5.7) $$\|d[f^\mu(u_0; t)]_{t_0}^{-1}\| \leqq \mu \, \|t\|.$$

This inequality implies (5.3) by [**12**, Lemma 6.2] since

$$[f^p(u_0)]_{t_0} = [F(u_0)]_{t_0} = [f(u_0)]_{t_0} = [\gamma(\theta)]_{t_0} = \theta.$$

COROLLARY TO LEMMA 5.1. *$u = u_0$ is an isolated zero of $[f(u)]_{t_0} = [f^0(u)]_{t_0}$ or, equivalently, θ is an isolated zero of $[\gamma(t)]_{t_0} = [\gamma^0(t)]_{t_0}$. Thus the Leray-Schauder index of this zero exists.*

DEFINITION 5.2. Under the assumptions of Lemma 5.1 the index $j_t(u_0, f_t)$ of the singularity u_0 of the vectorfield f_t is defined as the Leray-Schauder index of θ as zero of $[\gamma(t)]_{t_0}$.

THEOREM 5.1. *Suppose the nonvanishing vectorfield f of form (1.1) defined on $\dot{B}$ has the direction of the interior normal at a finite number of points $u_0^1, u_0^2, \ldots, u_0^l$. In addition f is supposed to satisfy the assumptions of Definition 5.1, and at each u_0^λ ($\lambda = 1, 2, \ldots, l$) to satisfy Assumption* (C). *Finally we assume that each u_0^λ has a neighborhood on which f satisfies a uniform Lipschitz condition. Then*

$$\chi(\dot{B}, f) = 1 - \sum_{\lambda=1}^{l} j_t^\lambda \tag{5.8}$$

where j_t^λ denotes the index of the singularity u_0^λ of the tangent field f_t.

PROOF. Let s_1 be an arbitrary positive number, and let M be a number such that

$$\|f(u)\| < M \quad \text{for all } u \in \dot{B}. \tag{5.9}$$

Finally let α be a number satisfying the inequality

$$\alpha > \max\{M/s_1, 1\}. \tag{5.10}$$

We now extend f from $\dot{B}$ to $\bar{B}' - B$ where B' is the ball concentric to B of radius $1 + s_1$: denoting the extension again by f we set

$$f(u) = f(\bar{u}) + \alpha s \bar{u} \quad \text{for } u \in \bar{B}' - B \tag{5.11}$$

where

$$\bar{u} = u/\|u\| \in \dot{B}, \qquad s = \|u - \bar{u}\|. \tag{5.12}$$

We note that f is of the form (2.6) since

$$f(u) = \frac{1 + \alpha s}{1 + s} u + F\left(\frac{u}{1 + s}\right). \tag{5.13}$$

We will need the following three lemmas concerning the extended field f.

LEMMA 5.2. *$(f(u), u) > 0$ for $u \in \dot{B}'$.*

PROOF. We see from (5.11), (5.12), (5.9) and (5.10) that for $u \in \dot{B}'$, i.e. for $\|u\| = 1 + s_1$

$$\begin{aligned}(f(u), u)/(1 + s_1) &= (f(u), \bar{u}) = (f(\bar{u}), \bar{u}) + \alpha s_1 \\ &\geqq \alpha s_1 - |(f(\bar{u}), \bar{u})| \geqq \alpha s_1 - \|f(\bar{u})\| > \alpha s_1 - M > 0.\end{aligned}$$

LEMMA 5.3. *The only zeros of $f(u)$ in $\bar{B}' - B$ are the points*

$$u_1^\lambda = u_0^\lambda + s_0^\lambda u_0^\lambda \quad \text{where} \quad s_0^\lambda = -(f(u_0^\lambda), u_0^\lambda)/\alpha. \tag{5.14}$$

PROOF. Let $u \in \bar{B}' - B$. By (5.12) the tangent space at u to the sphere (concentric to B) on which u lies coincides with the tangent space at $\bar{u}$ to B. Therefore

$$[f(u)]_t = [f(\bar{u})]_t \tag{5.15}$$

where the subscript t denotes projection on this tangent space (cf. Definition 5.1).

If now $\bar{u}$ is none of the u_0^λ then either the right, and therefore the left, member of (5.15) is $\neq \theta$ which implies $f(u) \neq \theta$, or $f(u) = \rho u$ for some positive ρ which by (5.11) again implies $f(u) \neq \theta$.

Let now $\bar{u} = u_0^\lambda$ for some λ, i.e. $u = (1 + s)u_0^\lambda$. Then by (5.15), $[f(u)]_t = \theta$, and therefore by (5.11) $f(u) = f(u) - [f(u)]_t = (f(u), u_0^\lambda)u_0^\lambda = [f(\bar{u}), u_0^\lambda) + \alpha s]u_0^\lambda$, and the bracket equals zero if and only if $s = s_0^\lambda$. Finally, that the points u_1^λ are actually contained in $B' - B_0$ follows from the fact that by (5.14), (5.9), and (5.10) the inequality $0 < s_0^\lambda < s_1$ holds.

LEMMA 5.4.

$$\chi(B, f) = 1 - \sum_{\lambda=1}^{l} j_0^\lambda \tag{5.16}$$

where j_0^λ denotes the Leray-Schauder index of u_1^λ as zero of f.

PROOF. From Lemmas 2.6 and 5.3 we see that

$$\chi(B', f) = \chi(B, f) + \sum_{i=1}^{l} j_0^\lambda.$$

But the left member of this equality equals 1 as follows from (5.13), (2.10), Lemma 5.2, and (1.2) (with B replaced by B').

From Lemma 5.4 just proved it is clear that for the proof of Theorem 5.1 it will be sufficient to show that

$$j_0^\lambda = j_t^\lambda, \qquad \lambda = 1, 2, \ldots, l, \tag{5.17}$$

where $j_t^\lambda = j(u_0^\lambda, f_t)$ (see Definition 5.2). For the proof of (5.17) we drop the index λ denoting by u_0 one of the u_0^λ. Correspondingly we write u_1, j_0, j_t, s_0 instead of u_1^λ, j_0^λ, j_t^λ, s_0^λ resp. For the proof we need several lemmas.

LEMMA 5.5. *Let ρ be a positive number $<\frac{1}{2}$, and*

$$Z_\rho = \{u \in (\bar{B}' - B) \mid \|t\| < \rho\} \tag{5.18}$$

(cf. Definition 5.1). We choose a ρ_0 in such a way that u_1 is the only zero of f in Z_{ρ_0}. This is possible by Lemma 5.3. It follows that

$$j_0 = A(\theta, Z_{\rho_0}, f). \tag{5.19}$$

Let now $\rho_1 \leqq \rho_0$ be such that the spherical neighborhood V_{ρ_1} of u_1 with radius ρ_1 is contained in Z_{ρ_0}. Then

$$j_0 = A(\theta, V_\rho, f) \quad \text{for } 0 < \rho \leqq \rho_1. \tag{5.20}$$

We then assert the existence of a $\mu > 0$ and of a positive $\rho_2 \leqq \rho_1$ such that

$$\|f(u)\| \geqq \|u - u_1\|(2\mu)^{-1} \quad \text{for } \|u - u_1\| \leqq \rho_2. \tag{5.21}$$

PROOF. If $\eta = \sigma u_0 + t$ with $t \in T(u_0)$ and if $m(\eta)$ is a bounded linear operator we use the following matrix notation

$$m(\eta) = \begin{pmatrix} [m(\sigma u_0)]_{u_0} & [m(t)]_{u_0} \\ [m(\sigma u_0)]_{t_0} & [m(t)]_{t_0} \end{pmatrix}.$$

A calculation which we omit shows that in this notation

$$df(u;\eta) = \begin{pmatrix} \sigma u_0(1 + (\alpha - 1)s_1) & [dF(u_0, t)]_{u_0} \\ \theta & \dfrac{t(1 + (\alpha - 1)s_1)}{1 + s_1} + [dF(u_0; t)]_{t_0} \end{pmatrix}. \tag{5.22}$$

By (5.10) and Assumption (C) the linear operator in t at the right lower corner of this matrix has a bounded inverse. Inspection of (5.22) shows that this implies that $df(u;\eta)$ has a bounded inverse. The assertion of our lemma follows now from [**12**, Lemma 6.2] since $f(u_1) = \theta$ (cf. the argument in the proof of Lemma 5.1.)

LEMMA 5.6. *We use the notations of the preceding lemma. Let $u \in Z_{\rho_1}$, and let*

$$v = u_0 + t \tag{5.23}$$

where t is the projection of $\bar{u} - u_0$ on $F(u_0)$. We then set

$$\tilde{f}(u) = t + u_0 + F(t + u_0) + \alpha s(t + u_0). \tag{5.24}$$

We assert the existence of a positive $\rho_3 \leqq \rho_2$ such that (i), u_1 is the only zero of $\tilde{f}$ in Z_{ρ_3}, and (ii)

$$A(\theta, Z_\rho, f) = A(\theta, Z_\rho, \tilde{f}) \quad \text{for } 0 < \rho \leqq \rho_3. \tag{5.25}$$

PROOF. By the definitions involved

$$f(u) = \tilde{f}(u) \quad \text{if } t = \theta. \tag{5.26}$$

It therefore follows from the definition of ρ_0 in Lemma 5.5 that u_1 is a zero of $\tilde{f}$ and the only one in $Z_{\rho_0} \cap \{t = \theta\}$.

Let now $u \in Z_{\rho_2} \in Z_{\rho_0}$ but $t \neq \theta$. We write

$$[\tilde{f}(u)]_{t_0} = [f(u)]_{t_0} + [\tilde{f}(u) - f(u)]_{t_0} \tag{5.27}$$

and estimate the two terms at the right member of (5.27). Now

$$[f([u]_{u_0})]_{t_0} = \theta, \tag{5.28}$$

for by (5.11), $f([u]_{u_0}) = f(u_0) + \alpha s u_0$, and since u_0 is orthogonal to $T(u_0)$ we see that $[f([u]_{u_0})]_{t_0} = [f(u_0)]_{t_0}$ which proves (5.28) since u_0 is a singularity of the tangent field. Since $[u]_{u_0}$ is of the form $(1 + s)u_0$ with $0 \leqq s \leqq s_1$ it follows easily from (5.28), (5.13) and Lemma 5.1 that there exists a positive constant μ such that

$$[f(u)]_{t_0} = [f([u]_{u_0} + [u]_{t_0})] \geqq \|[u]_{t_0}\| (2\mu)^{-1} \quad \text{for } 0 \leqq s \leqq s_1. \tag{5.29}$$

On the other hand we assert the existence of a constant C such that

$$\|[\tilde{f}(u) - f(u)]_{t_0}\| \leqq \|\tilde{f}(u) - f(u)\| \leqq C \|[u]_{t_0}\|^2. \tag{5.30}$$

Indeed by (5.23) and (4.7)

$$\bar{u} - v = u_0(\sqrt{1 - (t, t)} - 1). \tag{5.31}$$

From (5.31) one concludes easily the existence of a constant C_1 and of a positive $\rho_3' < \rho_2$ such that

$$\|\bar{u} - v\| \leqq \|t\|^2 C_1 \quad \text{for } u \in Z'_{\rho_3}.$$

From this inequality, from the definitions for f and $\tilde{f}$, and from the Lipschitz assumption in theorem 5.1 the existence of a constant C_2 is easily established for which

$$\|\tilde{f}(u) - f(u)\| \leqq C_2 \|t\|^2. \tag{5.32}$$

This proves (5.30) since $\|t\| \leqq \|[u]_{t_0}\|$.

From (5.27), (5.29) and (5.32) we see that

$$\|\tilde{f}(u)\| \geqq \|[u]_{t_0}\| ((2\mu)^{-1} - C_2 \|[u_{t_0}]\|).$$

The right member is greater than $\|[u]_{t_0}\|/2$ in Z_{ρ_3} if $\rho_3 = \min(\rho_3', 4/(\mu C))$. Thus $[\tilde{f}(u)]_{t_0}$, and a fortiori $\tilde{f}(u)$, is $\neq \theta$ in Z_{ρ_3} for $[u]_{t_0} \neq \theta$, and assertion (i) is proved.

For the proof of assertion (ii) it will, on account of (i), be sufficient to prove the existence of a positive $\rho_4 \leqq \rho_3$ such that

$$A(\theta, V_{\rho_4}, \tilde{f}) = A(\theta, V_{\rho_4}, f) \tag{5.33}$$

where V_ρ denoted the ball with center u_1 and radius ρ. Again for the proof of (5.33) it will, on account of Rouché's theorem [**10**, Theorem 4], be sufficient to show the existence of a ρ_4 such that

$$\|\tilde{f}(u) - f(u)\| < \|f(u)\| \quad \text{for } u \in \dot{V}_{\rho_4}. \tag{5.34}$$

Now for $\rho_4' < \min(\rho_3, \rho_2)$ we have from (5.32) and (5.21)

$$\|f(u) - \tilde{f}(u)\| < C_2 \|u - u_1\|^2 \leqq \|f(u)\| \, 2\mu C_1 \|u - u_1\| \quad \text{for } u \in \dot{V}'_{\rho_4}.$$

This proves (5.34) with $\rho_4 < \min(\rho_3', (2\mu C_1)^{-1})$.

LEMMA 5.7. *We set*

$$\tilde{\tilde{f}}(u) = \tilde{f}(u) - \alpha st. \tag{5.35}$$

Then there exists a positive $\rho_4 \leqq \rho_3$ *such that*

$$A(\theta, Z_{\rho_4}, \tilde{f}) = A(\theta, Z_{\rho_4}, \tilde{\tilde{f}}). \tag{5.36}$$

PROOF. For $0 \leqq \beta \leqq 1$, let

$$\tilde{f}_\beta(u) = \tilde{f}(u) - (1 - \beta)\alpha st. \tag{5.37}$$

Obviously the lemma will be proved once we show that

$$\tilde{f}_\beta(u) \neq \theta \quad \text{for } 0 \leqq \beta \leqq 1, \text{ and } u \in \dot{Z}_{\rho_4} \tag{5.38}$$

for some ρ_4. Now

$$[f_\beta(u)]_{t_0} = (1 + \beta\alpha s)t + [F(t + u_0)]_{t_0}$$

$$d[f_\beta(u_0; t)]_{t_0} = (1 + s)t + [dF(u_0; t)]_{t_0}.$$

By Assumption (C) the norm of the inverse of this linear operator has a bound μ independent of s for $0 \leqq s \leqq s_1$ (cf. the proof of Lemma 5.1). Moreover $[f_\beta(u_0)]_{t_0} = \theta$ the proof being similar to the one for (5.28). Therefore application of [**12**, Lemma 6.2] shows that there exists a ρ_4 such that $[f_\beta(u)]_{t_0} \geqq \|t\| \, (2\mu)^{-1}$ for $\|t\| < \rho_4$. This proves (5.38) for those points u of $\dot{Z}_{\rho_4}$ for which $t \neq \theta$. But $t = \theta$ on $\dot{Z}_{\rho_4}$ only for $u = u_0$ and $u = (1 + s_1)u_0$. But in these points $\tilde{f}_\beta(u) = f(u)$, and the normal component of f is there $\neq \theta$ (by assumption for $u = u_0$, and by Lemma 5.2 for $u = (1 + s_1)u_0$). This finishes the proof of Lemma 5.7.

Let now $u \in Z_\rho$, and let $\bar{u}$, s, t be as above (see Lemma 5.5 and (5.12)). We set

$$x = su_0 + t. \tag{5.39}$$

Then from (4.7)

$$\begin{aligned} u &= (1 + s)\bar{u} = (1 + s)(t + u_0\sqrt{1 - \|t\|^2}) \\ &= (1 + s)x + (1 + s)u_0(\sqrt{1 - \|t\|^2} - s). \end{aligned} \tag{5.40}$$

Since $s = (x, u_0)$, and $t = x - (x, u_0)u_0$ the right member of (5.40) is a function of x; this function which we denote by $\psi(x)$ obviously maps the "straight cylinder"

$$C_\rho = \{x \mid 0 \leqq s \leqq s_1, \ \|t\| \leqq \rho\} \tag{5.41}$$

in a one to one continuous manner onto Z_ρ.

Lemma 5.8. *Let*

$$g(x) = \tilde{\tilde{f}}(\psi(x)) \tag{5.42}$$

where x is given by (5.39) *and ψ by the right member of* (5.40). *Then*

$$A(\theta, Z_\rho, \tilde{\tilde{f}}) = A(\theta, C_\rho, g). \tag{5.43}$$

Proof. As is seen easily from the product theorem (see (2.20)) it will be sufficient to prove that $A(b, C_\rho, \psi) = +1$, for any $b \in \psi(C_\rho) = Z_\rho$. But this degree equals the Leray-Schauder index of the differential $d\psi(\psi^{-1}(b); \xi)$ [**5**, p. 56]. We choose $b = (1 + s)u_0$ where $0 \leqq s \leqq s_1$. Computation shows that then this differential equals $\xi + s\, dt$ where dt denotes the differential of t as function of x. Since s may be taken arbitrarily small it is clear that the index is $+1$.

LEMMA 5.9. *$g(x)$ has the following properties:*
(α) $(g(\theta), u_0) < 0$,
(β) $[g(\theta)]_{t_0} = \theta$,
(γ) *there exists a positive ρ such that*

$$[g(t)]_{t_0} \neq \theta \quad \text{for } t \in C_\rho \cap T(u_0) - \theta.$$

PROOF. By assumption (see the statement of Theorem 5.1) f has at u_0 the direction of the interior normal. Therefore $(f(u_0), u_0) < 0$. But $(g(\theta), u_0) = (f(u_0), u_0)$. This proves ($\alpha$). Again by assumption, $[f(u_0)]_{t_0} = \theta$. But $[g(\theta)]_{t_0} = [f(u_0)]_{t_0}$. This proves ($\beta$). Finally we see from (5.4) (with $p = 0$ and $u = u_0 + t$) that $[u_0 + t + F(u_0 + t)]_{t_0} = [t + F(u_0 + t)] \neq \theta$ for $\|t\|$ small enough. But $[t + F(u_0 + t)]_{t_0} = [g(t)]_{t_0}$. This proves ($\gamma$).

We recall that our goal is to prove (5.17). Now from (5.19), (5.25), (5.36), and (5.43) we see that for ρ small enough

$$j_0 = A(\theta, C_\rho, g). \tag{5.44}$$

On the other hand (β) and (γ) of Lemma 9 show that θ is an isolated singularity of the field $g_t = [g(t)]_{t_0}$. We denote the index of this singularity by $j(\theta, g_t)$.

LEMMA 5.10.

$$j(u_0, f_t) = j(\theta, g_t). \tag{5.45}$$

PROOF. The index of a singularity is determined by the index of the Fréchet differential at the singularity [**5**, p. 56]. Consequently (5.45) follows from Definition 5.2 in conjunction with the equalities

$$dg(\theta, t) = d\tilde{\tilde{f}}(u_0, t) = d\tilde{f}(u_0, t) = d\gamma(\theta, t),$$

which are easily verified from the definitions involved if one uses (5.2), the chain-rule for differentials, and the fact that $d\psi(\theta, t) = t$.

Since θ is an isolated singularity of g_t we have for ρ small enough

$$j(\theta, g_t) = A(\theta, C_\rho \cap T(u_0), g_t). \tag{5.46}$$

It follows from (5.44), (5.45), (5.46) that for the proof of (5.17) we have to show that for ρ small enough

$$A(\theta, C_\rho, g) = A(\theta, C_\rho \cap T(u_0), g_t). \tag{5.47}$$

The proof of (5.47) is based on the following decomposition

$$g(x) = g_1 \circ g_2(x), \tag{5.48}$$

where

$$\begin{aligned} g_1(x) &= x + [G(t)]_{t_0} \\ g_2(x) &= x + u_0\{(u_0, g(t)) + s(\alpha - 1)\} \\ &= t + u_0\{(u_0, g(t)) + s\alpha\}. \end{aligned}$$

(In verifying (5.48) observe that $g(t) = t + u_0 + G(t)$ and, therefore, $(u_0, g(t)) = (u_0, G(t)) + 1$.

We note that g_1 is a "layer map with respect to $T(u_0)$" (i.e. $g_1(x) - x \in T(u_0)$) and that g_2 is a layer map with respect to the one dimensional space E' spanned by u_0.

We want to apply Lemma 2.9 with $f = g_1$, $\psi = g_2$, and $a = \theta$. Let ρ' be a ρ-value for which Lemmas 5.9 and 5.10 are satisfied, and let ρ be a positive number $< \rho'$. Let $s_2 > s_1$ be such that $|(g_2(x), u_0)| = |(g(t), u_0) + s\alpha| \leqq s_2$ for $x \in C_\rho$, and let $H = \{x \mid -s_2 < s < s_2, |t| < \rho'\}$. Then $g_2(C_\rho) \subset H$ and (taking account of (α), (γ) in Lemma 5.9 and of $g(\theta) = u_0$) one easily verifies that the assumptions of Lemma 2.8 are satisfied. We therefore see from (2.20) and (5.48) that for $b \in g_2(C_\rho)$

(5.50) $$A(\theta, C_\rho, g) = A(\theta, g_2(C_\rho), g_1) A(b, C_\rho, g_2).$$

But since g_2 is a layer map with respect to E' the second factor at the right equals the degree of the map $s' = (u_0, g(\theta)) + (\alpha - 1)s$ of E' into E'. Since $\alpha - 1$ is positive this degree is $+1$, and (5.50) becomes

(5.51) $$A(\theta, C_\rho, g) = A(\theta, g_2(C_\rho), g_1).$$

Now θ is the only zero of g_1 for s arbitrary and $\|t\| \leqq \frac{1}{2}$. It therefore follows from (5.51) that

(5.52) $$A(\theta, g_2(C_\rho), g) = A(\theta, C'_\rho, g_1)$$

where

$$C'_\rho = \{x \mid -s' < s < s_1, \|t\| < \tfrac{1}{2}\}, \qquad s' > 0.$$

We now claim

(5.53) $$A(\theta, C'_\rho, g_1) = A(\theta, C'_\rho \cap T(u_0), g_1).$$

Indeed: $g_1(\dot{C}'_\rho) \neq \theta$. Therefore $g_1(\dot{C}'_\rho)$ has a positive distance ε from θ. Now $[G(t)]_{t_0}$ is a completely continuous map of $T(u_0) \cap C'_\rho$ into $T(u_0)$. Therefore there exists a finite dimensional subspace E^n of $T(u_0)$ and a continuous map $G^n : T(u_0) \cap C'_\rho \to E^n$ such that

$$\|[G(t)]_{t_0} - G^n(t)\| < \varepsilon.$$

We set $g^n(x) = x + G^n(t)$ for $x \in C'_\rho$. Then

(5.54) $$\|g_1(x) - g^n(x)\| < \varepsilon \quad \text{for} \quad x \in C'_\rho.$$

It follows from the definition of the Leray-Schauder degree that

(5.55) $$A(\theta, C'_\rho, g_1) = A(\theta, C'_\rho \cap E^n, g_1).$$

Now (5.54) holds a fortiori for $x \in C'_\rho \cap T(u_0)$. We therefore have also

(5.56) $$A(\theta, C'_\rho \cap T(u_0), g_1) = A(\theta, C'_\rho \cap T(u_0) \cap E^n, g_1).$$

But $T(u_0) \cap E^n = E^n$. Therefore the right members of (5.55) and (5.56) agree, and these two equalities imply (5.53).

Now on $T(u_0)$, $g_t(t) = g_1(t)$. If we replace g_1 by g_t in the right member of (5.53) and combine the equality thus obtained with (5.52) we see that

$$A(\theta, C_\rho, g) = A(\theta, C'_\rho \cap T(u_0), g_t).$$

But the right member of this equality equals the right member of (5.47) since θ is the only zero of g_t on C'_ρ and $\theta \in C_\rho \subset C'_\rho$. Thus (5.47) is shown to be valid and the proof of Theorem 5.1 is finished.

THEOREM 5.2. *Let $B_0, B_1, \ldots, B_q, f, x_1, \ldots, x_r, j_1, j_2, \ldots, j_r$ be defined as in the lines directly preceding* (1.3). *In addition we assume that on each B_k ($k = 0, 1, \ldots, q$), f satisfies the assumptions made for f on $\dot{B}$ in Theorem* 5.1. *Let J^k_t denote the sum of the indices of those singularities on $\dot{B}_k$ of the tangentfield f_t in which f has the direction of the interior normal* (*with respect to B_k*). *Then* (*in generalization of* (1.4)),

$$1 - q = \sum_{\rho=1}^{r} j_\rho + J^0_t - \sum_{i=1}^{q} J^i_t. \tag{5.57}$$

PROOF. From (5.8) applied to each B_k we see that

$$\chi(\dot{B}_k, f) = 1 - J^k_t, \qquad k = 0, 1, \ldots, q.$$

Substituting this equality in (1.3) we obtain (5.57).

REFERENCES

1. P. Alexandroff and H. Hopf, *Topologie*, Springer, Berlin, 1935.

2. J. Cronin, *Fixed points and topological degree in nonlinear analysis*, Math. Surveys No. 11, Amer. Math. Soc., Providence, R.I., 1964.

3. A. Granas, *The theory of compact vector fields and some of its applications to topology of functional spaces*. I, Rozprawy Mat. **30** (1962), 93 pp.

4. M. A. Krasnosel'skiĭ, *Topological methods in the theory of nonlinear integral equations*, Pergamon Press, New York, 1964.

5. J. Leray and J. Schauder, *Topologie et équations fonctionelles*, Ann. Sci. École Norm. Sup. **51** (1934), 45–78.

6. M. Morse, *Singular points of vector fields under general boundary conditions*, Amer. J. Math. **51** (1929), 165–178.

7. M. Nagumo, *A theory of degree of mapping based on infinitesimal analysis*, Amer. J. Math. **73** (1951), 485–496.

8. ———, *Degree of mapping in convex linear topological spaces*, Amer. J. Math. **73** (1951), 497–511.

9. E. H. Rothe, *Zur Theorie der topologischen Ordnung und der Vektorfelder in Banachschen Räumen*, Compositio Math. **5** (1937), 177–197.

10. ———, *The theory of the topological order in some linear topological spaces*, Iowa State J. Sci. XIII, No. **4** (1939), 373–390.

11. ———, *Critical points and gradient fields in Hilbert space*, Acta Math. **85** (1951), 73-98.

12. ———, *Critical point theory under general boundary conditions*, J. Math. Anal. Appl. **11** (1965), 357–409.

13. A. E. Taylor, *Functional analysis*, Wiley, New York, 1958.

THE UNIVERSITY OF MICHIGAN AND
WESTERN MICHIGAN UNIVERSITY

REGULARITY OF SOLUTIONS OF SOME VARIATIONAL INEQUALITIES

Guido Stampacchia

1. **Variational inequalities.** Let X be a reflexive Banach space over the reals and X' its dual space. Denote by $(\, , \,)$ the pairing between X and X'. Let A be a linear or nonlinear operator from X into X' and let $\boldsymbol{K}$ be a closed convex set of X. We say that u satisfies a variational inequality if

$$u \in \boldsymbol{K}: (Au, v - u) \geqslant 0 \quad \text{for all } v \in \boldsymbol{K}. \tag{1}$$

Note that when $\boldsymbol{K} = X$ or u is an interior point of Ω then the v's in $\boldsymbol{K}$ describe a neighborhood of u and thus the inequality (1) actually reduces to the equality

$$(Au, w) = 0 \quad \text{for all } w \in X$$

that is: $Au = 0$.

The theory of variational inequalities is also directly related to the calculus of variations. Let f be a function from X into the reals and assume that there exists u_0 in $\boldsymbol{K}$ such that

$$f(u_0) \leqslant f(v) \quad \text{for all } v \in \boldsymbol{K}.$$

Assume that f has a Fréchet differential Gu for all $u \in X$. It is well known that if u_0 is an interior point of $\boldsymbol{K}$ then: $Gu_0 = 0$. But if $u_0 \in \partial \boldsymbol{K}$ then we get the variational inequality

$$(Gu_0, v - u_0) \geqslant 0 \quad \text{for all } v \in \boldsymbol{K}.$$

Let A be a monotone hemicontinuous operator from X into X'. We recall that A is said to be monotone on X if

$$(Au - Av, u - v) \geqslant 0 \quad \text{for all } u, v \in X,$$

and that A is said to be hemicontinuous if the map

$$t \in [0, 1] \to (A(tu + (1 - t)v), w)$$

is continuous for all $u, v, w \in X$.

Let $\boldsymbol{K}$ be a bounded closed convex set; assuming that A is monotone and hemicontinuous, then a solution of the variational inequality (1) exists.

If $\boldsymbol{K}$ is unbounded, consider for $R > 0$ the bounded convex set $\boldsymbol{K}_R = \boldsymbol{K} \cap \Sigma_R$ where $\Sigma_R = \{v \in X: \|v\| \leqslant R\}$ and let u_R be a solution of the variational inequality

$$u_R \in \boldsymbol{K}_R: (Au_R, v - u_R) \geqslant 0 \quad \forall v \in \boldsymbol{K}_R. \tag{1_R}$$

Then a solution of the variational inequality (1) exists if and only if, for a suitable R, the following estimate holds: $\|u_R\| < R$. The following two conditions (of coerciveness) are sufficient for the existence of a solution of (1).

(a) There exists $R > 0$ and $\varphi_0 \in \boldsymbol{K}_R$ such that $(Au, \varphi_0 - u) < 0$ for all $u \in \boldsymbol{K} \cap \partial\Sigma_R$.

(b) There exists $\varphi_0 \in \boldsymbol{K}$ such that

$$\frac{(Au - A\varphi_0, u - \varphi_0)}{\|u - \varphi_0\|} \to +\infty \quad \text{as} \quad \|u\| \to +\infty,\ u \in \boldsymbol{K}.$$

In general the set of solutions of the variational inequality (1) is a closed convex subset of $\boldsymbol{K}$. This follows at once from a slight generalisation of a lemma by Minty [**9**], [**8**], [**5**]: namely,

"u is a solution of the variational inequality (1) if and only if

$$(1') \qquad u \in \boldsymbol{K}: (Av, v - u) \geqslant 0 \quad \text{for all } v \in \boldsymbol{K}."$$

The solution is unique if A is strictly monotone, i.e. equality holds in (2) only for $u = v$.

The results mentioned above have been proved independently by F. Browder [**5**] and P. Hartman-G. Stampacchia [**8**]. Previous results in the case of Hilbert space and linear operators were proved by G. Stampacchia [**15**] and J. L. Lions-G. Stampacchia [**14**]. Results for more general operators than those mentioned above have been given by H. Brezis [**3**]. U. Mosco [**10**] has shown that the existence of a minimum value for a convex lower semi-continuous function from $\boldsymbol{K}$ into $(-\infty, +\infty]$, not necessarily differentiable, is also included in the previous existence theorems for variational inequalities.

2. **Regularity of solutions.** When A is an elliptic differential operator (linear or nonlinear) the theory of variational inequalities generalises the variational theory of boundary value problems. For instance, consider, in a bounded open set Ω, the second order differential operator

$$(2) \qquad Av = -(a_{ij}(x)v_{x_i})_{x_j},$$

where the coefficients $a_{ij}(x)$ are bounded and satisfy the condition

$$a_{ij}(x)\xi_i\xi_j \geqslant \nu\,|\xi|^2, \quad \nu > 0,\ \xi \in \boldsymbol{R}^n.$$

Denote by $H^1(\Omega)$ the space of all real functions which are in $L^2(\Omega)$ together with their first derivatives. The closure in $H^1(\Omega)$ of all smooth functions vanishing near $\partial\Omega$ will be denoted by $H_0^1(\Omega)$. The dual space of $H_0^1(\Omega)$ will be denoted by $H^{-1}(\Omega)$. The operator A maps $H_0^1(\Omega)$ into $H^{-1}(\Omega)$. Let V be a linear space such that: $H_0^1(\Omega) \subset V \subset H^1(\Omega)$, and let f be an element of $H^{-1}(\Omega)$. An element $u \in V$ such that

$$\int_\Omega a_{ij}(x)u_{x_i}v_{x_j}\,dx = (f, v) \quad \text{for all } v \in V,$$

satisfies a boundary value problem for the operator A. The Dirichlet problem is obtained for $V = H_0^1(\Omega)$.

An example of a typical variational inequality is obtained by considering the closed convex set of $H_0^1(\Omega)$

$$K \equiv \{v \in H_0^1(\Omega): v \geqslant \psi \quad \text{on} \quad E\} \tag{3}$$

where E is a subset of Ω and the "$\geqslant$" is defined in a suitable way (see §4 below). An element u satisfying

$$u \in K: \int_\Omega a_{ij}(x) u_{x_i} (v - u)_{x_j} \, dx \geq 0 \quad \text{for all } v \in K \tag{4}$$

is a solution of a variational inequality. Note that in the case $a_{ij} = a_{ji}$, u is a solution of the variational problem:

$$\min_{v \in K} \int_\Omega a_{ij} v_{x_i} v_{x_j} \, dx.$$

As we have already mentioned the existence theory of boundary value problems is generalised in a natural way by the existence theory for variational inequalities. From this point of view problems of regularity for solutions of variational inequalities naturally arise. We shall describe here some recent results concerning this type of problems.

In a paper by Hans Lewy and G. Stampacchia [**13**] (see also [**12**]) problem (4) has been investigated. It turns out that it is basically an extension of investigations of the conductor potential in potential theory. We have proved that when ψ is a smooth function on all of $\bar{\Omega}$ and is negative on $\partial\Omega$, then the solution of problem (4) has Hölder-continuous first derivatives and second derivatives in $L^p(\Omega)$, for suitable p. Let us point out that, after all, we cannot expect a smoother solution for problem (4); for instance, we cannot expect to have solutions with continuous second derivatives. It suffices to consider $n = 1$, $\Omega \equiv (-2, 2)$ with $\psi = 1 - x^2$, where the solution is partially a parabola and partially its tangents.

In the same paper the question of the topological and analytic nature of the coincidence set where $u(x) = \psi(x)$ is treated for dimension 2 and for $A = -\Delta$. In this case the convexity of Ω and of the function $-\psi$ imply the simple connectivity of the coincidence set. Furthermore if ψ is an analytic function in Ω then the boundary of the coincidence set is a Jordan curve, which permits the representation of its coordinates as analytic functions of a real parameter.

In the paper [**4**], H. Brezis and G. Stampacchia have considered an abstract result of regularity for solutions of variational inequalities and have given several applications of it. From this point on, the discussion will be based on the joint paper [**4**].

3. **An abstract theorem of regularity.** We consider as in §1 the variational inequality (1) which we write now in the form

$$u \in K: (Au - f, v - u) \geqslant 0 \quad \text{for all } v \in K \tag{5}$$

with $f \in X'$, where K is a closed convex set of X and A is a monotone, hemicontinuous operator from X into X'.

Let W be a reflexive Banach space and denote by $\| \ \|_W$ its norm. Let W' be the dual of W. We assume that W is a subspace of X', dense in X', such that

$$\| \ \|_{X'} \leqslant \text{const} \ \| \ \|_W.$$

It follows that X may be identified with a subspace of W', dense in W', such that

$$\| \ \|_{W'} \leqslant \text{constant} \ \| \ \|_X.$$

We shall denote by the same notation (,) the pairing between X, X' and W', W.

Assume, that W and W' are strictly convex. Fix a gauge function $t \in [0, +\infty) \to \varphi(t)$, continuous and strictly increasing, such that $\varphi(0) = 0$ and $\varphi(r) \to +\infty$ as $r \to +\infty$. Let J be the duality mapping with the function φ and J^{-1} its inverse (see [**6**]).

The maps

$$J: W \to W', \qquad J^{-1}: W' \to W$$

are defined by the properties,

$$(Jw, w) = \varphi(\|w\|_W) \, \|w\|_W \quad \text{for all } w \in W,$$
$$\|Jw\|_{W'} = \varphi(\|w\|_W),$$
$$(J^{-1}w', w') = \varphi^{-1}(\|w'\|_{W'}) \cdot \|w'\|_{W'} \quad \text{for all } w' \in W',$$
$$\|J^{-1}w'\|_W = \varphi^{-1}(\|w\|_{W'}),$$

and they are monotone hemicontinuous.

We shall make use of the following

Definition. Let $u \in K$; we say that the tern $\{u, K, A\}$ is *J-compatible* if, for any $\varepsilon > 0$, there exists two maps

$$B_\varepsilon: X \to W, \qquad C_\varepsilon: X \to W'$$

uniformly bounded with respect to ε (i.e. $\|B_\varepsilon v\|_W \leqslant C$, $\|C_\varepsilon v\|_{W'} \leqslant C$ for all $\varepsilon > 0$ and for all $v \in K$, with C a constant independent of ε and v) such that the equation

$$u_\varepsilon + \varepsilon J(Au_\varepsilon + Bu_\varepsilon) = u + \varepsilon C_\varepsilon u_\varepsilon \tag{6}$$

has a solution $u_\varepsilon \in K$ with $Au_\varepsilon \in W$.

Note that the equation (3) has a meaning in W' since $u_\varepsilon, u \in V \subset W'$ and $J(Au_\varepsilon + B_\varepsilon u_\varepsilon)$, $C_\varepsilon u_\varepsilon \in W'$.

We have the following

Theorem 1. *Assume that K is bounded and that $f \in W$ and let u be a solution of the variational inequality*

$$u \in K: (f - Au, v - u) \leqslant 0, \quad \text{for all } v \in K. \tag{5}$$

If the tern $\{u, K, A\}$ is J-compatible, then $Au \in W$.

Note that the solution u of (5) exists because K is bounded and the operator $Au - f$ is monotone hemicontinuous. The last statement of Theorem 1 is obvious if u is an interior point of K since $Au = f$.

The assumption that the tern $\{u, K, A\}$ is J-compatible reduces the problem of regularity of a solution of the variational inequality (5) to a similar problem for equation (6). A fundamental role is played by the fact that it is possible to construct nonlinear operators which approximate the identity, and leave the convex set K invariant.

We have already mentioned that (5) is equivalent to

$$(5') \qquad u \in K : (f - Av, v - u) \leqslant 0, \quad \text{for all } v \in K.$$

Since the tern $\{u, K, A\}$ is J-compatible, we can set $v = u_\varepsilon$ in (5′)

$$(f - Au_\varepsilon, u_\varepsilon - u) \leqslant 0$$

and thus, because of (6), we get

$$(f - Au_\varepsilon, C_\varepsilon u_\varepsilon - J(Au_\varepsilon + B_\varepsilon u_\varepsilon)) \leqslant 0.$$

Consequently, setting $\xi_\varepsilon = Au_\varepsilon + B_\varepsilon u_\varepsilon$,

$$(\xi_\varepsilon, J\xi_\varepsilon) - (f + B_\varepsilon u_\varepsilon, J\xi_\varepsilon) \leqslant (\xi_\varepsilon - f - B_\varepsilon u_\varepsilon, C_\varepsilon u_\varepsilon)$$

or

$$\varphi(\|\xi_\varepsilon\|_W)\, \|\xi_\varepsilon\|_W - \varphi(\|\xi_\varepsilon\|_W)\, \|f + B_\varepsilon u_\varepsilon\|_W \leqslant C(\|\xi_\varepsilon\|_W + \|f + B_\varepsilon u_\varepsilon\|_W)$$

or

$$\varphi(\|\xi_\varepsilon\|)\left(1 - \frac{\|f + B_\varepsilon u_\varepsilon\|_W}{\|\xi_\varepsilon\|_W}\right) \leqslant C\left(1 + \frac{\|f + B_\varepsilon u_\varepsilon\|_W}{\|\xi_\varepsilon\|_W}\right).$$

This inequality implies, since $\varphi(r) \to +\infty$ as $r \to +\infty$, that ξ_ε, and thus Au_ε, is uniformly bounded with respect to ε.

On the other hand

$$\|u_\varepsilon - u\|_{W'} \leqslant \varepsilon(\varphi(\|\xi_\varepsilon\|_W) + C),$$

and thus

$$\lim u_\varepsilon = u \text{ strongly in } W'.$$

But since $u_\varepsilon \in K$ and K is bounded,

$$\lim u_\varepsilon = u \text{ weakly in } X.$$

Note, after all, that

$$|(Au_\varepsilon, u_\varepsilon - u)| \leqslant \|Au_\varepsilon\|_W \cdot \|u_\varepsilon - u\|_{W'},$$

i.e.,

$$\lim_{\varepsilon \to 0} (Au_\varepsilon, u_\varepsilon - u) = 0.$$

It follows, using a lemma by Brezis, that $Au_\varepsilon \to Au$ weakly in X'. This implies that $Au_\varepsilon \to Au$ weakly in W and thus $Au \in W$.

The lemma we have just used states that

"If $u_\varepsilon \to u$ weakly in X and $\lim_{\varepsilon\to 0} (Au_\varepsilon, u_\varepsilon - u) = 0$, then

$Au_\varepsilon \to Au$ weakly in X'".

The assumption that K is bounded can be removed easily if we assume that the operator A satisfies the following coerciveness condition,

"There exists $v_0 \in K$ such that

$$\frac{(Av, v - v_0)}{\|v\|} \to +\infty \quad \text{as} \quad \|v\| \to +\infty, \cdot v \in K\text{''}. \tag{7}$$

Then the conclusion of Theorem 1 holds true. In fact, under this assumption it is possible to prove that u and u_ε must lie in a bounded convex set $K_R = K \cap \Sigma_R$ for a suitable R and the tern $\{u, K_R, A\}$ is J-compatible. Thus the theorem is reduced to the previous one.

In the applications of the theorem of this section regularity of the solution will be reduced to the verification of the condition of J-compatibility. In some cases it is clear that the solution u_ε of (6) exists and $Au_\varepsilon \in W$; consequently the tern $\{u, K, A\}$ is J-compatible if $u_\varepsilon \in K$. This is the case, for instance, when $A + B_\varepsilon$ is monotone hemicontinuous and satisfies (7) and C_ε does not depend on v. In fact, equation (6) reduces to the equation

$$Au_\varepsilon + B_\varepsilon u_\varepsilon + J^{-1}\left(\frac{u_\varepsilon - u}{\varepsilon} - C_\varepsilon\right) = 0$$

and the operator on the left-hand side is monotone hemicontinuous from X into X'.

4. **Applications to a variational inequality.** In the paper [**4**] mentioned above we have considered applications of the previous theorem to variational inequalities for linear and nonlinear operators related to convex sets of the type:

$$K_1 = \{v \in H_0^1(\Omega);\ v \geqslant \psi \quad \text{in} \quad \Omega\},$$

$$K_2 = \{v \in H_0^1(\Omega);\ |\text{grad}\, v| \geqslant 1 \quad \text{a.e. in} \quad \Omega\}.$$

Here we want to consider variational inequalities related to a convex set of the type

$$K = \{v \in H_0^1(\Omega) : \psi_1 \leqslant v \leqslant \psi_2 \quad \text{in} \quad \Omega\}.$$

Let us recall some notation. Let Ω be an open set of $\boldsymbol{R}^n$, $\bar{\Omega}$ its closure and $\partial\Omega$ its boundary which we assume, for sake of simplicity, to be smooth. We denote by $C^m(\Omega)$ (or $C^m(\bar{\Omega})$) the space of all functions continuous in Ω (or $\bar{\Omega}$) together with their derivatives up to the order m.

$C^{m,\lambda}(\Omega)$ (or $C^{m,\lambda}(\bar{\Omega})$) denotes the space of all Hölder-continuous functions together with their derivatives up to the order m in any compact set of Ω (or in $\bar{\Omega}$).

We denote by $C_0^m(\Omega)$ the subspace of functions in $C^m(\Omega)$ vanishing near $\partial\Omega$.

For $\alpha \geqslant 1$ the $L^\alpha(\Omega)$ norm of $u \in L^\alpha(\Omega)$ will be denoted by $\|u\|_\alpha$. The completion of $C^m(\bar{\Omega})$ [or $C_0^m(\Omega)$] with respect to the norm

$$\|u\|_{m,\alpha} = \sum_{0 \leqslant |j| \leqslant m} \|D^j u\|_\alpha$$

will be denoted by $H^{m,\alpha}(\Omega)$ [or $H_0^{m,\alpha}(\Omega)$]. We have set

$$D^j u = \partial^{|j|} u / \partial x_1^{j_1} \partial x_2^{j_2} \cdots \partial x_n^{j_n}, \qquad |j| = j_1 + j_2 + \cdots + j_n.$$

For $\alpha = 2$ we shall write $H^m(\Omega)$ or $H_0^m(\Omega)$ in place of $H^{m,2}(\Omega)$ or $H_0^{m,2}(\Omega)$. $H_0^{m,\alpha}(\Omega)$ is a reflexive Banach space and its dual is denoted by $H^{-m,\alpha'}(\Omega)$ where $1/\alpha + 1/\alpha' = 1$.

For $1 < t < +\infty$, let ψ_1, ψ_2 be two functions of $H^{1,t}(\Omega)$ such that $\psi_1 \leqslant 0$, $\psi_2 \geqslant 0$ on $\partial\Omega$ and $\psi_1 \leqslant \psi_2$ in Ω in the sense of $H^{1,t}(\Omega)$. If u and v are elements of $H^{1,t}$ the inequality: $u \leqslant v$ on a set E of $\bar{\Omega}$ means that u and v are respectively limits in $H^{1,t}(\Omega)$ of two sequences $\{u_i\}$, $\{v_i\}$ of functions in $C^1(\bar{\Omega})$ such that: $u_i \leqslant v_i$ on E. Actually ψ_1 and ψ_2 are limits of two sequences $\{\psi_1^{(i)}\}$, $\{\psi_2^{(i)}\}$ of functions of $C^1(\bar{\Omega})$ such that: $\psi_1^{(i)} \leqslant 0 \leqslant \psi_2^{(i)}$ on $\partial\Omega$ and $\psi_1^{(i)} \leqslant \psi_2^{(i)}$ on Ω.

Let $\boldsymbol{K}$ be the convex set of the functions v defined by

$$\boldsymbol{K} = \{v \in H_0^{1,t}(\Omega) : \psi_1 \leqslant v \leqslant \psi_2 \quad \text{in} \quad \Omega\}.$$

This set is obviously closed and it is nonempty since

$$\max\{\psi_1, 0\} + \min\{\psi_2, 0\} \in \boldsymbol{K}.$$

DEFINITION. We shall say that an operator A from $H^{1,t}(\Omega)$ into $H^{-1,t'}(\Omega)$ is *T-monotone* if, for any $v_1, v_2 \in H^{1,t}(\Omega)$ such that $v_1 \leqslant v_2$ on $\partial\Omega$,

$$\int_{\{v_1 \geqslant v_2\}} (Av_1 - Av_2, v_1 - v_2)\, dx \geqslant 0.$$

The left-hand side stands for $\int_\Omega (Av_1 - Av_2, w)\, dx$ where $w = \max\{v_1 - v_2, 0\} \in H_0^{1,t}(\Omega)$.

If A is T-monotone from $H^{1,t}(\Omega)$ into $H^{-1,t'}(\Omega)$, then A is monotone from $H_0^{1,t}(\Omega)$ into $H^{-1,t'}(\Omega)$.

THEOREM 2. *Let A be T-monotone from $H^{1,t}(\Omega)$ into $H^{-1,t'}(\Omega)$ such that $A(v + k) = Av$ for $v \in H^{1,t}(\Omega)$ and k constant, and assume that its restriction to $H^{1,t}(\Omega)$ is hemicontinuous and satisfies the following condition of coerciveness,*

$$\frac{(Av, v - v_0)}{\|v\|_{1,t}} \to +\infty \quad \text{as} \quad \|v\|_{1,t} \to +\infty, \quad v \in \boldsymbol{K}$$

where v_0 is a point of $\boldsymbol{K}$. Let $f \in L^p(\Omega)$ with $p > t'n/(n - t')$. Then, there exists a solution u of the variational inequality

$$u \in \boldsymbol{K} : (Au - f, v - u) \geqslant 0, \quad \text{for all } v \in \boldsymbol{K}.$$

Moreover, assume that $A\psi_1$, $A\psi_2$ *are measures and*

$$\max\{A\psi_1, 0\}, \min\{A\psi_2, 0\} \in L^p(\Omega),$$

then $Au \in L^p(\Omega)$.

PROOF. Set $X = H_0^{1,t}(\Omega)$, $W = L^p(\Omega)$. Since $p > t'n/(n-t')$ it follows that $L^p(\Omega) \subset H^{-1,t'}(\Omega)$ and thus we are in the situation of §1, i.e.: $W \subset X'$.

Consider the gauge function $\varphi(r) = r^{p-1}$. The dual map J_p from $L^p(\Omega)$ into $L^{p'}(\Omega)$, associated with φ, is $J_p v = |v|^{p-2}\, v$ and $J_p^{-1} v = J_{p'} v = |v|^{p'-2} \cdot v$. In order to prove the statement of the theorem, according to the result of §3, it suffices to show that the tern $\{u, \boldsymbol{K}, A\}$ is J_p-compatible.

We are going to prove that, for any $\varepsilon > 0$, there exists an operator $B_\varepsilon v$ from $H_0^{1,t}(\Omega)$ into $L^p(\Omega)$, uniformly bounded in $L^p(\Omega)$, such that the equation $u_\varepsilon + \varepsilon J_p(Au_\varepsilon + B_\varepsilon u_\varepsilon) = u$, has a solution $u_\varepsilon \in \boldsymbol{K}$ with $Au_\varepsilon \in L^p(\Omega)$. To begin with, we consider the equation, depending on i (and ε),

$$u_i + J_p(Au_i - B_1^{(i)}u_i - B_2^{(i)}u_i) = u \tag{7}$$

where

$$B_1^{(i)}v = \max\{A\psi_1, 0\}\vartheta_1(v - \psi_1)$$

$$B_2^{(i)}v = \min\{A\psi_2, 0\}\vartheta_2(v - \psi_2),$$

and

$$\vartheta_1(t) = 1, \quad \text{for } t \leqslant -1/i; \qquad = -it \quad \text{for } -1/i \leqslant t \leqslant 0; \qquad = 0 \quad \text{for } t \geqslant 0$$

$$\vartheta_2(t) = 0, \quad \text{for } t \leqslant 0; \qquad = it, \quad \text{for } 0 \leqslant t \leqslant 1/i; \qquad = 1, \quad \text{for } t \geqslant 1/i.$$

The equation (7) may be written in the form

$$Au_i - B_1^{(i)}u_i - B_2^{(i)}u_i + J_p\left(\frac{u_i - u}{\varepsilon}\right) = 0 \tag{7'}$$

and, since the operator on the left-hand side is monotone, hemicontinuous and coercive from $H_0^{1,t}(\Omega)$ into $H^{-1,t'}(\Omega)$, it follows that a solution of (7), $u_i \in H_0^{1,t}(\Omega)$, exists and $Au_i \in L^p(\Omega)$.

Next we prove that

$$\psi_1 - 1/i \leqslant u_i \leqslant \psi_2 + 1/i. \tag{8}$$

Suppose that the inequality $u_i \leqslant \psi_1 - 1/i$ were satisfied in a set E of positive measure. Since $u_i \geqslant \psi_1$ on $\partial\Omega$, then $\bar{E} \subset \Omega$ and, by definition of $B_2^{(i)}$, $B_2^i u_i = 0$ on E.

From (7′) we get,

$$\int_E \left(Au_i - A\psi_1, u_i - \psi_1 - \frac{1}{i}\right) dx + \int_E \left(A\psi_1 - B_1^{(i)}u_i, u_i - \psi_1 + \frac{1}{i}\right) dx$$

$$+ \int_E \left(J_p\left(\frac{u_i - u}{\varepsilon}\right),\ u_i - \psi_i + \frac{1}{i}\right) dx = 0.$$

Note that on E,

$$J_{p'}\left(\frac{u_i - u}{\varepsilon}\right) \leqslant J_{p'}\left(\frac{u_i - \psi_1 + (1/i)}{\varepsilon}\right)$$

and therefore

$$\int_E \left(J_{p'}\left(\frac{u_i - u}{\varepsilon}\right), u_i - \psi_1 + \frac{1}{i}\right) dx \geqslant \int_E \left(J_{p'}\left(\frac{u_i - \psi_1 + (1/i)}{\varepsilon}\right), u_i - \psi_1 + (1/i)\right) dx \geqslant 0.$$

By the definition of $B_1^{(i)}$, we get

$$\int_E \left(A\psi_1 - B_1^{(i)}u_1, u_i - \psi_1 + \frac{1}{i}\right) dx \geqslant 0$$

and consequently, it follows that

$$\int_E \left(Au_i - A\psi_1, u_i - \psi_1 - \frac{1}{i}\right) dx = \int_E \left(Au_i - A\left(\psi_1 - \frac{1}{i}\right), u_i - \psi_1 + \frac{1}{i}\right) dx \leqslant 0.$$

The last inequality, since A is T-monotone, implies: $u_i \geqslant \psi_1 - 1/i$. In a similar way we prove that $u_i \leqslant \psi_2 + 1/i$.

It follows from (7′) that Au_i is bounded in $H^{-1,t'}(\Omega)$ and the assumption of coerciveness of A implies that the sequence $\{u_i\}$ is bounded in $H_0^{1,t}(\Omega)$. Then a subsequence, still called u_i, converges weakly in $H_0^{1,t}(\Omega)$ and strongly in $L^{p'}(\Omega)$ to a function $u_\varepsilon \in H_0^{1,t}(\Omega)$. It follows that: $\psi_1 \leqslant u_\varepsilon \leqslant \psi_2$ in Ω. On the other hand, the sequences $\vartheta_s^{(i)}(u_i - \psi_s)$ $(s = 1, 2)$ converge, weakly in any $L^r(\Omega)$, respectively to two functions $\vartheta_1(x)$, $\vartheta_2(x)$ such that $0 \leqslant \vartheta_s(x) \leqslant 1$ $(s = 1, 2)$.

Define now the operator

$$B_\varepsilon v = -\max\{A\psi_1, 0\}\vartheta_1(x) - \min\{A\psi_2, 0\}\vartheta_2(x)$$

which is uniformly bounded in $L^p(\Omega)$. Finally we prove that u_ε satisfies the equation

$$Au_\varepsilon + B_\varepsilon u_\varepsilon = J_{p'}\left(\frac{u - u_\varepsilon}{\varepsilon}\right). \tag{9}$$

In fact, from Minty's lemma, (7′) may be written

$$(Av, v - u_i) - \int_\Omega (B_1^{(i)}u_i + B_2^{(i)}u_i, v - u_i)\, dx \geqslant \int_\Omega J_{p'}\left(\frac{u - u_i}{\varepsilon}, v - u_i\right) dx$$

for all $v \in H_0^{1,t}(\Omega)$, and, letting i go to $+\infty$, we get

$$(Av, v - u_\varepsilon) + \int_\Omega (B_\varepsilon u_\varepsilon, v - u_\varepsilon)\, dx \geqslant \int_\Omega \left(J_{p'}\left(\frac{u - u_\varepsilon}{\varepsilon}\right), v - u_\varepsilon\right) dx$$

for all $v \in H_0^{1,t}(\Omega)$, and thus (9) follows. We have proved that the tern $\{u, K, A\}$ is J_p-compatible and thus the statement of the theorem.

When A is an elliptic second order differential operator, from the conclusion of the theorem: $Au \in L^p(\Omega)$, we can obtain, via "a priori" estimates for solutions of elliptic equations information about the function u.

Let A be the linear operator (2) with smooth coefficients (for instance: $(a_{ij} \in C^2(\bar{\Omega}))$. If $f \in L^p(\Omega)$ and $\max\{A\psi_1, 0\}$, $\min\{A\psi_2, 0\} \in L^p(\Omega)$ with $p > n$, the well-known estimates in $L^p(\Omega)$ for solutions of elliptic differential equation (see for instance [1]) imply that the solution u of the variational inequality related to the convex

$$K \equiv \{v \in H_0^1(\Omega): \psi_1 \leqslant v \leqslant \psi_2 \quad \text{in} \quad \Omega\},$$

belongs to $H^{2,p}(\Omega)$ and consequently $u \in C^{1,\alpha}(\bar{\Omega})$ with $\alpha = 1 - n/P$.

Consider the nonlinear operator

$$Av = -\{(1 + |\mathrm{grad}\, v|^2)^{(t-2)/2} v_{x_i}\}_{x_i}$$

for $1 < t < 2$, which is locally monotone from $H^{1,t}(\Omega)$ into $H^{-1,t'}(\Omega)$, hemicontinuous and coercive in $H_0^{1,t}(\Omega)$, and let u be the solution of the variational inequality of Theorem 2. Let $f \in L^\infty(\Omega)$ and $\max\{A\psi_1, 0\}$, $\min\{A\psi_2, 0\} \in L^\infty(\Omega)$. Then it can be proved that $Au \in L^\infty(\Omega)$ and, making use of the "a priori estimates" for solutions of the nonlinear equation $Au = f$ (see [8] for references), it follows that $u \in H^{2,p}$ for all $1 < p < +\infty$, and thus $u \in C^{1,\lambda}$ for $0 < \lambda < 1$.

Let us mention that the interesting case $t = 1$ of the minimal surfaces escapes from this result.

5. **Regularity of the solutions of some variational inequalities.** The problem considered in the previous section generalises the problem of §3 studied in [**13**], where only the lower constraint ψ_1 acts.

In [**4**] variational inequalities related to the convex set

$$K = \{v \in H_0^1(\Omega): |\mathrm{grad}\, v| \leqslant 1 \quad \text{in} \quad \Omega\},$$

have been considered. Such problems arise in the theory of elastic plastic torsion [**2**], [**11**], [**16**]; they turn out to be directly related to the problem considered in §4.

Some other problems have been considered in [**4**] and we refer to it.

Many other problems, (see for instance [**7**]) which have great interest in mechanics, still escape from the framework of the abstract theorem of regularity of §3. In fact the verification of J-compatibility makes essential use of comparison theorems for solutions or subsolutions of equations, and this is mainly the reason we are obliged to confine ourselves to investigate problems for second order operators.

BIBLIOGRAPHY

1. S. Agmon, A. Douglis and L. Nirenberg, *Estimates near the boundary for solutions of elliptic partial differential equations satisfying general boundary conditions*, Comm. Pure Appl. Math. **12** (1959), 623–727.

2. B. D. Annin, *Existence and uniqueness of the solution of the elastic-plastic torsion problem for a cylindrical bar of oval cross-section*, Prikl. Mat. Meh., J. Appl. Math. Mech. vol. **25** (1965), 879–887.

3. H. Brezis, *Equations et inéquations non-linéaires dans les espaces vectoriels en dualité*, Ann. Inst. Fourier (Grenoble) **18** (1968), 115–175.

4. H. Brezis and G. Stampacchia, *Sur la régularité de la solution d'inéquations elliptiques*, Bull. Soc. Math. France **96** (1968), 153–180.

5. F. Browder, *Nonlinear monotone operators and convex sets in Banach spaces*, Bull. Amer. Math. Soc. **71** (1965), 730–785.

6. ———, *On a theorem of Beurling and Livingston*, Canad. J. Math. **17** (1965), 367–372.

7. G. Fichera, *Problemi elastostatici con vincoli unilaterali: il problema di Signorini con ambigue condizioni al contorno*, Atti Accad. Lincei, Mem. Cl. Sci. Fis mat. natur Sez. I (8) VII (1963–1964), 91–140.

8. P. Hartman and G. Stampacchia, *On some non linear elliptic differential functional equations*, Acta Math. **115** (1966), 271–330.

9. G. J. Minty, *On the generalization of a direct method of the calculus of variations*, Bull. Amer. Math. Soc. **73** (1967), 315–321.

10. U. Mosco, *A remark on a theorem of F. E. Browder*, J. Math. Anal. Appl. **20** (1967), 90–93.

11. H. Lanchon and C. Duvant, *Sur la solution du problème de la torsion elastoplastique d'une barre cylindrique de section quelconque*, C. R. Acad. Sci. Paris **264** (1697), 520–523.

12. H. Lewy, *On a variational problem with inequalities on the boundary*, J. Math. **17** (1968), 861–884.

13. H. Lewy and G. Stampacchia, *On the regularity of the solutions of a variational inequality*, Comm. Pure Appl. Math. (to appear).

14. J. L. Lions and G. Stampacchia, *Variational inequalities*, Comm. Pure Appl. Math. **20** (1967), 493–519.

15. G. Stampacchia, *Formes bilinéaires coercitives sur les ensembles convexes*, C. R. Acad. Sci. Paris **258** (1964), 4413–4416.

16. T. W. Ting, *Elastic plastic torsion*, Arch Rational Mech. Anal. **25** (1967), 342–366.

Instituto di Matematica
Universita di Pisa

FURTHER APPLICATIONS OF MONOTONE METHODS TO PARTIAL DIFFERENTIAL EQUATIONS*

Walter A. Strauss

1. From the point of view of boundary value problems for nonlinear partial differential equations, monotone methods have been used primarily to prove the *existence of weak global solutions*. Thus the results are relatively superficial, although the range of applicability includes a very broad class of equations which possess *a priori* estimates of energy type. We shall give several examples of boundary value problems, followed by a discussion of a class of abstract equations of evolution. Further details will appear in [**11**].

EXAMPLE 1. Parabolic equations have been discussed in the present context by Višik [**12**], Browder [**3**] and Lions [**6**], and by Brezis at this symposium. A very special example is the system

$$u_t + F(u)_x = u_{xx}$$

in two independent variables $t \geqq 0$, $-\infty < x < \infty$, where the vector-valued solution u is prescribed initially. F is any irrotational C^1-vector field:

$$F(u) = (F_1(u), \ldots, F_N(u)), \qquad u = (u_1, \ldots, u_N),$$
$$\partial F_i/\partial u_j = \partial F_j/\partial u_i \qquad (i, j = 1, \ldots, N).$$

EXAMPLE 2. Equations of Schrödinger type. A typical example is

$$u_t = i\Delta u - |u|^{p-2}\, u$$

where $t \geqq 0$, Δ is the Laplacian in R^n, $p > 1$, $i = (-1)^{1/2}$ and u is prescribed initially.

EXAMPLE 3. Relevant examples of hyperbolic equations have been considered by Segal [**9**] and Lions and Strauss [**7**]. One example, which comes essentially from [**7**], is

$$u_{tt} - \Delta u + |u_t|^{p-2}\, u_t + F(u) = 0$$

in the domain $\{t \geqq 0, x \in \Omega\}$, Ω an open bounded subset of R^n, where u satisfies the boundary conditions

$$u = 0 \text{ on bdy } (\Omega),\ u \text{ and } u_t \text{ given at } t = 0.$$

* This work was supported in part by National Science Foundation Grant GP6721.

We require that $p > 1$ and that F be a continuous function such that

$$\int_0^u F(v)\,dv \geqq -\text{const}\,(1 + |u|^q), \qquad q < p;$$

$$|F(u)| = O(|u|^{(p-1)\theta}) \quad \text{as} \quad |u| \to \infty$$

where $\theta = \max\,(1, 2n[p(n-2)]^{-1})$; θ arbitrary if $n < 3$. We do not know whether this growth condition is essential.

EXAMPLE 4. Now we elaborate the preceding example to allow derivatives in x of arbitrarily high orders and fairly general kinds of perturbations. It is the latter which go beyond the results of [7]. The linear part of the equation is of order 2μ and the nonlinear part of order $2m + 1$, where m and μ are arbitrary non-negative integers. The equation is

$$u_{tt} + \mathscr{A}(t)u + \sum_{|\alpha| \leqq m} (-1)^{|\alpha|} D^\alpha b_\alpha(Du, Du_t) = f(x, t)$$

in the domain $t \geq 0$, $x \in \Omega$, Ω as above, with the boundary conditions, say,

$$D^\alpha u = 0 \quad \text{on} \quad \text{bdy}\,(\Omega) \quad \text{for} \quad |\alpha| < \max\,(m, \mu);$$

$$u \text{ and } u_t \text{ prescribed initially.}$$

The operators $\mathscr{A}(t)$ form a C^1-family of *linear* symmetric differential operators of divergence form in Ω of order 2μ which satisfy Gårding's inequality

$$\int_\Omega (\mathscr{A}(t)v)(v)\,dx \geqq c_0\,|v|_{\mu,2} - c_1\,|v|_{0,2}, \qquad c_0 > 0,$$

for all test functions v. The notation $|\ \ |_{\mu,p}$ indicates the norm in the Sobolev space $W^{\mu,p}(\Omega)$.

Next we describe the nonlinear term. By D is meant the collection of all derivatives in x of order $\leqq m$; and by ζ and ξ corresponding sets of independent variables:

$$D = \{D^\beta\}_{|\beta| \leqq m}, \qquad \zeta = \{\zeta^\beta\}_{|\beta| \leqq m}, \qquad \xi = \{\xi^\beta\}_{|\beta| \leqq m}.$$

The continuous functions $b_\alpha(\zeta, \xi)$ are assumed to satisfy:

(a) b_α is independent of ζ_β for $|\alpha| = |\beta| = m$ in case $m \geqq \mu$.

(b) (Monotonicity) If ζ, ξ and ξ^* are vectors such that $\xi_\beta = \xi^*_\beta$ for $|\beta| < m$, but $\xi \neq \xi^*$, then

$$\sum_{|\alpha| = m} [b_\alpha(\zeta, \xi) - b_\alpha(\zeta, \xi^*)][\xi_\alpha - \xi^*_\alpha] > 0.$$

(c) A "coercivity" condition as in Example 3.

(d) A growth condition as in Example 3.

Let us see how Example 4 (and thereby Example 3, which is a special case) can be considered as an evolution equation of the first order. Let us write the equation (with $f = 0$, say) as

$$(*) \qquad (d^2u/dt^2) + \mathscr{A}(t)u + \mathscr{B}(u, du/dt) = 0,$$

and, think of $u(t)$, for each t, as belonging to some space of functions of x. It is equivalent to the pair of equations

$$du_1/dt - u_2 = 0,$$
$$du_2/dt + \mathscr{A}(t)u_1 + \mathscr{B}(u_1, u_2) = 0.$$

Putting $\bar{u} = [u_1, u_2]$, we may write this system formally as

$$J(t)\, d\bar{u}/dt + A(t)\bar{u} + B(\bar{u}) = 0,$$

where

$$(^{**}) \qquad J(t) = \begin{pmatrix} \mathscr{A}(t) & 0 \\ 0 & 1 \end{pmatrix}, \qquad A(t) = \begin{pmatrix} 0 & -\mathscr{A}(t) \\ \mathscr{A}(t) & 0 \end{pmatrix}.$$

In Example 4, $A(t)$ is naturally definable as a skew-adjoint operator on the Hilbert space

$$H_t = W_0^{\mu,2}(\Omega) \oplus L^2(\Omega)$$

which is furnished with the inner product

$$(\bar{u}, \bar{v})_{H_t} = (J(t)\bar{u}, \bar{v}).$$

We may replace $J(t)$ by the unit operator I if we make the canonical identification of each H_t with its dual space H_t' (or of the direct integral of the H_t's with its dual).

2. We shall present a theorem on equations of evolution which applies neatly to this situation. It is closely related to theorems of Browder [**3**], Lions [**6**] and Bardos and Brezis [**2**]. First we make the

DEFINITION. Let X be the dual of a Banach space. An operator $B: X \to X'$ is called *semimonotone on a subset* $S \subset X$ if there exists another operator $\bar{B}: X \times X \to X'$ such that $Bu = \bar{B}(u, u)$ and

(i) for each $u, v \in X$, $(\bar{B}(u, u + \epsilon v), v)$ is a continuous function of real ϵ and

$$(Bu - \bar{B}(u, v), u - v) \geqq 0.$$

(ii) Suppose $u, v \in X$ and $\{u_\epsilon\}$ is a net such that $\{u_\epsilon\} \subset S$, $u_\epsilon \to u$ weakly* in X. In case $v \neq u$, suppose that

$$(Bu_\epsilon - \bar{B}(u_\epsilon, u), u_\epsilon - u) \to 0.$$

Then

$$\overline{\lim}\, (\bar{B}(u_\epsilon, v), u_\epsilon - v) \geqq (\bar{B}(u, v), u - v).$$

We also define B to be *coercive* if

$$(Bu, u)/|u|_X \to +\infty \quad \text{as } |u|_X \to \infty.$$

Condition (ii) holds, for instance, if the operator $u \to \bar{B}(u, v)$ is continuous from the weak* topology of X to strong topology of X'. The definition is motivated by the theorem of Leray and Lions [**5**], which asserts the existence of a solution of the equation $Bu = 0$ provided B is a semimonotone (on all of X), coercive, bounded operator which is weakly* continuous on finite-dimensional subsets.

THEOREM 1. *Let $u_0 \in H$, $f \in \tilde{X}'$ and $T > 0$. Then there exists a weak solution $u(t)$ in the interval $0 \leqq t \leqq T$ of the problem*

$$du/dt + Au + Bu = f, \qquad u(0) = u_0.$$

The hypotheses are as follows. H is a Hilbert space identified with its dual, X is a reflexive Banach space which is continuously embedded as a dense subset of H, $\tilde{X} = L^p(X) \cap L^2(H)$, $p > 1$. B is a bounded mapping of $\tilde{X}$ into its dual $\tilde{X}'$ which is weakly continuous on finite-dimensional sets. There exists $\lambda > 0$ such that the operator $e^{-\lambda t}(\lambda I + B)$ is coercive and is semimonotone on the set

$$S = \{u \in \tilde{X} \mid du/dt \text{ is bounded in the space } D(A^*)' + L^2(V')\}.$$

The operator A is closed and linear with dense domain $D(A)$ in X and values in X' such that $(Au, u) = 0$ for $u \in D(A)$. Furthermore, its adjoint A^* is the closure of its restriction to $D(A^*) \cap V$, where V is a Banach space with strictly convex dual, V is continuously embedded as a dense subset of H, $D(A) \cap V$ is dense in both X and V, and the restriction of A to $D(A) \cap V$ has a continuous extension from V to V'.

REMARKS. (i) $L^p(X)$ denotes the space of strongly measurable functions of $t \in (0, T)$ with values in X whose pth powers are integrable. $C(H)$ denotes the strongly continuous functions on $[0, T]$ with values in H. $\tilde{X}$ could be a more general space of functions of t with values in X, for instance an Orlicz space. The operator A^* may alternatively be considered as a closed operator from $\tilde{X}$ to $\tilde{X}'$. Its domain, a subset of $\tilde{X}$, may be given the graph norm; this is the space referred to in the above definition of S.

(ii) The meaning of the differential equation may be taken as follows: for all $v \in D(A^*)$ and $t \in [0, T]$,

$$(u(t), v) = (u(0), v) + \int_0^t \{(u(\tau), A^*v) + (Bu(\tau) - f(\tau), v)\}\, d\tau.$$

(iii) If $A = 0$ and $\lambda = 0$, this theorem and the next one reduce to those of Lions [**6**]. The space V is here at our disposal. For instance, we might take $V = D(A)$ with the graph norm (cf. [**3**]). However, it will be more convenient to take V as a certain Hilbert space larger than $D(A)$.

THEOREM 2. *If $\lambda = 0$ there exists a weak solution of the differential equation in the same class which is of period T: $u(0) = u(T)$.*

The proof of Theorem 1 takes off from an elliptic regularization. We may assume $f = 0$. Let Y be the space of functions $u \in L^2(V) \cap \tilde{X}$ such that $du/dt \in L^2(H)$. Let Λ be the duality mapping from V to V'. For any positive ϵ, define $(u' = du/dt)$:

$$\gamma_\epsilon(u, v) = \int_0^T \{\epsilon(u', v') + \epsilon(\Lambda u, v) + (u' + Au + Bu, v)\}e^{-\lambda t}\, dt + (u(0), v(0)).$$

By the theorem of Leray and Lions, there exists a solution u_ϵ in Y of

$$\gamma_\epsilon(u_\epsilon, v) = (u_0, v(0)) \quad \text{for all } v \in Y.$$

Bounds for u_ϵ, $u_\epsilon(0)$, $u_\epsilon(T)$, $\epsilon^{1/2}u_\epsilon$, $\epsilon^{1/2}u_\epsilon'$ in the spaces $\tilde{X}$, H, H, $L^2(V)$, $L^2(H)$, respectively, are easily obtained. From the differential equation

$$-\epsilon u_\epsilon'' + u_\epsilon' = -Bu_\epsilon - Au_\epsilon - \epsilon\Lambda u_\epsilon$$

we also find a bound for u_ϵ' in $D(A^*)' + L^2(V')$. The limit u of a weakly convergent subnet of $\{u_\epsilon\}$ in $\tilde{X}$ satisfies in an appropriate sense the equation

$$du/dt + Au + g = 0 \qquad \text{and} \qquad u(0) = u_0,$$

where g is a weak cluster point in $\tilde{X}'$ of $\{Bu_\epsilon\}$. Next we prove the lemma: $u \in C(H)$ and

$$\tfrac{1}{2}|u(T)|_H^2 - \tfrac{1}{2}|u(0)|_H^2 + \int_0^T (g(t), u(t))\,dt = 0.$$

It follows that

$$\int_0^T (g(t), u(t))\,dt \geqq \lim_{\epsilon\to 0} \int_0^T (Bu_\epsilon(t), u_\epsilon(t))\,dt,$$

whence $Bu = g$ by Minty's device. The lemma is proved by a regularization in t. The proof of Theorem 2 is very similar.

3. Returning to Example 1, we make the following choices: $A = 0$, $V = H$, $H = L^2(R)$, $X = W^{1,2}(R)$, where R denotes the real line, and

$$\bar{B}(u, v) = -v_{xx} + F(u)_x.$$

It follows from Sobolev's inequality in one dimension (trivial) that $\bar{B}$ maps $X \times X$ into X'. It follows from the hypothesis made on F that

$$\int_R F(u)_x u\,dx = 0, \qquad u \in X;$$

so that B is coercive. Therefore Theorems 1 and 2 apply. It is obvious that they also apply to Example 2.

Now consider Example 4, assuming that $\mathscr{B}$ depends only on Du_t (the x-derivatives of $\partial u/\partial t$) and that $\mathscr{A}$ is independent of t. Theorem 1 applies to this situation if we make the choices:

$$H = W_0^{\mu,2} \oplus L^2, \qquad V = W_0^{\mu,2} \oplus W_0^{\mu,2}, \qquad X = W_0^{\mu,2} \oplus (W_0^{m,p} \cap L^2).$$

These are spaces of pairs of functions defined on $\Omega \subset R^n$. We define A by (**) with domain

$$D(A) = \{\bar{u} \mid \bar{u} \in V \cap X, \mathscr{A}u_1 \in (W_0^{m,p} \cap L^2)' \quad \text{where} \quad \bar{u} = [u_1, u_2]\}.$$

It is not difficult to see that A is a skew-adjoint operator from X to X'. In particular,

$$D(A) = D(A^*) \subset V$$

so that all the conditions on A are valid. The operator B, defined by $B[u_1, u_2] = [0, \mathscr{B}(u_2)]$, is semimonotone on S, coercive, etc. The semimonotonicity is a consequence of Aubin's compactness theorem [**1**] and the method of [**5**].

This result was proved by Lions and Strauss [**7**] in case $\mathscr{B}$ depends in a purely monotone manner on Du_t, and by Strauss [**10**] in the case just discussed. In [**10**] $\mathscr{A}$ was allowed to depend on t. In the purely monotone case, theorems of Browder [**3**] and Kōmura [**4**] also are applicable, although extra conditions are required on the initial data.

The question of the existence of time-periodic solutions of the equations of Examples 3 and 4 is beyond the reach of Theorem 2. The reason is that $\lambda = 0$ and that B itself fails to be coercive on $\tilde{X}$. However, the introduction of an extra *a priori* estimate as in Prodi [**8**] can be used to bypass this problem in the majority of cases (see [**11**]).

What about the general assertions made in Examples 3 and 4 concerning the initial value problem? In the reduction of second-order equations (*) to first-order ones, the relation $du_1/dt = u_2$ becomes implicit rather than explicit. The coercivity condition may therefore fail as, for example, in the case of Example 3, where $\iint F(u_1)u_2\,dx\,dt$ enjoys no positivity property for general pairs of functions $[u_1, u_2]$. For this reason we are forced to be more direct.

Let us rewrite equation (*) formally as

$$\left(-\frac{d^3}{dt^3} - \frac{d}{dt}\mathscr{A}\right)u + \left(-\frac{d}{dt}\mathscr{B}\right)u = 0$$

with appropriate boundary conditions. Then $-d/dt\mathscr{B}$ is formally semimonotone. We "elliptically regularize" this equation by adding the terms

$$\epsilon\left(\frac{d^4}{dt^4} - \frac{d^2}{dt^2}\mathscr{A}\right)u$$

to the left-hand side, again with appropriate boundary conditions, and proceed to argue in analogy with the proof of Theorem 1.

In order to state a complete result for Example 4, we list sufficient "coercivity" and growth conditions (c) and (d) for the existence of weak solutions: (c) There exist positive constants k and δ and a function $E(\zeta) \geqq 0$ such that

$$\sum_{|\alpha| \leqq m}\left(b_\alpha(\zeta, \xi) - \frac{\partial E}{\partial \zeta_\alpha}\right)\xi_\alpha \geqq k \sum_{|\beta| = m} |\xi_\beta|^p - \text{const}\left\{1 + |\zeta|^{p-\delta} + \sum_{|\beta| < m} |\xi_\beta|^{p-\delta}\right\};$$

(d) There are bounds

$$\left|\frac{\partial E}{\partial \zeta_\alpha}\right| = O(|\zeta|^{\theta(p-1)}) \quad \text{as} \quad |\zeta| \to \infty,$$

$$\left|b_\alpha(\zeta, \xi) - \frac{\partial E}{\partial \zeta_\alpha}\right| = O(|\xi|^{p-1} + |\zeta|^{p-1}) \quad \text{as} \quad |\xi| + |\zeta| \to \infty,$$

where $\theta = \max\,(1, 2n[p(n + 2m - 2\mu)]^{-1})$, arbitrary if $n \leqq 2\mu - 2m$.

In conclusion, we have seen how all these examples may be put in the form

$$Lu + Bu = 0$$

in a domain, with boundary conditions, where B is a nonlinear degenerate-elliptic partial differential operator and L is a linear skew-symmetric operator, roughly speaking. There ought to exist an existence theorem which applies to a much broader situation of this type.

References

1. J. P. Aubin, *Un théoreme de compacité*, C.R. Acad. Sci. Paris **256** (1963), 5042–5044.

2. C. Bardos and H. Brezis, *Sur une classe de problèmes d'évolution non linéaires*, C.R. Acad. Sci. Paris **266** (1968), 56–59.

3. F. E. Browder, *Nonlinear initial value problems*, Ann. of Math. (2) **82** (1965), 51–87.

4. Y. Kōmura, *Nonlinear semigroups in Hilbert space*, J. Math. Soc. Japan **19** (1967), 493–507.

5. J. Leray and J. L. Lions, *Quelques résultats de Višik sur les problèmes elliptiques non linéaires par les méthodes de Minty-Browder*, Bull. Soc. Math. France **93** (1965), 97–107.

6. J. L. Lions, *Sur certaines équations paraboliques nonlinéaires*, Bull. Soc. Math. France **93** (1965), 155–175.

7. J. L. Lions and W. A. Strauss, *Some nonlinear evolution equations*, Bull. Soc. Math. France **93** (1965), 43–96.

8. G. Prodi, *Soluzioni periodiche dell'equazioni delle onde con termine dissipativo non lineare*, Rend. Sem. Mat. Univ. Padova **36** (1966), 37–49.

9. I. E. Segal, *The global Cauchy problem for a relativistic scalar field with power interaction*, Bull. Soc. Math. France **91** (1963), 129–135.

10. W. A. Strauss, *The initial value problem for certain nonlinear evolution equations*, Amer. J. Math. **89** (1967), 249–259.

11. ———, *The energy method in nonlinear partial differential equations*, 1967 Lecture notes, IMPA, Notas de Matemática, Rio de Janeiro (to appear).

12. M. I. Višik, *Solubility of boundary-value problems for quasi-linear parabolic equations of higher order*, Mat. Sb. **59** (101) (1962), 289–325. (Russian)

Brown University and
City University of New York

AUTHOR INDEX

Roman numbers refer to pages on which a reference is made to an author or a work of an author.

Italic numbers refer to pages on which a complete reference to a work by the author is given.

Boldface numbers indicate the first page of the articles in the book.

SUBJECT INDEX